REGULATION OF CELL MEMBRANE
ACTIVITIES IN PLANTS

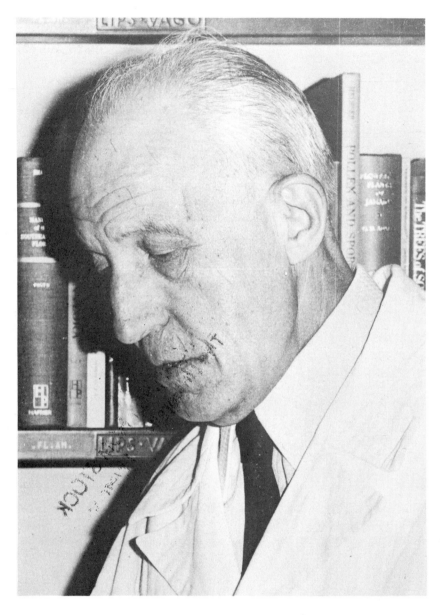

Dedicated to professor Sergio Tonzig on the occasion of his fiftieth year of scientific activity.

Cover design by Giancarlo Zompì

REGULATION OF CELL MEMBRANE ACTIVITIES IN PLANTS

Proceedings of the International Workshop held in Pallanza, Italy, August 26-29, 1976.

Editors
E. Marrè
and
O. Ciferri

1977

NORTH-HOLLAND PUBLISHING COMPANY
AMSTERDAM · OXFORD · NEW YORK

North-Holland ISBN: 0 7204 0615 3

Publishers:
North-Holland Publishing Company-Amsterdam
North-Holland Publishing Company, Ltd. - Oxford

Sole distributors for the U.S.A. and Canada:
Elsevier/North Holland Inc.
52 Vanderbilt Avenue
New York, N.Y. 10017

Library of Congress Cataloging in Publication Data

International Workshop on the Regulation of Cell Membrane
    Activities in Plants, Istituto Italiano di Idrobio-
    logia, 1976.
    Regulation of cell membrane activities in plants.

    Includes indexes.
    1. Plant cell membranes--Congresses. 2. Hormones
(Plants)--Congress. 3. Plant translocation--Congresses.
I. Marrè, Erasmo. II. Ciferri, O. III. Title.
QK725.I55 1976   581.8'75   76-57154
ISBN 0-7204-0615-3

PRINTED IN THE NETHERLANDS

PREFACE

In the last few years, the rapid developments in various fields of plant physiology have led to the recognition that the control of ion transport across the cell membrane plays a major role in the mechanism of action of plant hormones and thus on processes such as growth, seed germination and the control of transpiration by stomata. Indeed, it now appears that, while it is true that the control of cell membrane activity is a major point of hormone action, it is also true that the possibility of influencing experimentally ion transport by treatment with hormones represents a formidable tool in the study of the physiology of the plant membrane.

The "International Workshop on the Regulation of Cell Membrane Activities in Plants" held at the Istituto Italiano di Idrobiologia in Pallanza from August 26th to August 28th 1976, was a first attempt to integrate physiology of transport and hormonology, till now, two relatively separated fields of plant physiology. This book reports the results of such a joint effort and its layout closely reflects the programme of the workshop.

The workshop was organized by Professor Livia Tonolli, Professor Giorgio Forti and ourselves. The meeting was inaugurated by Professor Ernesto Quagliariello, President of the Consiglio Nazionale delle Ricerche that provided most of the financial support. The workshop was also aided by a grant from Montedison S.p.A. Doctors Roberta Colombo, Nicoletta Beffagna, Ida De Michelis and Raffaella Cerana did take care, with charm and efficiency, of the many organizational aspects while all the personnel of the Istituto Italiano di Idrobiologia made our stay both pleasant and fruitful. The success of the workshop was, however, due in good measure to the friendly attitude of all the participants. Indeed, we feel at Pallanza we met as colleagues and we parted as friends.

Erasmo Marrè

Orio Ciferri

Pallanza, October 1976

LIST OF PARTICIPANTS

A. Ballarin-Denti, C.N.R., Institute of Plant Sciences, University of Milan, Italy

A. Ballio, Institute of Biochemistry, University of Rome, Italy

R. Baudo, Istituto Italiano di Idrobiologia, Pallanza, Novara, Italy

N. Beffagna, C.N.R., Institute of Plant Sciences, University of Milan, Italy

F.W. Bentrup, Institut für Biologie 1, Universität Tübingen, Abteilung Biophysik, D-7400 Tübingen 1, Germany

P. Benveniste, Institut de Botanique, Laboratoire de Biochimie Végétale, Université de Strasbourg, France

R. Cerana, Institute of Plant Sciences, University of Milan, Italy

O. Ciferri, Institute of Microbiology and Plant Physiology, University of Pavia, Italy

R. Cleland, Department of Botany, University of Washington, Seattle, Washington, U.S.A.

M. Cocucci, C.N.R., Institute of Plant Sciences, University of Milan, Italy

M.C. Cocucci, C.N.R., Institute of Plant Sciences, University of Milan, Italy

S. Cocucci, Institute of Plant Physiology, University of Milan, Italy

R. Colombo, Institute of Plant Sciences, University of Milan, Italy

J. Dainty, Department of Botany, University of Toronto, Toronto, Ontario, Canada

M.I. De Michelis, Institute of Plant Sciences, University of Milan, Italy

G. Ducet, Laboratoire de Physiologie Cellulaire, Centre Universitaire Marseille-Luminy, France

D.C. Elliott, School of Biological Sciences, The Flinders University of South Australia, Bedford Park, Australia

D.S. Fensom, Department of Biology, Mount Allison University, Sackville, New Brunswick, Canada

G. Forti, Institute of Plant Sciences, Univeristy of Milan, Italy

A. Graniti, Istituto Sperimentale di Patologia Vegetale, Via Casal de' Pazzi, 250 Roma, Italy

J. Guern, Laboratoire de Physiologie Végétale et de Physiologie Végétale Appliquée, Université Paris VI, France

P. Guilizzoni, Istituto Italiano di Idrobiologia, Pallanza, Novara, Italy

M.A. Hall, Department of Botany and Microbiology, School of Biological Sciences, The University College of Wales, Penglais, Aberystwyth, Wales, U.K.

H.P. Haschke, Fachbereich Biologie (10) der Technischen Hochschule Darmstadt, Botanik, Darmstadt, Germany

W.D. Jeschke, Institut für Botanik, Universität Würzburg, Germany

S. Kuraishi, Department of Biology, University of Tokyo, Japan

P. Lado, Institute of Botany, University of Calabria, Cosenza, Italy

J. Levitt, Department of Horticultural Science, University of Minnesota, St. Paul, Minnesota, U.S.A.

T. Lomax, Department of Botany, University of Washington, Seattle, Washington, U.S.A.

G. Lucchini, C.N.R., Institute of Plant Sciences, University of Milan, Italy

E. Marrè, Institute of Plant Sciences, University of Milan, Italy

A. Melandri, Institute of Botany, University of Bologna, Italy

A. Novacky, Department of Plant Pathology, University of Missouri, Columbia, Missouri, U.S.A.

L.G. Paleg, Department of Plant Physiology, Waite Agricultural Research Institute, University of Adelaide, Glen Osmond, S.A. 5064, Australia

D. Penny, Massay University, Department of Botany and Zoology, Palmerston North, New Zealand

P. Pesci, Institute of Plant Sciences, University of Milan, Italy

M.G. Pitman, School of Biological Sciences, The University of Sydney, Australia

V.V. Polevoy, Department of Plant Physiology, Leningrad State University, Leningrad 199164, U.S.S.R.

M.C. Pugliarello, C.N.R., Institute of Plant Sciences, University of Milan, Italy

P. Pupillo, Institute of Botany, University of Bologna, Italy

M. Radice, Ricerca e Sviluppo, Montedison S.p.a., Milan, Italy

K. Raschke, MSU/ERDA Plant Research Laboratory, Michigan State University, East Lansing, Michigan, U.S.A.

F. Rasi-Caldogno, C.N.R., Institute of Plant Sciences, University of Milan, Italy

J.A. Raven, Department of Biological Sciences, University of Dundee, Scotland, U.K.

D. Rayle, Department of Botany, San Diego State University, San Diego, California, U.S.A.

F. Rollo, Institute of Microbiology and Plant Physiology, University of Pavia, Italy

B. Rubinstein, Department of Botany, University of Massachusetts, Amherst, Massachusetts, U.S.A.

A. Scacchi, Institute of Plant Sciences, University of Milan, Italy

E.J. Stadelmann, Department of Horticultural Science and Landscape Architecture, University of Minnesota, St. Paul, Minnesota, U.S.A.

W. Tanner, Fachbereich Biologie, Universität Regensburg, Germany

M. Thellier, L.A.T.P.E.I.C., Laboratoire associé au C.N.R.S., Faculté des Sciences, I.S.H.N., 76130 Mont-Saint-Aignan, France

L. Tonolli, Istituto Italiano di Idrobiologia, Pallanza, Novara, Italy

P.F. Wareing, Department of Botany and Microbiology, School of Biological Sciences, The University College of Wales, Penglais, Aberystwyth, Wales, U.K.

M.H. Weisenseel, Botanisches Institut der Universität Erlangen-Nürnberg, Erlangen, Germany

E.J. Williams, Department of Physics, Edinburgh, Scotland, U.K.

R.G. Wyn Jones, Department of Biochemistry and Soil Science, University College of North Wales, Bangor, Wales, U.K.

U. Zimmermann, Institut für Physikalische Chemie, Kernforschungsanlage, Jülich, Germany

CONTENTS

EFFECTS OF AUXIN AND FUSICOCCIN ON ION TRANSPORT

PHYSIOLOGICAL ASPECTS OF THE REGULATION OF MEMBRANE ACTIVITIES

GENERAL ASPECTS OF SOLUTE
TRANSPORT IN PLANTS

Regulation of Cell Membrane Activities in Plants
E. Marrè and O. Ciferri eds.
© *1977, Elsevier/North-Holland Biomedical Press, Amsterdam*

PASSIVE TRANSPORT PARAMETERS OF PLANT CELL MEMBRANES[*]

Eduard J. Stadelmann
Laboratory of Protoplasmatology
Department of Horticultural Science and Landscape Architecture
University of Minnesota
St. Paul, MN 55108 USA

I. INTRODUCTION

Uptake of materials, release of metabolites and the ability of the cell to retain or exclude specific substances was recognized early as central problem in cell physiology. The complex question of the material exchange of the cell with its environment was occasionally referred to as the permeability problem[1]. Naegeli[2] mentioned 1855 the existance of cytomembranes and Pfeffer[3] attribute to such membranes the control of the material exchange. He visualized the possibility that these membranes may be of bimolecular thickness[4]. An important step for conceptual clarification was made by Overton[3] with the distinction between active (adenoid) and passive transport. From his work came the first inference that the transport characteristics of the cell membrane suggest that this membrane is of lipoid nature. Later work gave a more detailed insight into the role of cell membranes in cell transport processes which are now conveniently classified into three groups[5,6,7,8]:

1) Non-mediated transport (passive transport; e.g., diffusion or osmosis)

2) Mediated transport (e.g., active transport)

3) Transport by membrane flow and membrane transformation (e.g., pinocytosis)

Passive transport processes take place along an electro-chemical potential gradient. When such transport proceeds through a membrane this membrane must exhibit permeability, i.e., the capacity to allow substances pass through it.

Passive membrane permeability is determined by membrane structure and composition. Determination of membrane permeability by absolute measurements or in a comparative manner for test substances with specific characteristics is a powerful tool to probe structure and composition of cytomembranes under conditions close to normal. Such an approach proved most instructive in earlier work, that led to the recognition of the lipid nature of the cell membrane (at least of the portion responsible for the passive permeation) and the enhanced permeability of these membranes for small molecular non-electrolytes[9].

---

*Part of this text was prepared during tenure of a Humboldt Award at the University of Hohenheim (Stuttgart, Germany). Paper No. 9623, Scientific Journal Series, Minnesota Agricultural Experiment Station, University of Minnesota, St. Paul, MN 55108, USA.

The discovery of the great variability of the permeability of plant cell cytoplasmic membranes[10] (i.e., plasmalemma and tontoplast in series) with cell type and cell history impressively underlines the differences and alterations which are possible in membrane structure and membrane composition[11,12].

Membrane permeability can not be measured directly. Rather a relationship must be postulated between the permeability measure (permeability constant, also called permeability coefficient) and such measurable quantities as, for instance, time, cell volume, cell surface, and driving force.

Passive permeation can be considered as a flow process with the gradient of the electrochemical potential as the driving force. For non-electrolytes as permeators the driving force can be expressed as the concentration difference. In a typical permeability experiment with plant cells an activity (concentration) gradient for the permeator is established between external solution and vacuole; the permeability measured is the one resulting from the permeation resistance of plasmalemma and tonoplast in series (with negligible permeation resistance of the mesoplasm[13].

A formula similar to Fick's First Law was introduced by Runstroem[14] to relate the permeability measure with the magnitude of measurable quantities in a permeability experiment and can be written as:

$$\frac{dm}{dt} \cdot \frac{1}{A} = F = K . \Delta C \tag{1}$$

where dm is the amount in moles of the permeator passing through the membrane during the time differential dt, in sec, A area of the membrane in cm^2, F flux in $cm^{-2} . sec^{-1} . mol$, ΔC concentration difference between the two sides of the membrane in $cm^{-3} . mol$, and K is the measure for permeability (permeability constant) for the membrane in $cm . sec^{-1}$.

A variety of experimental designs was developed for specific cell types to measure permeability[15] (See Chapter VI). In many of them ΔC as well as A varies with time and formula (1) has to be modified to substitute for one dependent variable thus allowing integration[16].

Passive transport through cytomembranes takes place in the liquid phase. Since water is the common solvent in living organisms, solid or gaseous substances must be present as aqueous solutions. Therefore, solute and solvent (water) permeability must be distinguished although the mode of interaction between membrane and solute as well as membrane and solvent molecules may be similar.

In the following chapters some details on the permeability constants, on the theoretical basis for their derivation, and on the underlaying membrane structure should be discussed. Finally conclusions are derived on the necessary parameters for measuring membrane permeability and a few experimental methods are indicated which are frequently used for permeability determinations of plant cells.

II. WATER PERMEABILITY CONSTANT

Water permeability (i.e., permeability for water) of cytomembranes was first measured by producing transport of water through the membrane by osmotic pressure differences[17]. Correspondingly, formula (1) was modified:

$$\frac{dm}{dt} \cdot \frac{1}{A} = K_{wo} \cdot \Delta P \tag{2}$$

where dm, dt, and A, as in (1), ΔP is the osmotic pressure difference in atm, K_{wo} is the osmotic water permeability constant in $cm.sec^{-1}.atm^{-1}$.

Since hydrostatic pressure acts for water transport like osmotic pressure[18], equation (2) can also be applied to experiments where hydrostatic pressure is the driving force[19].

To allow comparison between solute and water permeability constants, K_{wo} must be expressed in the unit $cm.sec^{-1}$. This can be accomplished by multiplying the numerical value of K_{wo} by 1.358×10^3 atm (for $25°C$). Some conversion factors for K_{wo} expressed in other units are given in the appendix (Table 2[20]).

Soon after its discovery, deuterium[21] was introduced by Hevesey et al. and Lucké and Harvey[22] for water permeability studies in animal cells and later by Wartiovaara[23] in plant cells. These "diffusion permeability experiments" allowed direct application of equation (1) thus enabling the calculation of a "diffusional water permeability constant", K_{wd}.

When K_{wo} and K_{wd} are determined for the same cell material in most cases the latter was found to be smaller than the former[24]. This discrepancy gave rise to much discussion and to the assumption that aqueous pores are present in the cell membranes[25] (cf. Chapter V). A re-evaluation, however, of the experimental conditions for the diffusion experiment[26] lead to the conclusion that in diffusion experiments unstirred layers develop. These layers (already recognized by Noyes and Whitney 1897[27]) appear on both sides of the membrane and result in an additional resistance which becomes significant for the exchange of fast permeators (such as water) through the membrane. Several approaches are now available for estimation[28] or calculation[29] of the contribution of the unstirred layer to the total permeation resistance measured in the diffusion experiments. When this contribution is taken into account the true diffusional water permeability constant of the membrane can be calculated. Its value comes close to or is almost the same as that of K_{wo}[30]. Such agreement of K_{wd} and K_{wo} after consideration of the unstirred layer has also been proven for artificial phospholipid membranes[31]. The water permeability of the plant cell, therefore, can be expressed by a single water permeability constant K_w, which may be determined by a diffusion method or by osmotic or hydrostatic pressure methods. From an experimental point of view the latter two methods seem to be more reliable, because disturbing factors ("sweeping away effect"[32]) are less significant than in diffusion experiments[33].

In tissues composed of normal sized plant cells the effect of unstirred layers on the value of the water permeability determined by the diffusion methods will be significant, but it still allows the appearance of considerable differences in the water permeability, e.g., between dead and living carrot discs[34]. For well stirred cell monolayers (onion scale epidermis) the unstirred layer may contribute to the total permeation resistance for water in the same order of magnitude as the water permeability of cell membrane and tonoplast additively (unpublished).

Values for K_w, determined for a great variety of different plant cells[35] (i.e., plasmalemma and tonoplast in series) are mostly between 1 to 15 x 10^{-4} cm.sec^{-1}. Artificial phospholipid membranes have comparable values[36]. Characean cells exhibit a water permeability upto about 100 times higher[37].

III. SOLUTE PERMEABILITY CONSTANT

Equation (1) can be used directly to determine solute permeability when the concentration gradient of the permeator is the driving force, i.e., when the permeator is a non-electrolyte[38].

This formula for the permeation process was proven to be valid by Szuecs[39]. Collander and Baerlund[38] tested the concept that the protoplasm layer, as membrane, is the transport controlling structure.

Solute permeability values (K_s) vary greatly for the different permeators depending upon their lipid solubility and upon the cell species[40] and cell condition. Small molecular substances (molecular weight below 50[41]) permeate generally faster than their corresponding lipid solubilities would suggest ("sieve effect"), an enhancement which led to the lipoid-filter hypothesis[42] and may be explained by considering the molecular mass of the permeator[43,44] or kinks in the hydrocarbon chain[45,46]. The former relationship may reflect involvement of the molecular moment of inertia of the solute[43].

IV. REFLECTION COEFFICIENT

Already in 1911, in his discussion on leaky osmometers Antropoff[47] proposed a "Durchlaessigkeitskoeffizient" $\mathcal{S}$ as measure for the semipermeability of the membrane, which he derives from the relation $P_m = P. \mathcal{S}$ (P_m osmotic pressure for the dynamic equilibrium, P theoretical osmotic pressure for an ideal semipermeable membrane). From a similar concept Lepeschkin[48] postulated a proportionality factor μ for the membrane permeability for a solute ($1 - \mu = \mathcal{S}$) and a factor σ for the solvent[49]. None of these measures received wide usage.

The coefficient $\mathcal{S}$ gained attention when Staverman[50] derived the same quantity from thermodynamics and introduced it as the reflection coefficient σ.

The magnitude of σ depends on the water permeability and the solute permeability of the membrane. The maximum height attained in the vertical glass tube of an osmometer cell with a leaky membrane and filled with a permeator is only a transient one, a factor which is often not clearly stated in papers discussing σ (cf.[50] p. 350: "a state with constant height of liquid"). The final equilibrium in such an osmometer is reached when the concentrations of the permeator on both sides of the membrane are equalized and hydrostatic pressure differences disappear. At the time of the transient maximum height of the solution in the vertical glass tube of the osmometer cell, the concentration $(1 - \sigma).C$ of the permeator has leaked out from the osmometer. Thus, the concentration difference generating the transient maximum hydrostatic pressure in a leaky osmometer is $\sigma.C$.

When solute (non-electrolyte) and solvent (water) permeation do not interact with each other, equation (1) and (2) correctly describe the solute and water permeation through a permeable membrane of an osmometer and σ can be calculated from K_w and K_s.

In such a leaky osmometer the resulting total volume flux $\frac{dV}{dt.A}$ during the water intake phase (rising of the meniscus in the vertical glass tube) can be construed as composed of two components. One component results from the initial osmotic potential of the solution in the osmometer. The other component is brought about by the leakage of the solute into the external solution leading to a water movement in the opposite direction.

The first component can be expressed as $K_w.(P_o - P_h).\frac{\bar{V}_w}{R.T}$ and the second one as $K_s.(C_o - C_a).\bar{V}_s$, where P_o is the osmotic pressure of the solution as determined e.g., by the freezing point depression, in atm; P_o is related to the concentration inside the osmometer at the beginning of the experiment by the expression $P_o = R.T.C_o$, K_w and K_s, and $\bar{V}_w$ and $\bar{V}_s$ are the permeability constants and the partial molar volumes for water and solutes in cm.sec^{-1} and cm^3.mol^{-1}, respectively; P_h is the actual hydrostatic pressure resulting from the rise of the water column in the vertical tube of the osmometer at the time t in atm; and C_o and C_a are the concentrations of the solute inside and outside of the osmometer respectively. For the moment of maximum transient hydrostatic pressure in the vertical tube the total volume flux becomes zero, and therefore $P_h = \sigma.P_o$, corresponding to the definition of σ. When the amount of the solute permeated through the leaky membrane is continuously removed $C_a = 0$ and for this condition we can write:

$$\frac{dV}{dt.A} = J_v = K_w . (P_o - .P_o) . \frac{\bar{V}_w}{R.T} + K_s.C_o.\bar{V}_s = 0 \qquad (3)$$

and therefore[51]
$$\sigma = 1 - \frac{K_s.\bar{V}_s}{K_w.\bar{V}_w} \qquad (4)$$

When water and solute permeation processes interact inside the membrane, equations (1) and (2) are not applicable and σ must be derived as shown in the next chapter.

The reflection coefficient σ as calculated from formula (4) contains the same information about the solute permeability as K_s when K_w is known, as is the case in most experiments. Equation (4), however reveals a serious shortcoming of σ: for most of the non-electrolytes which are frequently used in permeability experiments the value of σ will be very close to 1. Therefore, within the limits of experimental observations it will not be possible to recognize the differences in the σ values for those permeators[52]. The insensitivity of σ has recently led workers to return to the use of K_s or relative permeability values for measuring solute permeation[46,53,54].

V. INTERACTION OF SOLUTE AND WATER TRANSPORT

The influence of water transport on a simultaneous transport of solute through a cytomembrane and vice versa is of fundamental importance for the determination of the correct value of the membrane permeability for water and for the solute, since such permeability determinations are based on the measurement of the permeation rate. When this permeation rate is altered by the simultaneous occurrence of other permeation processes these alterations must be considered in deriving the permeability parameters.

Two types of interaction between solute (e.g., non-electrolyte) and solvent (e.g., water) permeation should be distinguished:

1) Interaction caused by alteration of the driving force for one transport process (e.g., water permeation) by the other transport process (e.g., solute permeation).

2) Interaction of the permeation processes inside the membrane when solute and solvent use the same pathway.

The first type of interaction can be considered to be an indirect one only, since equal changes in driving force can be produced by completely different means (e.g., addition of a non-permeant solute to the opposite compartment, application of hydrostatic pressure). Thus, the effects of these types of interactions can be accounted for by appropriately considering the alterations of the driving force in the calculations of the permeability constants using formulas (1) and (2).

When the transport processes interfere with each other inside the membrane, a third parameter has to be introduced to measure their interaction. Equations relating the three parameters (for water permeation, solute permeation, and interaction of these permeation processes) can be derived from the concept of irreversible thermodynamics[55]:

$$J_v = L_p \cdot \Delta p + L_{pD} \cdot \Delta \pi \qquad (5)$$

$$J_D = L_{Dp}.\Delta p + L_D.\Delta \pi \tag{6}$$

where J_v is the total volume flow in $cm^3.cm^{-2}.sec^{-1}$, L_p is the hydraulic conductivity in $cm.sec^{-1}.atm^{-1}$, Δp hydrostatic pressure difference in atm, $L_{pD} = L_{Dp}$ are the cross coefficients[55] in $cm.sec^{-1}.atm^{-1}$, $\Delta \pi$ osmotic pressure difference due to the permeator in atm, and L_D coefficient for the exchange flow in $cm.sec^{-1}.atm^{-1}$. In the absence of an interaction between water and solute flow inside the membrane $L_{pD} = L_{Dp} = 0$ and equations (5) and (6) reflect the same relationships as in equations (1) and (2)[56].

The reflection coefficient σ here is derived from the equation (5) for the total volume flow $J_v = 0$ as:

$$L_p.\Delta p = -L_{pD}.\Delta \pi \tag{7}$$

For the transient state of maximum height the maximum value of Δp in a leaky osmometer is given by the relationship $\Delta p = \sigma.\Delta C.R.T$, where σ is again defined as the coefficient by which the osmotic pressure of a solution (as determined, e.g., by freezing point depression) must be multiplied to obtain the observed transient maximum hydrostatic pressure in a leaky osmometer.

This reflection coefficient (derived for the case of an interaction of the permeation of the solute and of the water inside the membrane) can be related to other membrane parameters by the equation[57]:

$$\sigma = 1 - \frac{\omega.\bar{V}_s}{L_p} - \frac{K.f_{sw}}{\mathscr{Y}.(f_{sw} + f_{sm})} \tag{8}$$

where ω is the proportionality factor between the solute flow and the concentration difference in $mol.cm^{-2}.sec^{-1}.atm^{-1}.mol$ when there is no volume flow; $\bar{V}_s$ is the partial molar volume of the solute in $cm^3.mol^{-1}$; f_{sw} and f_{sm} and the frictional coefficients solute/water and solute/membrane; K, distribution coefficient membrane/water; and $\mathscr{Y}$, volume fraction of water in the membrane.

When there is no interaction between solute and water transport $f_{sw} = 0$ and for the transient maximum height in the osmometer (volume flow zero)[58] $\omega = \frac{K_s}{R.T.}$. Now when $\frac{Kw.\bar{V}w}{R.T}$ is substituted for L_p in equation (8)[59] σ results as given in (4).

The introduction of equations (5) and (6) into cell permeability research[60] tacitly implies the presence of aqueous pores in the membrane. This is generally not very clearly stated as an assumption in the papers applying irreversible thermodynamics in biological transport. This lack of clarity may relate to the concept of the membrane as a "black box"[61] for the purpose of transport studies, whereby only the transport processes and their interactions are investigated, without the consideration of the underlying structures.

Re-evaluation of earlier findings and new data now available strongly suggest that there is no or only a negligible interaction between water and passive solute transport. This conclusion is based on experiments where water permeation was tested with simultaneous permeation of a rapid permeator of small molecular size in the opposite direction and no interference between those permeation processes was found[62]. Furthermore, a possible interaction between water and solute permeation has to take place in aqueous membrane pores. Recent results and concepts, however, indicate the absence of aqueous pores in cell membranes[30,31,36,46,53].Therefore the passive permeation of solute and water through a cytomembrane is now considered to take place by the dissolution of the individual permeator molecule into the membrane lipid region correspondingly to the distribution coefficient of water/lipid and by molecular diffusion through the membrane following a concentration gradient inside the membrane[56,63,64].

Since water and solvent permeation through cytomembranes do not interfere with each other, the equations (1) and (2) must be used to derive the water and solute permeability constants for the membrane. These formulas, of course, do not apply for permeation through porous membranes (as e.g., used for technical processes) for which equations (5) and (6) are valid.

To consider the various types of permeation processes a general scheme can be prepared to characterize every kind of passive membrane transport by its permeation parameters which will describe the membrane passability. The kind and number required of these parameters depends on the type of the permeation process and the mechanism (Table 1).

TABLE 1.   NUMBER OF PARAMETERS NEEDED TO DESCRIBE MEMBRANE PERMEABILITY

FOR A GIVEN PERMEATOR

Membrane quality transport mechanism	Number of parameters needed	Symbols	Example of such membrane
Semipermeability	1	K_w	copper ferrocyanide membrane (idealized)
Permeable for solute and solvent (solution-diffusion)	2	K_w, K_s	cell membranes
Permeable for solute and solvent (through pores)	3	L_p, σ, ω	dialysis membranes

K_w = water permeability constant $(cm.sec^{-1})$; K_s: solute permeability constant $(cm.sec^{-1})$;

$L_p = \dfrac{K_w . \bar{V}w}{R.T}$ coefficient of hydraulic conductivity $(cm.sec^{-1}.atm^{-1})$;

ω = $R.T.K_s$ $(J_v = 0)$ solute permeability coefficient at zero volume flow $(mol.cm^{-2}.sec^{-1}.atm^{-1})$; $\sigma = -\dfrac{L_{pD}}{L_p}$ reflection coefficient (no dimension).

VI. SOME METHODS FOR MEASURING PASSIVE PERMEABILITY OF PLANT CELLS

To determine passive permeability three quantities have to be measured: (1) permeator (non-electrolyte or water) flux, (2) surface area, and (3) driving force. The methods must be capable of obtaining these quantities for the individual cell. In the giant Characean cells, chemical analysis of the cell sap[65], observation of water flow through the cell (transcellular osmosis)[66], or measurement of the cell wall extension[67] can be used to calculate permeator flux. Surface area can be measured directly and the driving force must be obtained from the parameters of the experiment.

A method which most directly allows the determination of the permeator flux in normal sized cells of higher plants is the elution method. The tissue is first loaded with the labeled permeator and next transferred into an efflux chamber with a constant flow of water for elution[68]. The elution samples are collected at suitable time intervals and used to establish an efflux curve from which the permeability constant can be calculated. Tissue geometry is important since it may alter the efflux rates. Using cell monolayers (e.g., inner epidermis of Allium cepa bulb scales) and a sufficiently high flow rate of the elution liquid, the error due to additional diffusion resistance outside the protoplasm layer can be minimized. Other methods of measuring permeator flux are seldom used because they are applicable only to cells and permeators with specific qualities (e.g., stains, electrolytes)[69].

The most important and widely used methods for measuring permeability of tissues or individual cells of higher plants to water or solutes are indirect ones making use of the osmotic qualities of the cell. Changes in volume (of tissue, cell, or protoplast) with time are evaluated to indicate the permeator flux.

Depending upon the osmoticum applied in the external solution and the cell material, the protoplasm layer exhibits almost ideal semipermeability (e.g., for sugar as osmoticum, at least during the useful time of the experiment) or an appreciable permeability of a certain degree (e.g., for specific harmless non-electrolytes). In the first case an osmotic driving force will be brought about by such external solutions. This driving force will cause water movement out of the cell or protoplast. The rate of this water flux can be used to determine the water permeability of the protoplasm layer (i.e., essentially of the plasmalemma[70]).

When, on the other hand, the cell is transferred into a solution of a permeator, a diffusion gradient is established for the permeator in addition to the osmotic driving force. The diffusion of the permeator into the cell will modify the osmotic driving force. Therefore for a non-electrolyte which permeates relatively slowly compared with water, its diffusion into the cell will continue even after reaching an osmotic equilibrium, and lead to a water transport[71] to maintain the osmotic equilibrium.

The contact of the cell membranes with an artificial milieu of appreciable osmotic activity is often considered to cause alterations in their permeability properties. Even harmless solutes (e.g., mannitol for water permeability studies) may change permeability by the withdrawal of water from within the membrane by the osmotic action of the solution[72]. No such effect on water permeability[73], however, was found in artificial phospholipid bilayer membranes. The considerable alteration (3 to 10 fold increased)[74] of water permeability in _Valonia_ _utricularis_ and Characean cells under conditions of low turgor pressure could not be verified for _Allium_ _cepa_ inner epidermal cells, where water permeability at normal and zero turgor pressure remained the same[68].

The concentration strength of the external solution will determine whether the cells undergo changes in turgor only when hypotonic concentrations are used (turgescence methods), or whether the protoplast separates partially from the cell wall for a more or less extended time period (plasmolytic methods).

Turgescence methods are of limited usefulness in higher plant cells and generally only applicable for detection of permeability changes or differences (e.g., by determination of rates of bending or changes of length of a tissue strip or organ)[69]. This is because it is difficult to measure accurately small volume changes involved in turgor alterations. Furthermore, the relation between cell volume and cell wall pressure must be known to quantify the driving force, since wall pressure acts as a driving force or is a component of it. Only recently some of these data on wall pressure have become available[75].

Plasmolytic methods evaluate volume changes of a protoplast which is partially separated from the cell wall. These methods are the only ones available for general use with higher plant cells and are most widely applied. The exposure of free protoplasm surfaces was shown not to alter the permeability qualities[76]. In most cells the plasmalemma separates from the cell wall without difficulties. Local new formation of plasmalemma from the cytoplasm or resorption of plasmalemma into the cytoplasm generally must occur very rapidly considering the often observed rapid changes in shape and area of protoplast surfaces.

When a plant cell comes in contact with the hypertonic solution of a slow permeator first the cell relaxes (initial phase) and then plasmolyzes (contraction phase). The contraction of the protoplast is maximum when osmotic equilibrium is reached between the vacuole and external solution. During the subsequent deplasmolysis the permeator continues to enter into the vacuole. This can be monitored by the corresponding volume increase of the protoplast due to the simultaneous diffusion of water (dilatation phase[77]) into the vacuole.

In cylindrical cells the sufficiently plasmolyzed protoplast takes the shape of a cylinder with two hemispheric ends (plasmometric method[78]). In this case surface area, driving force and flux can be quantitatively expressed by the protoplast length as the only variable and equation (1) or (2) can be integrated yielding formulas to calculate K_w and K_s from two protoplast length measurements at different times[79].

Used with proper precautions and experience the plasmolytic methods, especially the plasmometric method, gives reliable values for water and solute permeability which agree well with others obtained by different methods on a similar material[80].

OUTLOOK

Investigations of the passive permeability of cytomembranes of plants as well as animals can be expected to contribute to our knowledge on structure and function of these cell membranes in the future as significantly as they have done in the past.

Recent emphasis on membrane proteins led to some underestimation of the function of membrane lipids. It is well established now, however, that the membrane proteins do not function properly in the absence of lipids. The vital importance of membrane lipids in cell transport is clearly shown by their involvement in the transport of three of the most important substances for cell life; water, O_2, and CO_2. All of these move passively through the lipid portion of the membrane[81]. Furthermore, the osmotic qualities of the cytomembranes, which are so essential for plant cells, are also derived from the lipid component of these membranes.

Along with better data on the lipid species composing the cell membranes, permeability measurements can be expected to contribute considerably to a clearer understanding of the permeation process[46]. This will also help to quantitatively determine the forces of molecular interaction between the permeator and the membrane. Parallel experiments with better controlled and less complicated artificial phospholipid membranes and with reconstituted membranes will help to interpret the data from cytomembranes.

Another field of great potentiality, as initiated by Hoefler[82], is comparative permeability research with plant cells. Establishing the permeability series for specific cell types of plants adapted to extreme environmental conditions (e.g., drought, frost, heat) will reveal the magnitude of the variability of cell permeability and lead to important conclusions concerning the function of the membrane lipid components.

Finally, permeability measurements with cytomembranes which have undergone alterations by external or internal factors will enhance our understanding of the permeation process and allow better interpretation of accompanying changes in membrane structure and/or composition. Besides its value for understanding the permeation process, this information may be of great practical interest in pathology, pharmacology, toxicology and for understanding of uptake of herbicides in plant cells.

The topics indicated above are only the more urgent ones expected to advance by future work on membrane permeability. Not mentioned are such aspects as the effect of the lipid component on the functioning of the active transport system and problems such as the mobility of phospholipid molecules in the membrane, and lipid-protein interactions within the membrane.

Inspite of all the impressive progress in our understanding of cell membranes, many major problems still remain unsolved today. They await the intense efforts of future workers whose noble aim of enhancing our knowledge of the fascinating variety of transport processes through cytomembranes will reaffirm the vital importance of membranes for the living cell.

APPENDIX: TABLE II. FACTORS TO CONVERT WATER PERMEABILITY CONSTANTS EXPRESSED IN VARIOUS UNITS TO CM.SEC^{-1}. Values given in the units indicated in (I) must be multipled by the factor in (II) to obtain the value in cm.sec^{-1}.

Column I	Column II
$cm.sec^{-1}.atm^{-1}$	1.358×10^3 atm*
$cm^4.sec^{-1}.mol^{-1}$	$5.556 \times 1\text{-}^{-2}$ $cm^{-3}.mol$*
$cm.lit.h^{-1}.mol^{-1}$	1.543×10^{-2} $lit^{-1}.h.sec^{-1}.mol$*
$\mu.sec^{-1}$	10^{-4} $cm.\mu^{-1}$
$\mu.min^{-1}$	1.667×10^{-2} $cm.\mu^{-1}.min.sec^{-1}$
$\mu.min^{-1}.atm^{-1}$	2.264×10^{-3} $cm.\mu^{-1}.min.sec^{-1}.atm$*
$cm^3.dyn^{-1}.sec^{-1}$	1.376×10^9 $cm^{-2}.dyn$*
$sec^{-1}.dyn^{-1}.mol$	2.477×10^{10} $dyn.cm.mol^{-1}$*
$m.sec^{-1}.Pa^{-1}$	1.376×10^{10} $cm.m^{-1}.Pa$*
$m.sec^{-1}.(N.m^{-2})^{-1}$	1.376×10^{10} $cm.m^{-3}.N^{-1}$*
$m.sec^{-1}.bar^{-1}$	1.376×10^5 $cm.m^{-1}.bar$*

*Calculated for 25 °C. R.T = 8.314×10^7 erg.°K^{-1}.mol^{-1} x 298 °K;
= 0.08205 lit.atm.°K^{-1}.mol^{-1} x 298 °K.

REFERENCES

1. Gellhorn, E. (1929) Das Permeabilitaetsproblem. Seine physiologische und allgemein pathologische Bedeutung. Berlin: Springer. 441 p.

2. Naegeli, C. (1855) In: Naegeli, C., and C. Crammer: Pflanzenphysiologische Untersuchungen (1), 1-20. Zuerich: F. Schulthess. 120 p., see p. 7.

3. Pfeffer, W. (1877) Osmotische Untersuchungen. Leipzig: W. Engelmann. 236 p., see p. 123. Overton, E. (1899) Vierteljahrschr. Naturf. Ges. Zuerich 44, 88-135, see p. 102.

4. Pfeffer. W. (1897) Pflanzenphysiologie. Vol. 1. Leipzig: Engelmann. 620 p., see p. 93.

5. Netter, H. (1961) 12th Coll. Ges. Physiol. Chem. in Mosbach, 15-53. Berlin: Springer. 270 p., see p. 29.

6. Sitte, P (1969) Ber. Deutsch. Bot. Ges. 82, 329-383, see p. 337.

7. Davis, B. D. (1961) In: Goodwin, T. W., and O. Lindberg (Eds.): Biological structure and function. p. 571-580. London: Academic Press. 665 p., see p. 573.

8. Stadelmann, Ed. (1970) What's New in Plant Physiol. 2 (7):1-5. Dept. of Botany, University of Florida, Gainsville, FL.

9. Collander, R., and Baerlund, H. (1933) Acta Bot. Fenn. 11, 1-114, see p. 89f.

10. Bennet, H. S. (1963) In: Seno, S., and Cowdry, E. V. (Eds.): Intercellular Membraneous Structure. p. 7-13. Okayama (Japan): Japan. Soc. Cell Biol. 588 p., see p. 7.

11. Hoefler, K. (1965) Protoplasma 60, 150-158.

12. Url, W. (1968) Abh. Deutsch. Akad. Wiss. Berlin 4a, 17-27.

13. Url, W. (1971) Protoplasma 72, 427-447.

14. Runnstroem, J. (1911) Arkiv Zool. 7, 1-17, see p. 14.

15. Stadelmann, Ed. (1956) Protoplasma 46 (Weber-Festschrift), 692-710.

16. Stadelmann, Ed. (1951) Sitzsgber. Oesterr. Akad. Wiss. Math.-naturwiss. Kl., Abt. I, 160, 761-787.

17. Lillie, R. S. (1916) Amer. J. Physiol. 40, 249-266, see p. 260f. Northrop, J. H. (1928) J. Gen. Physiol. 11, 43-56, see p. 45.

18. Dainty, J. (1963) Adv. Bot. Res. 1, 271-326, see p. 291f. Stein, W. D. (1967) The Movement of Molecules across Cell Membranes. New York: Academic Press. 369 p., see p. 45. Forster, R. E. (1971) Current Topics in Membrane Transport 2, 41-98, see p. 43.

19. Steudle E., and Zimmermann, U. (1971) Z. Naturf. 26b, 1302-1311.

20. Stadelmann, Ed. (1963) Protoplasma 57, 660-718, see p. 675.

21. Urey, H. C., Brickwedde, F. G., and Murphy, G. M. (1932) Physical. Rev. 39, 164-165.

22. v.Hevesey, G., Hofer, E., and Krogh, A. (1935) Skand. Arch. Physiol. 72, 199-214. Lucke, B., and E. N. Harvey (1935) J. Cell. Comp. Physiol. 5, 473-482.

23. Wartiovaara, V. (1944) Acta Bot. Fenn. 34, 1-22.

24. Stein, W. D. (1967) The Movement of Molecules across Cell Membranes. New York: Academic Press. 369 p., see p. 110.

25. Stein, W. D., and Danielli, J. F. (1956) Disc. Faraday Soc. 21, 238-251. Paganelli: C. V., and Solomon, A. K. (1957) J. Gen. Physiol 41, 259-277.

26. Dainty, J. (1963) Adv. Bot. Res. 1, 279-326, see p. 300f.

27. Noyes, A. A., and Whitney, W. R. (1897) Z. Physikal. Chem. 23, 689-692, see p. 690. Osterhout, W. J. V. (1936) Botan. Rev. 2, 283-315, see p. 297.

28. Wartiovaara, V. (1944) Acta Bot. Fenn. 34, 1-22, see p. 10f. Collander, R. (1954) Physiol. Plant. 7, 420-455, see p. 423. Gutknecht, J. (1968) Biochim. Biophys. Acta 163, 20-29, see p. 24.

29. Andreoli, T. E., and Troutman, S. L. (1971) J. Gen. Physiol. 57, 464-478, see p. 468f.

30. Gutknecht, J. (1968) Biochim. Biophys. Acta 163, 20-29, see p. 24.

31. Cass, A., and Finkelstein, A. (1967) J. Gen. Physiol. 50, 1765-1784, see p. 1778f. Andreoli, T. E., and Troutman, S. L. (1971) J. Gen. Physiol. 57, 464-478, see p. 472.

32. Dainty, J., and Ginzburg, B. Z. (1964) Biochim. Biophys. Acta 79, 102-111, see p. 103.

33. Everitt, C. T., and Haydon, D. A. (1969) J. Theoret. Biol. 22, 9-19, see p. 16, Table 1.

34. Glinka, Z., and Reinhold, L. (1972) Plant. Physiol. 49, 602-606, see p. 603.

35. Stadelmann, Ed. (1963) Protoplasma 57, 660-718, see p. 698f.

36. Oschman, J. L., Wall, B. J., and Gupta, B. L. (1974) Sym. Soc. Exp. Biol. 28, 305-350, see p. 310.

37. Dainty, J., and Ginzburg, B. Z. (1964) Biochim. Biophys. Acta 79, 129-137, see p. 133. Steudle, E., and Zimmermann, U. (1974) Biochim. Biophys. Acta 332, 399-412, see p. 409.

38. Collander, R., and Baerlund, H. (1933) Acta Bot. Fenn. 11, 1-114, see p. 25f.

39. Szuecs, J. (1910) Sitzgsber. Kaiserl. Akad. Wiss. Mathem.-naturwiss. Kl. Abt. I, 119, 737-773.

40. Url, W. (1968) Abh. Deutsch. Akad. Wiss. Berlin 4a, 17-27, see p. 25.

41. Collander, R. (1959) In: Steward, F. C. (Ed.) : Plant Physiology, 2, 3-102. New York: Academic Press. 758 p., see p. 81.

42. Poijarvi, L. A. P. (1928) Acta Bot. Fenn. 4, 1-102, see p. 94.

43. Collander, R. (1959) In: Steward, F. C. (Ed.) : Plant Physiology, 2. New York: Academic Press. 758 p., see p. 83.

44. Stein, W. D. (1962) The Movement of Molecules across Cell Membranes. New York: Academic Press. 369 p., see p. 72f. Stein, W. D. (1969) Nature 224, 240-243. Diamond. J. M. and Wright, E. M. (1969) Proc. Roy. Soc. B 172, 273-316.

45. Traeuble, H. (1971) J. Membrane Biol. 4, 193-208.

46. Wright, E. M., and Bindslev, N. (1976) J. Membrane Biol. (In Press).

47. Antropff, A. V. (1911) Z. Physikal. Chem. 76, 721-731. .

48. Lepeschkin, W. (1908) Ber. Deutsch. Bot. Ges. 26a, 198-214, see p. 204.

49. Lepeschkin, W. W. (1932) J. Phys. Chem. 36, 2625-2638, see p. 2631.

50. Staverman, A. J. (1951) Rec. Trav. Chim. Pays-Bas 70, 344-352.

51. See also the following paragraphs and equation (8).

52. Dainty, J., and Ginzburg. B. Z. (1964) Biochim. Biophys. Acta 79, 129-137, see p. 133.

53. Bindslev, N., and Wright, E. M. (1976) J. Membrane Biol. (In Press).

54. Smulders, A. P., and Wright, E. M. (1971) J. Membrane Biol. 5, 297-318, see p. 315f. Cohen, B. E. (1974) J. Membrane Biol. 20, 205-234. J. Membrane Biol. 20, 235-268. Poznansky, M., Tong, S., White, P. C., Milgram, J. M., and Solomon, A. K. (1976) J. Gen. Physiol. 67, 45-66.

55. Katchalsky. A. (1960) In: Keinzeller, A., and Kotyk A. (Eds.) 1961. Membrane Transport and Metabolism. (Proceed. Symp. Prague, 1960). London: Academic Press 608 p., see p. 745. Dainty, (1963) Adv. Bot. Res. 1, 279-326, see p. 285f. House, C. R. (1974) Water Transport in Cells and Tissues. London: E. Arnold Ltd. 562 p., see p. 36f.

56. Laeuger, P. (1969) Angew. Chemie. Internat. Ed. 8, 42-54. see p. 46f.

57. Kedem, O., and Katchalsky, A. (1961) J. Gen. Physiol. 45, 143-179, see p . 156.

58. Dainty, J. (1963) Adv. Bot. Res. 1, 279-326, see p. 291.

59. Stadelmann, Ed. (1963) Protoplasma 57, 660-718, see p. 675.

60. Kedem, O., and Katchalsky, A. (1958) Biochim. Biophys. Acta. 27, 229-246, see p. 245.

61. Paterson, R. (1970) In: Bittar, E. E. (Ed.). Membranes and Ion Transport. 1. London: Wiley-Interscience. 483 p., see p. 133.

62. Klocke, R. A., Andersson, K. K., Rotman, H. H., and Forster, R. E. (1972) Amer. J. Physiol. 222, 1004-1013, see p. 1009. Forster, R. E. (1971) Curr. Topics Membrane Transport 2, 41-98, see p. 87.

63. Collander. R., and Baerlund, H. (1933) Acta Bot. Fenn. 11, 1-114, see p. 81.

64. Hanai, T., and Haydon, D. A. (1966) J. Theoret. Biol. 11, 370-382, see p. 380f.

65. Collander R., and Baerlund, H. (1933) Acta Bot. Fenn. 11, 1-114, see p. 7f.

66. Osterhout, W. J. V. (1948) J. Gen. Physiol. 32, 553-557, see p. 554. Kamiya N., and Tazawa, M. (1956) Protoplasma 46, 394-422, see p. 396f.

67. Steudle, E., and Zimmermann, U. (1974) Biochim. Biophy. Acta 332, 399-412, see p. 402.

18

68. Palta, J. P., and Stadelmann, E. J. (1976) Plant Physiol. 57, (5). Ann. Meeting Suppl. p. 79.

69. Stadelmann, E. J. (1956) Protoplasma 46, p. 692-710, see p. 693f. Wieringa, K. T. (1930) Protoplasma 8, 522-584.

70. Url, W. G. (1971) Protoplasma 72, 427-447, see p. 439. Huber, B. and Hoefler, K. (1930) Jahrb. Wiss. Bot. 73, 351-511, see p. 448.

71. Diamond, J. D. (1962) J. Physiol. 161, 503-527, see co-diffusion, p. 510.

72. Tazawa, M., and Kamiya, N. (1966) Austral. J. Biol. Sci. 19, 399-419, see p. 410. Dainty, J., and Ginzburg, B. Z. (1964) Biochim. Biophys. Acta 79, 79, 102-111, see p. 107f. Kiyosawa, K., and Tazawa, M. (1972) Protoplasma 74, p. 257-270.

73. Hanai, T., and Haydon, D. A. (1966) J. Theoret. Biol. 11, 370-382, see p. 376, Table 1.

74. Zimmermann U. and Steudle, E. (1974) J. Membrane Biol. 16, 331-352, see p. 340f. Zimmermann, U. and Steudle, E. (1975) Austral. J. Plant Physiol. 2, 1-13.

75. Dainty, J. (1976) In: Luettge, U., and Pitman, M. G. (Eds.): Encyclopedia of Plant Physiology, New Series. Vol. II, Transport in Plants, part A, p. 12-35. Berlin: Springer-Verlag. 400 p.

76. Huber, B., and Hoefler K. (1930) Jahrb. Wiss. Bot. 73, 351-511, see p. 375f., 447. Schmidt, H. (1936) Jahrb. Wiss. Bot. 83, 470-512.

77. Stadelmann, Ed. (1951) Sitzsber. Oesterr. Akad. Wiss., Mathem.-naturw. Kl., Abt. I, 160, 761-787, see p. 770f.

78. Hoefler, K. (1917) Ber. Deutsch. Bot. Ges. 35, 706-726.

79. Stadelmann, Ed. (1951) Sitzsber. Oesterr. Akad. Wiss. Mathem.-naturw. Kl., Abt. I, 160, 761-787, see p. 779. Stadelmann, Ed. (1963) Protoplasm 57, 660-718, see p. 682. Stadelmann, Ed. (1966) In: Prescott, D. M. (Ed.): Methods in Cell Physiology. 2. New York: Akademic Press. 426 p., see p. 189f.

80. Collander, R., and Baerlund, H. (1933) Acta Bot. Fenn. 11, 1-114, see p. 62. Vreeman, H. J. (1966) Proc. Acad. Wetensch. B, 69, 564-577, see p. 576. Poznansky, M., Tong, S., White, P. C. Milgram, J. M., and Solomon, A. K. (1976) J. Gen. Physiol. 67, 45-66, see p. 50.

81. Noggle, G. R., and Fritz, G. J. (1976) Introductory Plant Physiology. Englewood Cliffs, NJ. Prentice-Hall Inc. 688 p., see p. 338.

82. Hoefler, K. (1942) Ber. Deutsch. Bot. Ges. 60, 179-200.

Regulation of Cell Membrane Activities in Plants
E. Marrè and O. Ciferri eds.
© *1977, Elsevier/North-Holland Biomedical Press, Amsterdam*

PROTON TRANSLOCATION IN FACULTATIVE PHOTOSYNTHETIC BACTERIA

A. Baccarini Melandri and B.A. Melandri
Institute of Botany, University of Bologna, Bologna
Italy

Increasing evidence is accumulating that ion gradients across biological membranes represent a major energy storage, which can be used
by cells to perform several endoergonic processes; in particular protonic gradients are believed to represent a primary driving force for
transport of ions or metabolites in bacteria[1,2], organelles from Eu-
kariotic cells[3] and possibly unicellular green algae[4].

According to Mitchell's original proposal[5], which is supported by
extensive experimental evidence, transmembrane proton gradients can
be generated by several independent primary active transport systems
present in the membrane and coupled to ATP hydrolysis or to redox
reactions of respiration, photosynthesis or chemiosynthesis. The proposal of the presence on the same membrane of multiple reversible
proton translocators, driven by ATP or redox reactions, leads also
to the essential concept of Mitchell's chemiosmotic coupling hypothesis[6] of membrane associated phosphorylation, namely the obligatory
and reversible coupling of redox reactions and ATP synthesis only
through the transmembrane gradient of protons.

Among the different prokariotic systems, some facultative photosynthetic bacteria of the Rhodospirillaceae family offer special ad-
vantages as experimental material, due to their unique metabolic flexibility. In fact they can grow indifferently autotrophically in the
light or heterotrophically in the dark and possess therefore on the
same membrane proton translocators driven by photosynthesis or respiration as well as by ATP hydrolysis. In this system the multiplicity
and independence of the proton pumps present with parallell polarity
on the membrane can be readily demonstrated, utilizing membrane fragments prepared from cells grown under different conditions.

In Fig.1 proton translocation coupled to respiration in aerobic
membranes from Rhodopseudomonas capsulata is demonstrated by the use
of a fluorescent probe for transmembrane pH differences. The experiments show that a proton gradient can be formed in response to NADH
oxidation either by oxygen or, in presence of cyanide, by an exogenous electron acceptor(Q_1), indicating in the former case the operativity in proton translocation of the whole respiratory chain or in

the latter of only the first site of energy conservation associated with NADH dehydrogenase. By means of similar experiments the formation of proton gradients by respiration of succinate or by oxidation of artificial electron donors to cytochrome oxidase (the terminal energy conserving site of the respiratory chain) can also be demonstrated.

Membrane fragments from photosynthetically grown cells, on the other hand, have been extensively utilized for studies on light induced proton translocation: in these preparations a very extensive proton concentration difference can be established, following illumination of the membranes, as shown in Fig.2. In the same preparation moreover ATP hydrolysis in the dark also promotes the formation of proton gradients, although of lower extent.

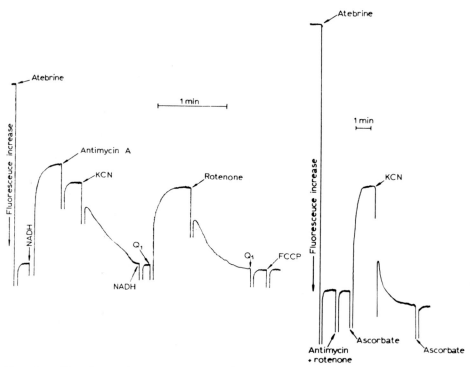

Fig. 1 - The formation, across respiring membranes from Rps.capsulata, of a protonic gradient, driven by the overall aerobic respiration or by partial reactions of the respiratory pathway and its response to specific electron transfer inhibitors. (Taken by Baccarini Melandri et al.[16]).

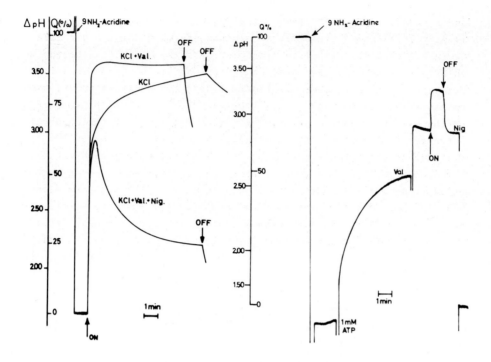

Fig. 2 - The formation of a proton gradient across photosynthetic
membranes from Rps. capsulata, induced by light dependent
electron transport or by ATP hydrolysis. Additions: KCl,
50 mM; valinomycin, 2 µg/ml; nigericin, 0.2 µg/ml.

The independency of these two mechanisms for proton transport can
be clearly demonstrated when the ATPase enzyme is detatched from the
membrane by EDTA washings[7,8]. In decoupled membrane preparations
only light can drive proton transport, being the formation of ATP induced proton gradients strictly associated with the operativity of
the membrane bound ATPase, which, after purification, can be rever-
sibly reassociated to the decoupled membranes[8].

In a closely similar bacterial system, R. rubrum, Isaev et al.[9]
have utilized permeant synthetic anions as indicators of the onset
of a membrane potential upon membrane energization. Their results
clearly show an electrophoretic accumulation of anions in the membrane compartment into which protons, driven by respiration, photosynthesis or ATP hydrolysis, accumulate, demonstrating the electrogenic nature of all these proton transport mechanisms.

The evaluation of the amount of energy stored across the membrane

under the form of an electrochemical potential difference of protons
is of particular interest for an experimental test of the bioenerge-
tic parameters of metabolite transport systems, as well as of the mechanism of membrane associated phosphorylation. In view of the electrogenic characteristics of primary proton transport, however, such
quantitative extimations require the measurements of both components
of the electrochemical proton gradient, namely of both the chemical
potential difference of protons (ΔpH) and of the electrostatic potential difference ($\Delta\psi$) between the inner and the outer compartment of
a vesicular system. Although such measurements in microscopic systems
present considerable experimental difficulties, attempts have been
made in this direction, by means of exogenous or endogenous spectro-
scopic probes.[10] In bacterial photosynthetic membrane fragments the
use of a fluorescence monoamine, 9-aminoacridine, as a quantitative
indicator of ΔpH, can be associated to an evaluation of $\Delta\psi$ following
the electrochromic response of endogenous pigments[11] or of an exogenous dye[12].

For these measurements we have used the spectral red shift of the
endogenous carotenoids: this spectral signal can be calibrated
against transmembrane potentials, artificially induced in the dark

TABLE 1

THE EXTENT OF ELECTROCHEMICAL POTENTIAL DIFFERENCE OF PROTONS INDUCED
BY MEMBRANE ENERGIZATION IN CHROMATOPHORES FROM RHODOPSEUDOMONAS CAP-
SULATA

Energy Source	Additions	$\Delta\psi$ (mV)	$-\dfrac{RT}{F}\,\Delta$pH (mV)	$\Delta\bar{\mu}_{H^+}$ (mV)
light	none	206	198	404
	2 μg/ml valinomycin	71	214	285
	2 μg/ml nigericin	272	$\leqslant$ 36	$\leqslant$ 308
	6 μM FCCP	103	106	209
ATP	none	12	153	165
	2 μg/ml valinomycin	0	178	178
	2 μg/ml nigericin	30	$\leqslant$ 36	$\leqslant$ 66

Chromatophores from Rps.capsulata, grown photoheterotrophically in
anaerobiosis were used. The assays contained: glycylglycine buffer,
pH 8.5, 40 mM; KCl, 50 mM; Na-succinate, 0.02 mM; MgCl$_2$, 5 mM; chromatophores corresponding to about 20 μg/ml of Bacteriochlorophyll.
ΔpH and $\Delta\psi$ were measured using 9-aminoacridine (4 μM) and carotenoid
electrochromic shift respectively.

by K^+ pulses in the presence of valinomycin[11]. If the assumption is made that the light induced signals can be quantitized by this procedure, an accurate and continuous monitoring of $\Delta\psi$ during membrane energization can be obtained[13].

Quantitative data obtained by these methods are presented in Table 1.

To the formation of a considerably large electrochemical potential difference induced by light ($\sim$ 400 mV in electrical units) $\Delta\psi$ and ΔpH contribute in approximately equal amounts.

Additions of specific ionophorous antibiotics can however drastically reduce the overall $_H^+$ by decreasing one of the two components: thus the electroneutral H^+/K^+ exchanger nigericin will eliminate the concentration component of $_H^+$, while the electrophoretic K^+ carrier valinomycin will affect specifically $\Delta\psi$. None of these two antibiotics alone will be able, however, to completely dissipate $\Delta\bar{\mu}_H^+$ also if added at high concentrations: complete uncoupling can however be obtained with the protonophorous uncoupler FCCP, which can influence both ΔpH and $\Delta\psi$. These results can be taken as a further proof of the electrogenic nature of the light induced proton translocation in bacterial photosynthesis membranes and are also consistent with independent observations on light induced ATP synthesis. This reaction, in fact, is resistant to nigericin or valinomycin added alone but is completely inhibited by FCCP or by a combination of these two ionophores[14].

The extent of $\Delta\bar{\mu}_H^+$ induced by ATP hydrolysis in the dark is also shown in Table 1; under this condition the predominant component of the proton gradient is represented by ΔpH and accordingly practically complete deenergization can be caused by addition of nigericin alone. The overall ATP induced $\Delta\bar{\mu}_H^+$ is therefore considerably lower than the light induced one.

This experimental approach offers the possibility of comparing the extent of $\Delta\bar{\mu}_H^+$ with the free energy change required for photophosphorylation under extreme conditions, i.e. at the maximal ATP/ADP.P_i ratio at which net ATP synthesis can be observed. In this type of experiments a maximal free energy change of about 15 kcal/mole has been obtained, which corresponds to about 330 mV, if a chemiosmotic coupling mechanism is assumed to operate and a stechiometry of $2H^+$/ATP is postulated. These results are therefore self consistent, since the experimental $\Delta\bar{\mu}_H^+$ is well about the energy required for ATP synthesis; this consistency generally holds also upon addition of low concentrations of different uncouplers[15].

REFERENCES

1. Harold, F.M.(1972) Bacteriological Reviews 36, 172.

2. Hamilton, W.A.(1975) Adv.Microb.Physiol. 12, 1.

3. Klingenberg, M.(1970) Essays in Biochemistry 6, 119.

4. Komor, E., Haass, D., Komor, B. and Tanner, W.(1973) Eur.Journ.
    Biochem. 39, 139.

5. Mitchell, P.(1970) in "Membrane and Ion Transport"(Bittar, E.E.
    eds.) vol.1, p.192, Wiley, New York.

6. Mitchell, P.(1968) Chemiosmotic Coupling and Energy Transduction
    Glynn Res. Ltd., Bodmin.

7. Melandri, B.A., Baccarini Melandri, A., Gest, H. and San Pietro,
    A.(1970) Proc. Natl. Acad. Sci., USA, 67, 477.

8. Melandri, B.A., Baccarini Melandri, A., Crofts, A.R. and Codgell,
    R.J.(1972) FEBS Lett. 24, 141.

9. Isaev, P.I., Liberman, E.A., Samuilov, V.D., Skulachev, V.P. and
    Tsofina, L.M. (1970) Biochim.Biophys.Acta 216, 22.

10. Rottenberg, H.(1975) J.Bioenerg. 7, 63.

11. Jackson, J.B. and Crofts, A.R. (1969) FEBS Lett. 4, 185.

12. Pick, U. and Avron, M.(1976) Biochim.Biophys.Acta 440, 189.

13. Casadio, R., Baccarini Melandri, A. and Melandri, B.A.(1974)Eur.
    J.Biochem. 47, 121.

14. Casadio, R., Baccarini Melandri, A.,Zannoni, D. and Melandri, B.A
    (1974), FEBS Lett. 49, 203.

15. Casadio, R., Baccarini Melandri, A. and Melandri, B.A. (1975) in
    "Electron Transfer Chains and Oxidative Phosphorylation" (Qua-
    gliarello, E., Papa, S., Palmieri, F., Slater, E.C. and Silipran-
    di, N. eds.) North Holland Pub. Co, Amsterdam, p.407.

16. Baccarini Melandri, A., Zannoni, D. and Melandri, B.A.(1973)
    Biochim.Biophys.Acta 314, 298.

Regulation of Cell Membrane Activities in Plants
E. Marrè and O. Ciferri eds.
© *1977, Elsevier/North-Holland Biomedical Press, Amsterdam*

CHARACTERISTICS, FUNCTIONS AND REGULATION
OF ACTIVE PROTON EXTRUSION

J.A. Raven
Department of Biological Sciences,
University of Dundee,
Dundee DD1 4HN,
Scotland

and

F.A. Smith
Department of Botany,
University of Adelaide,
Adelaide, S.A. 5001,
Australia

I. INTRODUCTION

The H^+ pump will be discussed here mainly in relation to the eukaryote plasmalemma. The main thesis which is developed pertains to the primary function of H^+ pumps in regulating pH in the cytoplasm.

II. CHARACTERISTICS

(1) Definition: Active H^+ transport will be defined as a net movement of H^+ across a membrane in the opposite direction to that predicted from the prevailing passive driving forces of H^+ concentration gradient and electrical potential difference; such a movement requires the input of energy from some other reaction[1]. This coupled reaction could be an exergenic biochemical reaction, or a downhill solute flux. It appears (Section II (3)) that H^+ pumps are powered by biochemical reactions, i.e. are examples of primary active transport; the biochemical reaction can be either a redox process (Section II (4)) or ATP use (Section II (5)). The characteristics described under (2) and (3) below are common to H^+ pumps with either of the energy sources.

(2) Topology and sidedness: Some membranes in which active H^+ transport is thought to occur in photosynthetic eukaryotic cells are indicated in Figure 1. When the H^+ pump is working in the 'forward' direction, i.e. converting chemical energy into an H^+ free energy gradient, the H^+ flux is from the side of the membrane at which adenine nucleotides and nicotinamide adenine dinucleotides interact with the pump[1]. Thus at the plasmalemma the active H^+ flux is from the cytoplasm to the medium. The direction of pumping across internal membranes can be rationalised in terms of

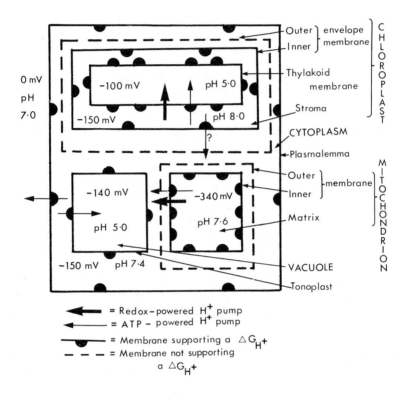

Fig. 1. pH values of compartments, electrical potential differences between compartments, and directions of H+ pumping, in a plant cell. All electrical potentials are referred to the external solution = 0 mV. Based on data from references 1 – 8.

the derivation of membranes such as the tonoplast, and the outermost membranes of chloroplasts and mitochondria, by invagination of the plasmamembrane. The inner membranes of chloroplasts and mitochondria would then have arisen either by further invagination of these membranes[9] or from endosymbionts[10].

(3) Electrogenic uniport: The H+ pump is widely held to be an example of uniport[1,5,11], i.e. H+ is the only transported species. Uniport of an ion involves net charge transfer and hence alters the electrical potential difference (E_m) across the membrane, i.e. is electrogenic. Such a pump is best recognised by a pump-dependent increment of E_m such that the resulting E_m cannot be explained as a

diffusion potential.   In the case of the eukaryote plasmalemma the most negative value which a diffusion potential can reach is E_K, the K^+ diffusion potential.   A value of E_m more negative than E_K is taken as evidence for an electrogenic pump either involving anion influx or cation efflux[3].   In glycophytes the hyperpolarising electrogenic pump does not appear to involve anion influx[3,12]. The involvement of H^+ efflux in the hyperpolarisation cannot be shown by removing cytoplasmic H^+;   the best evidence for implicating this active ion flux in electrogenesis comes from experiments in which there is a correlation between net H^+ efflux and hyper- polarisation of the membrane potential[3,11] (Section III (5) (a)). The involvement of either the $\triangle E$ or the $\triangle \left[H^+ \right]$ components of the H^+ electrochemical potential gradient in co-transport phenomena can destroy the correlation between net H^+ efflux and membrane hyperpolarisation[3] (Section III (5)).   Some electrical data suggest an electro-neutral extrusion (e.g. K^+-H^+ antiport) under some conditions[13-15].

    (4) Redox H^+ pumps: Active H^+ transport with membrane- associated redox reactions as the energy source occurs in the plasma membrane of many prokaryotes, and in the thylakoid membrane of chloroplasts and the crista membranes of mitochondria in eukaryotes (Figure 1).   These pumps can be associated with photo- chemical or thermochemical redox reactions.   Transfer of one reducing equivalent in the respiratory chain from $NADH_2$ to O_2 probably pumps three H^+ while in photosynthesis both the non-cyclic pathway from H_2O to NADP and the cyclic pathway probably pumps two H^+ [1,16].   The maximum transmembrane H^+ free energy gradient which these redox chains can generate is some 35 kJ mole^{-1}. Purified extracts from various biological membranes are functional as H^+ pumps when reincorporated into lipid bilayers[17].

    The evidence for the occurrence of redox-driven H^+ pumps in membranes not normally mediating the interconversion of redox energy and adenylate energy is equivocal.   While the chloroplast envelope[18] does not appear to contain any redox agents which could be involved in active H^+ transport, the plasmalemma [19,20], and the lysosomes[21] may well contain such redox components. Some in vivo evidence[22,23] is consistent with the occurrence of a redox-coupled H^+ efflux at the eukaryote plasmalemma.

    (5) ATP-driven H^+ pumps: Active H^+ transport with ATP as the energy source is probably more widespread than is redox-driven H^+

transport (Figure 1). The most direct evidence, including
extraction and re-incorporation studies, has been obtained for
membranes which can couple redox reactions to ATP synthesis; less
direct evidence has been obtained for the other membranes shown in
Figure 1 as having ATP-powered H^+ transport[1,3,17,18,24-27].

The most widely accepted value[1,11,28-30] for the H^+/ATP ratio
of the ATP-driven H^+ pump is 2. However, values of 3 have been
suggested[4,5,31] on various grounds for H^+-ATPases from membranes
which can interconvert redox and adenylate energy, while $1H^+$/ATP
seems likely[32,33] for at least some eukaryote plasma membrane
ATPases.

III. FUNCTION

(1) Regulation of cytoplasmic pH: It has been argued[2,3,34-35]
that the primary selective pressure which can lead to the widespread
occurrence of active H^+ transport is the requirement for regulation
of cytoplasmic pH in the face of metabolic production or consumption
of H^+, and of passive H^+ fluxes at the plasmalemma. Regulation of
pH (as opposed to the total electrochemical potential for H^+)
demands close interaction between H^+ transport and passive cation
(K^+) influx[33]. If this did not occur, the major effect[3,33] of
changes in H^+ active transport would be large changes in ΔE rather
than ΔpH. In a medium with a pH below about pH 9, the passive
driving forces on H^+ at the plasmalemma tend to move H^+ inwards[2,3],
i.e. cause cytoplasmic acidification. The other major sources of
cytoplasmic pH perturbation are essential biosyntheses required for
cell growth. Growth with CO_2 or carbohydrate as C source, and NH_4^+
as N source, leads to H^+ production in the cytoplasm in excess of
that required for maintenance of constant intracellular pH of more
than 1.0 H^+ per NH_4^+ assimilated. Conversely, when NO_3^- is the N-
source[35], up to 1.0 excess OH^- are generated in the cytoplasm per
N assimilated.

The observed long-term constancy[2,3] of cytoplasmic pH in the
range 6.5 - 8.0 cannot be explained in terms of pre-existing
cytoplasmic buffers, since these are themselves made by pH-
perturbing processes.

The two dynamic methods of countering these pH changes are a
biochemical production of H^+ (or OH^-), and active H^+ transport (and
regulated passive H^+ transport) at the plasmalemma[3]. The biochemical production of H^+ (OH^-) by secondary metabolism as a counter
to excess OH^- (H^+) production in primary metabolism has been

eloquently advocated by Davies[39,40]. Here excess OH^- in the cytoplasm, by raising the pH, increases the rate at which a strong acid (malic acid) is produced from such neutral and weak acid substrates as carbohydrate and CO_2; the major pH-sensitive reaction here is PEP carboxylation. NH_4^+ assimilation, or passive H^+ influx, by contrast, lower the cytoplasmic pH. This inhibits malic acid synthesis, and stimulates malic enzyme which converts $malate^{2-}$ into $pyruvate^-$, CO_2 and OH^-. In this way the biochemical pH stat can counter both alkalinisation and acidification of the cytoplasm. Excess OH^- can then be countered by this biochemical mechanism in the long term (i.e. more than one cell generation time), since the malic acid is produced from exogenous substrates. However, for a cell growing on CO_2 or carbohydrate, excess H^+ in the cytoplasm cannot be corrected over long periods by this biochemical pH stat. OH^- can only be generated from $malate^{2-}$, and this $malate^{2-}$ can only have arisen by malic acid synthesis from the neutral substrates, followed by extrusion of excess H^+ in exchange for inorganic cations, e.g. K^+ (see below). Thus over times greater than a cell generation time, disposal of the excess H^+ generated in NH_4^+ assimilation cannot be by the biochemical pH stat alone[3]. All of this excess H^+ generated in the cytoplasm during NH_4^+ assimilation is actively transported out of the cytoplasm into the medium[3,35]. Since it is likely that NH_4^+ was the earliest inorganic N-source for cell growth[36], pH regulation by active H^+ extrusion was probably an important metabolic requirement in primitive prokaryotic cells [36,41,42]. NO_3^- is the major N-source for plants to-day[35], so that the majority of growing plants have an overall excess OH^- production in their cells. In most algae this is disposed of entirely by excretion; in algae such as Acetabularia, and in both aquatic and terrestial angiosperms, some of the excess OH^- is neutralised by the biochemical pH stat[35]. The extent to which the biochemical pH stat is involved is rarely sufficient to prevent the NO_3^- assimilating plant from having a net OH^- efflux (H^+ influx). In those aquatic plants in which HCO_3^- assimilation occurs during photosynthesis all of the excess OH^- generated in HCO_3^- assimilation (almost 1 OH^- per HCO_3^-) is excreted[2,3,44-47]. Teleologically the absence of OH^- neutralisation by the biochemical pH stat is reasonable in that organic acid synthesis used at least 1 C per OH^- neutralised (i.e. HCO_3^- assimilated).

Despite the fact that the net OH^- efflux during HCO_3^- and NO_3^- assimilation is downhill (Section II (1)), 'biophysical' pH control is not solely by a controlled passive OH^- efflux since there is evidence[43,45-48] for active H^+ efflux at the same time as the net OH^- efflux.

It would appear that active H^+ extrusion is a common feature of plants, even when there is a net passive efflux of OH^-, and that this is related to the regulation of cytoplasmic pH. Unless the external pH is very high, an inwardly-directed electrochemical potential gradient for H^+ is maintained. The role of the H^+ pump during net OH^- efflux is considered in Section III (2).

(2) <u>pH regulation in multicellular land plants</u>: The discussion up till now has considered a cell in an extensive aquatic milieu (e.g. aquatic algae) which provides a large sink for excreted H^+ or OH^-. The shoot cells of land plants are isolated from such an extracellular sink other than the very small cell wall (apoplast) in the shoot[3,35,40]. There is no mechanism whereby excess H^+ produced by such cells can be transported to subterranean parts of the plant and then excreted to the soil solution. It is likely that excessive acidification of the small apoplast ("free space") volume[3] of the shoot is prevented by inhibition of net H^+ efflux at low extracellular pH values[50]. However, this prevents net H^+ disposal from the cytoplasm during NH_4^+ assimilation, and during K^+ organate$^-$ accumulation with <u>in situ</u> organic acid synthesis. Raven and Smith[3] point out the artificiality of experiments in which higher plant shoot tissues are exposed to large volumes of extracellular solution; this allows large <u>net</u> H^+ fluxes, and hence K^+ organate$^-$ accumulation involving <u>in situ</u> organic acid synthesis in the absence of intracellular sinks for H^+ (e.g. NO_3^- assimilation).

For an intact plant with NH_4^+ as N-source, it is likely that all the organic N in the shoot, as well as the K^+ salts of organic acids which are involved in turgor generation, have been transported there from the root, where their synthesis does not produce a large pH stress: the excess H^+ is excreted into the rooting medium[3]. Thus the nutrients supplied to the shoot are of such a composition as to not involve net H^+ production during shoot growth[3,35].

NO_3^- assimilation is not so spatially constrained, since the excess OH^- generated in NO_3^- assimilation can be neutralised by the biochemical stat as well as by excretion. When, as frequently occurs, there is a significant assimilation of NO_3^- in shoot tissue, organic acid synthesis via the 'biochemical pH stat' can occur in shoot cells as a means of neutralising the excess OH^- generated in

NO_3^- assimilation, and turgor can be generated by accumulation of K^+ (from the K^+ NO_3^- transported in from the roots) and organate$^-$ (generated in the biochemical pH stat)[3,35,51,52]. However, this can lead to osmotic problems during NO_3^- assimilation in the shoot; how the plant deals with these is discussed by Raven and Smith[35].

Despite the absence of large net H^+ fluxes out of shoot cells of higher plants, it is still possible that processes which depend on H^+ electrochemical potential gradients in the absence of large net H^+ effluxes could still occur; these processes include various kinds of co-transport (e.g. K^+ Cl^- accumulation), and cell wall extension induced by low pH. It would appear that further restrictions are placed on cells which have a net OH^- efflux during NO_3^- or HCO_3^- assimilation; here the H^+ electrochemical potential gradient can be maintained (Section II (1)), so that co-transport can occur, although cell wall extension due to a lowering of extracellular pH presumably cannot. In the case of roots, where acid-induced growth also occurs[53], it is possible that OH^- excretion related to NO_3^- assimilation occurs in the mature cell zone, with net H^+ excretion in the expanding zone (cf. 54). This hypothesis could be tested by experiments with indicator dyes[55] similar to those carried out on charophytes.

(3) Intracellular pH regulation: The location of the intra-cellular H^+ pumps in a photosynthetic plant cell, and the pH values which they produce in various cell compartments, are shown in Figure 1. The pH differences across the coupling membranes of mitochondria and chloroplasts are related to both the mechanism of energy coupling (Section III (4)), and in metabolic regulation[6,56]. The low pH of lysosomes and vacuoles is related to the pH optima of the hydrolytic enzymes they contain[3,57].

(4) Coupling of membrane-associated redox reactions to ATP synthesis and breakdown: Membranes which have both a redox-powered H^+ pump and an ATP-powered H^+ pump oriented in the same direction (Section II (2)) can have the redox reactions and the adenylate reactions coupled stoichiometrically via their common intermediate, the H^+ electrochemical potential gradient. This is termed chemiosmotic coupling[1-3,34].

Raven and Smith[36] have proposed an evolutionary scheme for the evolution of chemiosmotic coupling from pH-regulating ATP-driven and photoredox-driven H^+ pumps, employing plausible evolutionary pressures in the Precambrian.

(5) <u>Coupling to membrane transport of other solutes</u>: Another major use of the H^+ electrochemical potential gradient generated by the electrogenic H^+ pump is in bringing about the transmembrane transport of other solutes[3,11]. This, like chemiosmotic coupling of redox reactions to ATP synthesis (Section III (4)) is held to be a secondary role of the H^+ gradients set up by pH-regulating H^+ pumps[3,36].

The coupling of the H^+ electrochemical potential gradient to solute transport can involve the $\triangle E$ component alone, the $\triangle \left[H^+ \right]$ component alone, or both components.

(a) <u>Passive cation influx</u>: Here a net passive cation influx occurs in response to an hyperpolarisation of the membrane potential by the active electrogenic H^+ efflux pump. The most easily analysed cases involve a substantial net H^+ efflux, i.e. net H^+ production within the tissue. This occurs during organic acid synthesis and $cation^+$ $organate^-$ accumulation in fungi and in higher plants[11,58,59], and during NH_4^+ assimilation[35]. During $cation^+$ $organate^-$ accumulation there is frequently a large hyper-polarisation of E_m. Since (at a constant internal concentration of the cation) the net passive influx of a cation is a function of P_{cation}, E_m and $\left[cation \right]_o$ a low value of P_{cation}. $\left[cation \right]_o$ means that a very negative value of E_m is required to produce a net passive cation influx equal to the net active H^+ efflux. This case provides a clear correlation between hyperpolarisation and net active H^+ efflux, i.e. <u>prima facie</u> evidence for an electrogenic H^+ efflux.

During NH_4^+ assimilation it appears from plasmalemma conductance measurements[60] that $P_{NH_4^+}$ is high, so that the product P_{cation} $\left[cation \right]_o$ is large, and only a small hyperpolarisation is needed to make net passive cation influx equal net H^+ efflux; this may account[3] for the depolarising effect of NH_4^+.

Two further cases of passive cation entry related to electrogenic H^+ efflux are of interest; neither of them involve a <u>net</u> H^+ efflux. As an alternative to K^+ $organate^-$ accumulation, many plants (especially vacuolate algae) accumulate K^+ Cl^-. Smith[55] has suggested that this can be achieved by using the $\triangle E$ component of the H^+ electrochemical potential gradient generated by the H^+ pump to drive K^+ influx, while Cl^- influx is driven by an electroneutral Cl^--OH^- antiport, using the $\triangle \left[H^+ \right]$ component of the gradient (see (c)).

Finally, if $P_{H^+} \cdot \left[H^+\right]_o$ is high, e.g. at low external pH, there appears to be a large passive H^+ entry so that both the ΔE and the $\Delta\left[H^+\right]$ components of the gradient are short-circuited[3]. One possibility which would make the apparent high passive H^+ entry less of an energetic liability is that much of the apparent high passive H^+ influx is not by uniport, but is part of an electro-genic cotransport system (see (b) below).

(b) <u>Passive electrogenic H^+ symport</u>: Here passive entry of H^+ is coupled to the influx of some other neutral or anionic solute in such a way that the overall charge on the complex crossing the membrane is positive[3,12,61]. Active transport (secondary) occurs in such a system if the free energy available from H^+ entry exceeds that required to bring about influx of the cotransported solute.

As with passive H^+ uniport, both the ΔE and the $\Delta\left[H^+\right]$ components of the H^+ electrochemical potential gradient are 'short-circuited' by this symport mechanism.

(c) <u>Passive electroneutral H^+ symport (OH^- antiport)</u>: This is a special case of (b) in which there is a 1:1 stoichiometry of $anion^-$ influx with H^+ influx (OH^- efflux). This means that only the $\Delta\left[H^+\right]$ component of the H^+ electrochemical potential gradient can be used to energise the influx of the anion and an additional (biochemical) energy input may be required[43,55,74,75].

(d) <u>Weak electrolyte distribution</u>: It is possible that the polar transport of IAA in plant tissues is energised by both the $\Delta\left[H^+\right]$ and the ΔE components of the H^+ electrochemical potential gradient[78-80].

IV. REGULATION OF THE H^+ PUMP

General aspects of the regulation of solute transport in plants are discussed by Cram[51,52]. Figure 2 represents an idealised relationship between the processes which dispose of H^+ and OH^- in plant cytoplasm, and the pH of the cytoplasm, for a system showing negative feedback[51,52]. Here the 'normal' cytoplasmic pH corresponds to the pH at which the processes disposing of H^+ and of OH^- are proceeding at equal rates. A <u>decrease</u> in cytoplasmic pH causes an increase in the rate of the processes which dispose of H^+ and/or a decrease in the rate of processes which dispose of OH^-; an <u>increase</u> in cytoplasmic pH causes the opposite changes in the rates of these processes.

The <u>in vivo</u> and <u>in vitro</u> properties of the metabolic sequences

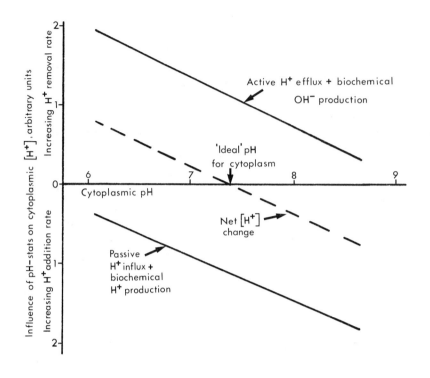

Fig. 2. The combined effect on cytoplasmic H^+ content at various cytoplasmic pH values of the operation of biochemical and bio-physical pH-stat mechanisms.

involved in the two portions of the 'biochemical pH stat' are consistent with this relationship (Fig. 2). The 'biophysical pH stat' is less satisfactory. In animal cells[68.71] there is evidence consistent with a variation of active and passive H^+ fluxes with (measured), cytoplasmic pH such as is indicated in Figure 2, while in plants no such relationship with cytoplasmic pH has yet been established. In Chara corallina there is evidence that the H^+ pump is close to thermodynamic equilibrium with its biochemical energy source over a range of external pH values, assuming that the pump is a $2H^+$/ATP ATP-ase, and making plausible assumptions about the free energy of hydrolysis of ATP within the cells. If this is generally true it imposes considerable restrictions on the feedback regulation of the active H^+ extrusion pump[3,33,51-52] (Section III (2)). For example, increases in total ΔG_{H^+} produced by signals such as growth regulators would be impossible in the absence of increases in ΔG_{ATP} or (if appropriate) ΔG_{redox} at the pump site (cf. Marre, 72).

Regulation in the scheme shown in Figure 2 is dynamic, in that both the biochemical and biophysical pH state show bidirectional fluxes even when the net flux is negligible (Section III (1) [39,40]). This 'cycling' may amplify the sensitivity of control mechanisms [40,73] and, in the case of the biophysical pH stat, allows the maintenance of a large electrochemical potential gradient even when there is no net H^+ efflux (Section III (1),[2]). One further problem relates to the relationship between the rate of intra-cellular H^+ production and the rate of H^+ extrusion. Perhaps the best investigated case is that of organic acid synthesis and K^+-H^+ exchange in higher plant tissue treated with indoleacetic acid or fusicoccin. Kinetic data suggest that, following the addition of indoeleacetic acid, electrogenic H^+ efflux is stimulated before organic acid synthesis[74-77]. This is consistent with an activation of the H^+ extrusion pump: this increases cytoplasmic pH, which in turn increases the rate of the 'OH$^-$ disposal' process, which in these tissues is organic acid synthesis. Cleland[13,14] suggests the reverse scheme for fusicoccin, in which activation of organic acid synthesis decreases cytoplasmic pH, and thus increases the rate of H^+ extrusion.

Analogous possibilities for control exist for the pH regulation reactions related to the N assimilation reactions, which are quantitatively very important during growth[5] (Section III (1)) particularly in relation to the evidence that active H^+ extrusion still occurs even when there is a net downhill OH$^-$ efflux during NO_3^- (and HCO_3^- assimilation): Section III (1). Again, measurements of cytoplasmic pH are needed. If the active H^+ efflux is due to excess OH$^-$ efflux (e.g. by 1:1 antiport with the metabolised anion[43]), then a decreases cytoplasmic pH would be predicted, while if a direct activation of the H^+ pump is the primary event, then a cytoplasmic pH increase would be expected, leading to a stimulation of OH$^-$ efflux. It is clear that progress in distinguishing the possibilities discussed above requires further measurements of cytoplasmic pH; the techniques for cyto-plasmic pH measurement which are currently available do not permit such measurements to be made on the majority of mature, vacuolate higher plant cells[2,3,39].

V. CONCLUSIONS

An active H^+ efflux is of very widespread occurrence in the plasma membrane of both prokaryote and eukaryote cells. Active H^+ transport is also common at internal membranes (mitochondria, chloroplasts, lysosomes, vacuoles) of eukaryote cells. These pumps are generally electrogenic H^+ uniporters. In those membranes (prokaryote plasma membrane; chloroplast thylakoid membrane, mitochondrial crista membrane) which can couple redox reactions to ATP synthesis, redox-driven H^+ pumps occur. ATP-driven H^+ pumps are apparently found in all membranes which can pump H^+.

The H^+ pump probably had its evolutionary origin in the plasmalemma of a prokaryote as a means of regulating cytoplasmic pH when essential metabolic processes (NH_4^+ assimilation, fermentation) caused cytoplasmic acidification. This pump was probably ATP-powered; evolution of a photoredox H^+ pump allowed pH regulation by the direct use of light energy; these two pumps together could bring about a chemiosmotic coupling of exergonic redox reactions and ATP synthesis[36]. Another secondary role of the H^+ electrochemical potential gradient generated by the H^+ pump(s) was coupled transport of other solutes.

In prokaryotes all three major functions are carried out by the plasmalemma H^+ pumps. In eukaryotes chemiosmotic ATP synthesis has been 'internalised' in mitochondria and chloroplasts, and the plasmalemma H^+ pump functions only in pH regulation and cotransport. Later metabolic evolution involved such OH^--generating processes as NO_3^- assimilation and HCO_3^- assimilation: despite the net downhill OH^- efflux which accompanies these processes, there is evidence that the H^+ efflux pump is still operating. In higher eukaryotic plants there is evidence that the H^+ pump is involved in the action of growth-regulating chemicals and photoreceptors.

In the above-ground parts of terrestial plants there is very limited scope for pH regulation processes which involve net H^+ or OH^- efflux. pH regulation in these plants involves 'biophysical' (H^+ or OH^- efflux) regulation in the below-ground portions, interacting with 'biochemical' pH regulation (synthesis and breakdown of organic acids) leading to restrictions on the siting of N-assimilation processes, and long-distance transport of carboxylates.

VI. ACKNOWLEDGEMENTS: We should like to thank Drs. S. M. Glidewell and H.D. Jayasuriya for their helpful comments on the manuscript.

REFERENCES:

1. Mitchell, P. (1976). Biochem. Soc. Transactions 4, 399-430.
2. Raven, J.A. and Smith, F.A. (1974). Can. J. Bot. 52, 1035-1048.
3. Raven, J.A. and Smith, F.A. (1976) Current Advances in Plant Science 9, 649-660.
4. Rottenberg, H. (1975) Bioenergetics 7, 61-74.
5. Papa, S. (1976) Biochim. Biophys. Acta 456, 39-84.
6. Werdan, K., Heldt, H.W. and Milovancev, M. (1975) Biochim. Biophys. Acta 396, 276-292.
7. Heber, U. (1974) Annu. Rev. Plant Physiol. 25, 393-421.
8. Davis, R.F. (1974) in Membrane Transport in Plants (Zimmermann, U. and Dainty, J., eds.), pp. 197-202. Springer, Berlin.
9. Bell, P.R. (1970) Symp. Soc. Exp. Biol. 24, 109-127.
10. Margulis, L. (1974) Evolutionary Biology 7, 45-78.
11. Slayman, C.L. (1974) in Membrane Transport in Plants (Zimmermann, U. and Dainty, J., eds.), pp. 107-119. Springer, Berlin.
12. Anderson, W.P. and Robertson, R.N. (1976) In Proceedings of the International Workshop on Transmembrane Ionic Exchanges in Plants (Ducet, R., Heller, R., and Thellier, M., eds.), in press C.N.R.S., Paris.
13. Cleland, R.E. (1976) Planta 128, 201-206.
14. Cleland, R.E. (1976) Biochem. Biophys. Res. Commun. 69, 333-338
15. Prins, H.B.A., Harper, J.R., Higinbotham, N. and Cleland, R.E. (1976) Plant Physiol. 57, 3s.
16. Raven, J.A. (1976) in The Intact Chloroplast (Barber, J., ed.), in press. North-Holland, Amsterdam.
17. Racker, E. (1975) Biochem. Soc. Transactions 3, 785-802.
18. Poincellot, P. and Day, P.R. (1974) Plant Physiol. 54, 780-783.
19. Briggs, W.R. (1975) Carnegie Institution of Washington Yearbook, 74, 807-809.
20. Brain, R.D., Freeberg, J., Weiss, C.V. and Briggs, W.R. (1976) Plant Physiol. 57, 19a.
21. d'Auzac, J., Dupont, J., Jacob, J.L., Lance, C., Marin, B. and Moreau, F. (1976) in Proceedings of the International Workshop on Transmembrane Ionic Exchanges in Plants (Ducet, R., Heller, R., and Thellier, M., eds.) in press. C.N.R.S., Paris.

22. Lüttge, U., Pallaghy, C.K. and Osmond, C.B. (1970) J. membrane Biol. 2, 17-30.

23. Polevoi, V.V. and Salamatova, T.S. (1975) Fiziol. Rastenii. 22, 519-526.

24. Mego, J.L. (1975) Biochem. biophys. Res. Commun. 67, 571-575.

25. Semano, R., Kanner, B.I. and Racker, E. (1976) J. Biol. Chem. 251, 2453-2461.

26. Slayman, C.L., Long, W.S. and Lu, C. Y-H. (1975) J. Membrane Biol. 14, 305-338.

27. Scarborough, G.A. (1976) Proc. Natl. Acad. Sci. U.S. 73, 1485-1488.

28. Thayer, W.S. and Hinkle, P. (1975) J. Biol. Chem. 248, 5395-5402.

29. Walker, N.A. and Smith, F.A. (1975) Pl. Sci. Lett. 4, 125-132.

30. Hall, D.O. (1976) in The Intact Chloroplast (Barber, J., ed.), in press. North Holland, Amsterdam.

31. Portis, A.R., Jr. and McCartey, R.E. (1976) J. Biol. Chem., 251, 1610-1617.

32. Raven, J.A. (1976) in Encyclopedia of Plant Physiology, New Series, (Lüttge, U., and Pitman, M.G., eds.) Volume 2A, 129-188. Springer, Berlin.

33. Walker, N.A. and Smith, F.A. (1976) in Proceedings of the International Workshop on Transmembrane Ionic Exchanges in Plants (Ducet, R., Heller, R., and Thellier, M., eds.) in press. C.N.R.S., Paris.

34. Raven, J.A. and Smith, F.A. (1973) in Ion Transport in Plant Cells (Anderson, P., ed.) pp. 271-278. Academic Press, London.

35. Raven, J.A. and Smith, F.A. (1976) New Phytol. 76, 415-431.

36. Raven, J.A. and Smith, F.A. (1976) J. Theoret. Biol. 57, 301-312.

37. Smith, F.A. and Raven, J.A. (1974) in Membrane Transport in Plants (Zimmermann, U., and Dainty, J., eds.) pp. 380-385. Springer, Berlin.

38. Smith, F.A. and Raven, J.A. (1976) in Encyclopedia of Plant Physiology, New Series (Lüttge, U., and Pitman, M.G., eds.), Volume 2A, 317-346. Springer, Berlin.

39. Davies, D.D. (1973) in Biosynthesis and its control (Millborrow, B.V. ed.) pp. 1-20. Academic Press, London.

40. Davies, D.D. (1973) Symp. Soc. Exp. Biol. 27, 531-539.

41. Clarke, D.J. and Morris, J.G. (1976) Biochem. J. 154, 725-729.

39

42. Riebeling, U., Thauer, B.K. and Jungermann, K. (1975) Eur. J. Biochem. 55, 445-454.

43. Raven, J.A. and Jayasuriya, H.D. (1976) in Proceedings of the International Workshop on Transmembrane Ionic Exchanges in Plants (Ducet, R., Heller, R., and Thellier, M., eds.) in press, C.N.R.S., Paris.

44. Raven, J.A. (1970) Biol. Revs. 45, 167-225.

45. Lucas, W.J. and Smith, F.A. (1973) J. Exp. Bot. 24, 1-14.

46. Lucas, W.J. (1975) J. Exp. Bot. 26, 271-286.

47. Lucas, W.J. (1976) J. Exp. Bot. 27, 19-31.

48. Richards, J.L. and Hope, A.B. (1974) J. Membrane Biol. 16, 121-144.

49. Raven, J.A. (1977) Adv. Bot. Res. 5, in press.

50. Cleland, R.E. (1975) Planta 127, 233.

51. Cram, W.J. (1976) in Encyclopedia of Plant Physiology, New Series, (Lüttge, U., and Pitman, M.G., eds.), Vol. 2A, pp. 284-316. Springer, Berlin.

52. Cram. W.J. (1976) in Transport and Transfer Processes in Plants. (Wardlaw, I.F., Passioura, J., and Evans, L.T., eds.), in press. Academic Press, New York.

53. Lado, P., de Micheli, H.I., Cerano, R. and Marre, E. (1976) Pl. Sci. Letts. 6, 5-20.

54. Fensom, D.S. (1959) Can. J. Bot. 37, 1003-1024.

55. Smith, F.A. (1970) New Phytol. 69, 903-917.

56. Barber, J. (1976) in The Intact Chloroplast (Barber, J., ed.), in press. North-Holland, Amsterdam.

57. ap Rees, T. (1975) in M.T.P. International Reviews of Science, Biochemistry, Series 1 (Northcote, D.H., ed.), Vol. 11, pp. 89-129. Butterworths, London.

58. Poole, R.J. (1974) Can. J. Bot. 52, 1023-1028.

59. Marre, E., Lado, P., Rosi-Caldagro, F., Colombo, R., Coccuci, M. and de Michelis, M.I. (1975) Phys. Veg. 13, 797-811.

60. Smith, F.A., Walker, N.A. and Raven, J.A. (1976) in Proceedings of the International Workshop on Transmembrane Ionic Exchanges in Plants (Ducet, R., Heller, R., and Thellier, M., eds.), in press. C.N.R.S., Paris.

61. Smith, F.A. and Walker, N.A. (1976) J. Exp. Bot. 27, 451-459.

62. Smith, F.A. (1972) New Phytol. 71, 595-601.

63. MacRobbie, E.A.C. (1975) Curr. Top. Membranes and Transport 7, 1-48.

64. Raven, J.A. (1974) in Membrane Transport in Plants (Zimmermann, U., and Dainty, J., eds.), 167-172. Springer, Berlin.
65. Raven, J.A. (1976) in Perspectives in Experimental Biology, (Sunderland, N. ed.), Vol. 2, pp. 381-389. Pergamon Press, Oxford.
66. Raven, J.A. (1975) New Phytol. 74, 163-172.
67. Conde, W.Z. and Ray, P.H. (1976) Planta 129, 43-52.
68. Goldsmith, M.H.M., Breault, D. and Rieur, E. (1976) Plant Physiol. 57, 52s.
69. Boron, W.F. and de Weer, P. (1976) J. Gen. Physiol. 67, 91-112.
70. Menard, M.R. and Hinke, J.A. (1976) Biophys. J. 16, 29a.
71. Thomas, R.C. (1976) J. Physiol. 255, 715-736.
72. Marre, E. (1976) in Proceedings of the International Workshop on Transmembrane Ionic Exchanges in Plants (Ducet, R., Heller, R., and Thellier, M., eds.), in press. C.N.R.S., Paris.
73. Newsholme, E.A. and Crabtree, B. (1974) Symp. Soc. Exp. Biol. 27, 429-460.
74. Haschke, H-P. and Lüttge, U. (1975) Plant Physiol. 56, 696-698.
75. Bayer, M.H. and Sonka, J. (1976) Plant Physiol. 57, 1s.
76. Johnson, K.D. and Rayle, D.L. (1976) Plant Physiol. 57, 2s.
77. Stout, R., Johnson, K.D. and Rayle, D.L. (1976) Plant Physiol. 57, 2s.

ADDENDUM

Treatments[78] which decrease intracellular pH hyperpolarise E_m in Nitella and Trianea cells, while increased intracellular pH depolarises E_m.   Provided the changes in E_m reflect the rate of electrogenic H^+ extrusion, and cytoplasmic pH changes in the same directions as overall intracellular pH, these results support the hypothesis of 'biophysical' pH regulation shown in Figure 2.

The increase in intracellular pH (correlated with net H^+ efflux) which follows the fertilization of sea urchin eggs may activate a range of biosyntheses[79]; this possibility is worth investigating in relation to treatments (e.g. IAA) which activate H^+ extrusion in plant cells.

78. Lyalin, O.O., and Kritorova, I.N. (1976).   Fisiologiya Rastenii 23, 305-314.
79. Johnson, J.D., Epel, D., and Paul, M. (1976) Nature 262, 661-664.

Regulation of Cell Membrane Activities in Plants
E. Marrè and O. Ciferri eds.
© *1977, Elsevier/North-Holland Biomedical Press, Amsterdam*

SYMMETRY OF THE TRANSMEMBRANE PHENOMENOLOGICAL CROSS-COEFFICIENTS,

AS A MEANS OF TESTING THE ACTIVE OR PASSIVE CHARACTER OF A CELLULAR TRANSPORT

Michel Thellier and Jean-Paul Lassalles

L.A.T.P.E.I.C., Faculté des Sciences, I.S.H.N., 76130 Mont-Saint-Aignan, France
(C.N.R.S.: L.A. 203 and R.C.P. 285; D.G.R.S.T. 74 7 0194)

SUMMARY

Growth promoting substances can perturb the ionic exchanges of the target cells. It would be important to know whether these effects are directly related to some active ionic pumping; but it is generally difficult to achieve the measurements which would enable one to apply the flux-ratio equation in this case. Thanks to Onsager's symmetry relations, we try here to develop a new criterion of active and passive transport (the "symmetry-criterion"), such that the required measurements would be easier to perform.

INTRODUCTION

Growth stimulating substances can exert a quick effect on the transport systems of the treated cells[1]. Thus it would be important to know which of these effects interfere on truly active systems (possibly on the induction of a new pump), and which are mere passive consequences of the modification of the electrochemical forces.

The direct use of metabolic inhibitors is not very helpful for such a problem: it is true that they will perturb the active processes; but, in turn, this will modify the electrochemical forces, hence the passive transports themselves...

The flux-ratio equation is written[2]

$$RT \ln(J_j^{ei}/J_j^{ie}) = RT \ln(\gamma_j^e c_j^e / \gamma_j^i c_j^i) - z_j F(\psi^i - \psi^e) \qquad (1)$$

with

j: the transported substance

e and i: the respective indices for the "exterior" and "interior" media

J^{ei} and J^{ie}: respectively influx and outflux

γ: activity coefficient

c: concentration

Ψ: electric potential

Being verified only if the transport is a purely passive, electrochemical one, this equation gives the wanted criterion. It has proved very efficient indeed for the study of the transepithelial fluxes[2]; but its practical use might be much more difficult in the case of a cellular transport, because the measurement of some of the variables (especially unidirectionnal fluxes and internal activity-coefficients) might become cumbersome or even impossible.

Starting from an approach of non-equilibrium thermodynamics, which we had called the "electrokinetic formulation" of the transmembrane fluxes[3-4], we try to develop an alternative criterion whose practical application to the cellular problem would be easier.

OUR "SYMMETRY-CRITERION"

Consider different substances (j, k,...), the cellular transport of which might interact. Let us write the net transmembrane fluxes

$$\begin{cases} J_j = L_{jj} \, X_j + L_{jk} \, X_k + \ldots \\ J_k = L_{kj} \, X_j + L_{kk} \, X_k + \ldots \end{cases} \qquad (2)$$

where the Xs are the electrochemical forces, the Ls the transmembrane phenomenological coefficients, and the dots hold for the possible interference of other forces (active coupling, etc.). These equations can always be written

$$\begin{cases} J_j = L_{jk} \, \ln c_k^e + K_j \\ J_k = L_{kj} \, \ln c_j^e + K_k \end{cases} \qquad (3)$$

where the Ks summarize a great number of terms (activity coefficients, internal concentrations, electric terms, possible active couplings, etc.).

Formerly, we asserted that it was possible to find experimental situations (quasi stationary conditions, small net fluxes, etc.) such that the Ks and the Ls would remain constant[5]; and we assumed that we should then obtain parallel linear graphs[6-7], for J_j as a function of c_k^e and J_k as a function of c_j^e, if coefficients L_{jk} and L_{kj} obeyed Onsager's symmetry [8]. Following the advices of R. Caplan and O. Kedem, a more thorough examination of the problem has now shown us that, in fact, this is impossible if an active coupling is involved in the transfer of j or k.

The theoretical reasoning, which needs to go back to the local Onsager's relations, will be detailed somewhere else. But, as it is, this approach already gives us the principle of our new criterion: $\underline{J_j}$ will be drawn in terms of $\underline{c_k^e}$ and $\underline{J_k}$ in terms of $\underline{c_j^e}$: if the two graphs are parallel for the small net fluxes, then it excludes the possibility for either j or k or both to be actively transported.

REFERENCES

1. Lado, P., Rasi-Caldogno, F., Colombo, R., de Michelis, M.I. and Marré, E. (1976) Plant Science Letters Vol. 7, 199-209.

2. Ussing, H.H. (1960) in: Handbuch der experimentellen Pharmakologie. Eichler, O. and Farah, A. eds., Springer-Verlag, Berlin, Vol. 13, 1-195.

3. Thellier M. (1968) C.R. Acad. Sci. Paris, série D, Vol. 266, 826-829

4. Thellier M. (1971) J. Theoret. Biol. Vol. 31, 389-393.

5. Thellier M. (1973) in: Ion transport in plants. Anderson, W.P. ed., Acad. Press, London and New-York, 47-63.

6. Thellier, M. Stelz, T. and Ayadi, A., C.R. Acad. Sci. Paris, Vol. 273, 2346-2349

7. Lassalles, J.P., Ayadi, A., Monnier, A., Stelz, T. and Thellier, M. (1973) C.R. Acad. Sci. Paris, série D, Vol. 276, 1053-1056.

8. Onsager, L. (1931) Phys. Rev. Vol. 37, 405-426.

Regulation of Cell Membrane Activities in Plants
E. Marrè and O. Ciferri eds.
© *1977, Elsevier/North-Holland Biomedical Press, Amsterdam*

BICARBONATE ASSIMILATION AND HCO_3^-/OH^- EXCHANGE
IN CHARA CORALLINA

W.J. Lucas and J. Dainty
Department of Botany, University of Toronto,
Toronto, Ontario, Canada.

INTRODUCTION

Numerous plant species form $CaCO_3$ incrustations on the outer surfaces of their photosynthetic organs (Lewin[1]). In the species Chara corallina, a fresh water member of the Characeae, these $CaCO_3$ deposits are spaced in a regular pattern along the whorl and internodal cells. This calcification banding phenomenon was first noted by Hanstein[2] and since then numerous hypotheses have been proposed in an attempt to explain this interesting feature (Dahm[3], Arens[4], Spear et al.[5], Smith[6]). The basic characteristics of this system, for Chara corallina, will be reviewed in this paper.

In general, $CaCO_3$ precipitation was considered to be related to the ability of a particular species to photosynthetically assimilate exogenous HCO_3^- (Ruttner[7], Steemann Nielsen[8]). (The reader is referred to the work of Raven[9] for a detailed account of the criteria employed in ascertaining whether a particular species can assimilate HCO_3^-.) However, it was not until the work of Smith[10,11] that it was conclusively shown that Characean cells can photosynthetically utilize exogenous HCO_3^-. Smith found that several species of the Characeae could fix $H^{14}CO_3^-$ at rates of 1-9 pmol $cm^{-2}s^{-1}$.

In the following year Spear et al.[5] reported, qualitatively, that in the light, the surface of Nitella clavata became partitioned into acid and alkaline zones. Smith[6] also observed this pH banding pattern on the surface of Chara corallina, and he concluded that the alkalinity and concomitant $CaCO_3$ precipitation was probably related to HCO_3^- assimilation.

To test this hypothesis, a technique for mapping the pH pattern established at the cell surface was developed by Lucas and Smith[12]. These workers showed that the acidification and alkalinization processes were indeed light-mediated. Furthermore, they were also able to demonstrate that the alkaline regions, or bands, were totally dependent upon the presence of exogenous HCO_3^- in the experimental medium. It appeared, therefore, that Smith's[6] proposal was correct, and it was consequently assumed that there would be a close relationship between HCO_3^- influx and OH^- efflux across the plasmalemma of this species. It should be stressed that photosynthetic assimilation of the carbon supplied by HCO_3^- influx, results in the production of one OH^- for each CO_2 molecule fixed. It is the efflux of these ions that generates the observed alkalinity.

[14]CARBON ASSIMILATION

Since the work of Smith[6] and Lucas and Smith[12], considerable progress has been made in elucidating some of the operational characteristics of HCO_3^- and OH^- transport in this species. A study of the photosynthetic capacity of these giant internodal cells revealed that they could assimilate [14]CO_2 and $H^{14}CO_3^-$ at rates higher than previously reported (Fig. 1). Maximum fixation rates, under saturating exogenous [14]CO_2 were found to be approximately twice those obtained under saturating $H^{14}CO_3^-$ conditions. Lower rates of photosynthesis, during HCO_3^- assimilation, were similarly reported by Steemann Nielsen[8] and Raven[13], using Myriophyllum and Hydrodictyon, respectively. This implies that fixation, in the presence of HCO_3^-, is limited by a process other than the carboxylating reactions. This limiting process may be the supply of energy from the chloroplasts to the HCO_3^- transport system, or, alternatively, the kinetics of this transport system may be rate-limiting. These possibilities remain to be investigated.

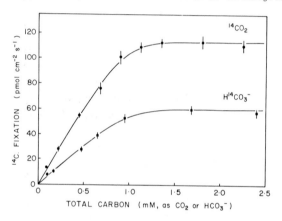

Fig. 1. Effect of substrate concentration on fixation. Light intensity 25 Wm^{-2}, [14]CO_2 and $H^{14}CO_3^-$ experiments conducted at pH 5.3 and 9.2 respectively.

The carbon source available for photosynthesis can be manipulated from predominantly CO_2 (pH 5.0) to HCO_3^- (pH 9.0) simply by adjusting the solution pH value. Cellular response to this change in [14]C. substrate was investigated (Fig. 2). An analysis of the data (Lucas[14]) revealed that under sub-saturating [14]CO_2 levels, both gaseous diffusion and HCO_3^- transport operate simultaneously, presumably in an attempt to maximize the rate of photosynthesis (c.f. Raven[9,13]). This maximization, governed by the prevailing external conditions, is demonstrated by the decline in fixation from 120 to 60 pmol $cm^{-2}s^{-1}$, over the pH range 7.0 to 8.4. Over this pH range the CO_2 level, in this experimental sequence, fell from 0.54 to 0.03 mM.

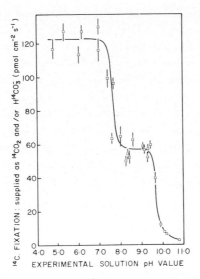

Fig. 2. Effect of increasing pH on ^{14}C. fixation. A total carbon concentration of 3 mM and 25 Wm^{-2} light intensity were used for each experiment.

The observed inhibition of H^{14}CO$_3^-$ transport at pH values above 9.5 (see Fig. 2) may have been due to either the influence of high pH _per se_ on the membrane or to the presence of CO$_3^{2-}$, which could competitively inhibit the HCO$_3^-$ transport system. Because cells grow in culture tanks in which the pH value often exceeds 9.5, it would seem that CO$_3^{2-}$ inhibition is the more likely explanation. This would be in accord with the inhibitory effect of CO$_3^{2-}$ on HCO$_3^-$ assimilation reported by Österlind[15]. Since hydroxyl bands can develop localized pH regions having values greater than 10.0 (see Fig. 3), CO$_3^{2-}$ in these regions may inhibit the HCO$_3^-$ carriers. This must also be considered as a possible explanation for the lower rates of HCO$_3^-$ fixation compared with those obtained under CO$_2$ assimilating conditions.

HYDROXYL EFFLUX PATTERN

Typical alkaline (OH$^-$) banding patterns established on the internodal cell surface, under different light intensities, are shown in Fig. 3. (For details of the techniques employed in mapping these pH profiles, see Lucas and Smith[12] and Lucas[16].) Under a light intensity of 10 Wm^{-2} and 0.5 mM total carbon, four OH$^-$ bands developed; the pH values at the centres of these bands were raised significantly above that of the background solution. The regions between the OH$^-$ bands were only weakly acidic, i.e. the pH values were only slightly depressed below background. Lower light intensities resulted in a reduction in the number of operational bands, i.e. the four bands did not reactivate and operate at lower OH$^-$ efflux rates. A full report of this hierarchical status in OH$^-$ efflux activity has been presented by Lucas[16]; based on this study, band C would be ascribed primary OH^- band efflux status. Upon illuminating the cell, this primary band would activate first and be the last efflux site to deactivate upon reduction of either light intensity or HCO_3^- substrate. The second band which remained operational under 3 Wm^{-2} (band B) would have the status of a sub-primary OH^- efflux site and bands A and D subsidiary band status. We have found that, following illumination at a particular light intensity and HCO_3^- substrate level, the primary and sub-primary bands generally activate. If these two efflux sites cannot cope with the rate of OH^- generation, subsidiary bands are activated until the combined efflux through the bands balances this OH^- production.

The presence of this hierarchy, in terms of OH^- efflux sites, indicates that the efflux of this ion must be controlled by an intricate cytoplasmic mechanism.

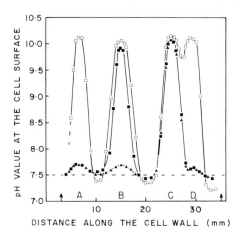

Fig. 3. Cell wall pH values obtained after 2h illumination. Light intensities were as follows: □ , 10 Wm-2; ■ , 3 Wm-2; ▲ , 2 Wm-2. A 1h dark period was employed between different intensities.

LIGHT ACTIVATION OF OH^- EFFLUX

A study on the influence of light intensity on the activation of the primary efflux site revealed an interesting feature (Fig. 4). As the light intensity was decreased, the lag period prior to activation increased. Lucas[16] showed that a reciprocal plot of these OH^- lag periods against light intensity gave two linear regions (Fig. 5). For most cells, light intensities greater than 5 Wm^{-2} did not cause an increase in the reciprocal of the OH^- lag period; the lag period has a minimum value determined by factors other than light intensity. (Fig. 5 indicates that cell chlorophyll concentration influences this minimum value.) The linear region observed at low light intensities (3 Wm^{-2}) was informative. It implied that light produced a substrate (i.e. for the OH^- efflux system) at a rate prop-

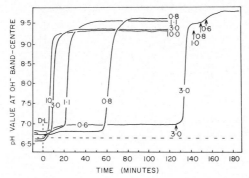

Fig. 4. Hydroxyl band response to illumination. Light intensities
(Wm^{-2}) are indicated on the diagram. Bathing solution pH indicated by
the broken line. 'Light on' indicated by the vertical broken line.

ortional to the light intensity and that this substrate was required at a specific
level before the primary band could activate. The light intensity at the intercept of the abscissa is the value that would take an infinite time to establish
the activating condition. This intercept is termed the critical light intensity
(Lucas[16]), and values below it never facilitate activation. The experimental cell
of Fig. 4 had a critical light intensity of 0.7 ± 0.05 Wm^{-2}, hence the primary band
remained inactive at 0.6 Wm^{-2}. However, once activated, efflux activity could be
maintained by a sub-critical intensity. This suggested that there was not only a
critical activating substrate level, but that there was also a deactivating
substrate level, below which OH$^-$ ions could not be transported (see Lucas[16] for
full details).

Figure 4 also demonstrates that, with primary bands, the band centre pH value
often increased as the light intensity was reduced. It is considered that this
increase in efflux activity, under lower light intensities, reflects the transfer
of OH$^-$ efflux function from inactive subsidiary bands to the primary site.

HCO$_3^-$/OH$^-$ EXCHANGE: POSSIBLE ELECTRICAL INVOLVEMENT?

The electrical response of the Characean plasmalemma to illumination has
received considerable attention (Nishizaki[17], Andrianov et al.[18], Spanswick[19],
Volkov[20], Saito and Senda[21,22], Hansen[23]). The electrical changes produced by
light have an action spectrum similar to that of photosynthesis (Walker[24]), and
photosynthetic inhibitors reduce or eliminate the photo-response of the membrane
(Volkov[20], Saito and Senda[21,22]). At present, the nature of the coupling between
photosynthesis and this electrical response, at the plasmalemma, remains to be
fully elucidated.

Since the work of Kitasato[25], many workers in this field consider that an
electrogenic H$^+$ transport system serves as this coupling factor. This hypothesis

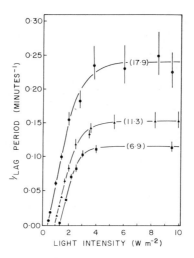

Fig. 5. Reciprocal plot of lag period against light intensity. Values in parenthesis are cell chlorophyll concentrations in μg chl. cm^{-2}.

is probably valid for the physiological pH range 5.0 to 7.0. However, other ions may also be involved. Lucas[26] stressed that since the nature of the HCO_3^- and OH^- transporters remains unknown, their contribution to the electrical properties of the plasmalemma should not be discounted (c.f. Richards and Hope[27]).

Electrical neutrality at the plasmalemma, during HCO_3^- assimilation, could be maintained by transporting HCO_3^- and OH^- on an antiporter system (Lucas and Smith[12]). Such a system could not contribute to either the electrical potential or conductance of the plasmalemma. This HCO_3^-/OH^- antiporter hypothesis was tested experimentally by isolating the cell's primary band within a special chamber (see Fig. 1 of Lucas[28]). (The primary band was selected since it would always be the last OH^- efflux site to deactivate.) Based on the above hypothesis, cessation of OH^- efflux would be expected when the isolating chamber was flushed with a solution 'totally' free of inorganic carbon. Removal of HCO_3^- from the isolated band had little or no effect on its efflux activity (Fig. 6). Activity could be maintained over prolonged periods, provided the outer cell segment was in contact with exogenous HCO_3^-. When the outer cell segment was immersed in liquid paraffin, the OH^- efflux activity was immediately influenced and the site eventually deactivated (see Fig. 6). Replacement of the liquid paraffin by a 0.2 mM $NaHCO_3$ solution resulted in reactivation of the isolated primary band. Loss of OH^- efflux activity could also be achieved by replacing the outer bathing solution with a solution containing 0.2 mM CO_2 (pH 5.0) (Lucas[28]).

These results demonstrate that OH^- ions can be effluxed in the absence of exogenous HCO_3^- at the actual OH^- efflux site. Hence, the HCO_3^-/OH^- antiporter hypothesis is invalid, and it can be assumed that the transports of HCO_3^- and OH^- are independent, insofar as the two processes can be spatially separated. It seems probable that quite distinct carriers are involved in the transport of these two ions.

This 'independent' operation of HCO_3^- and OH^- transport should establish a situation in the plasmalemma whereby, during the initial illumination of the cell, any imbalance between the influx of HCO_3^- and the efflux of OH^- should result in fluctuations in the membrane potential. The main electrical influence of these fluxes is likely, however, to be _via_ their contribution to the electrical conductance of the plasmalemma. This would be particularly so at pH values where a supply of exogenous HCO_3^- exists. The high HCO_3^- and OH^- effluxes (Fig. 1) may account for a large proportion of the observed rise in electrical conductance which has been observed at alkaline pH values (Nishizaki[17], Spanswick[19], Walker[24], Hope[29]).

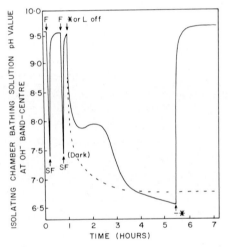

Fig. 6. Hydroxyl band activity in the presence of a 'carbon-free' bathing solution. A _primary_ band was located in a chamber that isolated its cell wall segment from the rest of the cell. Chamber was flushed with 'carbon-free' solution (F); cessation of flushing indicated by (SF). Symbol (*) indicates sealing of outer cell segment in liquid paraffin, (-*) is the replacement of this seal by 0.2 mM $NaHCO_3$ solution.

NUMERICAL ANALYSIS OF OH^- EFFLUX ACTIVATION

One method of testing the electrical involvement of HCO_3^- assimilation would be to compare the time course of photo-activated HCO_3^- and OH^- transport with the time course of membrane conductance. A close parallel between these respective time courses would be strong evidence for the electrical involvement of these ions. Using numerical analysis (Lucas et al.[30]), we were able to analyse the time-

dependent rise in the pH value at the centre of an alkaline band, to obtain the
dynamics of OH⁻ efflux activation (Fig. 7). This analysis revealed that the initial stage of activation followed an exponential form. Aside from its obvious
importance as an expression of a fundamental characteristic of the OH⁻ transport
system, this exponential form will be an extremely useful marker in elucidating
the electrical involvement of this ion.

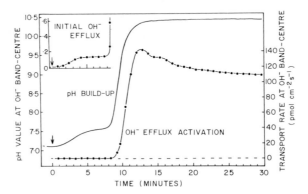

Fig. 7. pH Build-up and OH⁻ efflux activation curves. The symbol (●) represents the OH⁻ efflux activity required to give an exact fit of the experimental
pH build-up. Scale on the ordinate of the insert represents OH⁻ efflux in
pmol cm^{-2}s^{-1}, and the arrows indicate the commencement of illumination.

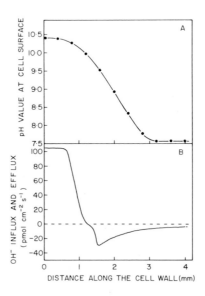

Fig. 8. Steady-state pH and OH⁻ efflux values along the cell surface.
A: experimental pH plot commencing at the OH⁻ band centre.

Another important characteristic that was discovered, as a result of this numerical analysis, was that OH^- bands are formed by a relatively wide uniform OH^- efflux region with a ramp decline in activity at the boundaries (Fig. 8). Of equal importance was the discovery that an OH^- sink exists in the lateral regions adjacent to the OH^- efflux zone (see Fig. 8B).

HCO_3^-, OH^- TRANSPORT MODEL

Several hypotheses have been proposed to account for the observed features of this HCO_3^- assimilating system (Lucas[16], Lucas et al.[30], Lucas and Dainty[31]). A brief description of the basic features of the present model will be given, but we wish to stress that because the control of HCO_3^- and OH^- transport, at the molecular level, is a complex process, this model will inevitably require significant modification. Also, our studies have concentrated upon the OH^- efflux system, and consequently information relating to the HCO_3^- transport system is sparse. Hence, our model is heavily biased towards the OH^- transport aspect of HCO_3^- assimilation.

Let us begin by discussing the small flux associated with the lag period prior to full efflux activity (see Figs. 4 & 7). We consider that the linkage between cell illumination and this small OH^- efflux process is _via_ the photosynthetic reactions occurring in the chloroplasts. The chloroplasts may provide energy to the transport system, or alternatively, these reactions may be required to generate a compound that is essential for transport activity. This 'compound' may also be involved in establishing the OH^- efflux band centres (Lucas et al.[30]). Our model assumes that activation of the transporters occurs in such a way that they will actively transport OH^- in both directions across the plasmalemma. Thus, in the presence of a higher pH value inside the cell than outside, a small net efflux results.

The activation of undirected active transport spreads out along the membrane in both directions from the band centre. The mechanism that imparts the location of the band centre is unknown, but by experimentation we know that, once selected, the band centre location is stable over a series of activations and deactivations.

The main exponential efflux build-up is considered to be due to the cooperative effect of two events. Firstly, the imposition of directionality on the active transport process, and secondly, the establishment of a critical concentration of a cytoplasmic substrate that is generated by photosynthesis (see Fig. 5). Obviously there are many ways in which directionality could be imparted to this membrane-bound transport system. One tentative proposal is that a conformational change, at the cytoplasm-plasmalemma interface, increases the binding affinity of the enzyme transport site for OH^- at this surface. Similarly, a conformational change within the OH^- transport 'complex' may increase the availability of energy to the reactions associated with the efflux process. Whatever the molecular mechanism

52

involved, we consider that the activation of directionality does not extend as far along the plasmalemma surface, from the band centre, as does the activation of undirected transport. This lack of complete matching between the regions covered by the two activation processes results in the observed phenomenon of the lateral OH^- sink.

Radioactive $H^{14}CO_3^-$ time course experiments have indicated that there is not a discernible lag prior to activation of HCO_3^- influx (Lucas[14]). However, we have found that a concentration gradient is required across the plasmalemma before full HCO_3^- transport activity can be obtained (Lucas[28]). Under a light intensity of 10 Wm^{-2} this gradient must be established very quickly, however, it is obvious that HCO_3^- transport cannot operate at high rates before the OH^- transport system is activated, because this situation would generate large membrane potentials. The nature of the coupling between these two transport processes still remains to be resolved. We do know, however, that there is a limit to the number of OH^- bands that can operate along any one cell. This limit is related to the inhibition of HCO_3^- transport by CO_3^{2-} at pH values above 9.5; cell surfaces having pH values above 9.5 would not be available for HCO_3^- transport. Hence, the more bands operating, the smaller the area available for HCO_3^- influx. Thus, in the steady state, there exists a balance between maximizing HCO_3^- influx and the area of plasmalemma required to dispose of the OH^- generated by this influx.

The limitations of our simple model, at the molecular level, are all too obvious. Nevertheless, we are now in a position to investigate the involvement of HCO_3^- and OH^- transport with respect to the electrical properties of the plasmalemma. This work should certainly facilitate the further refinement of the present model.

REFERENCES

1. Lewin, J.C., 1962. In physiology and Biochemistry of Algae. Ed. R.A. Lewin, New York, Academic Press, pp. 457-65.
2. Hanstein, J., 1873. Bot. Ztg 31, 694-97.
3. Dahm, P., 1926. Jahrb. wiss. Bot. 65, 314-51.
4. Arens, K., 1933. Planta 20, 621-58.
5. Spear, D.G., Barr J.K., and Barr, C.E., 1969. J. gen. Physiol. 54, 397-414.
6. Smith, F.A., 1970. New Phytol. 69, 903-17.
7. Ruttner, F., 1921. Sbor. Akad. Wiss. Wien. math-nat. Kl. I, 130, 71-108.
8. Steemann Nielsen, E., 1947. Dansk Bot. Ark. 12, 1-71.
9. Raven, J.A., 1970. Biol. Rev. 45, 167-221.
10. Smith, F.A., 1967. J. exp. Bot. 18, 509-17.
11. Smith, F.A., 1968. J. exp. Bot. 19, 207-17.
12. Lucas, W.J., and Smith, F.A., 1973. J. exp. Bot. 24, 1-14.

13. Raven, J.A., 1968. J. exp. Bot. $\underline{19}$, 193-206.

14. Lucas, W.J., 1975. J. exp. Bot. $\underline{26}$, 331-46.

15. Österlind, S., 1949. Symb. bot. upsal. $\underline{10}$, 1-137.

16. Lucas, W.J., 1975. J. exp. Bot. $\underline{26}$, 347-60.

17. Nishizaki, Y., 1968. Pl. Cell Physiol. $\underline{9}$, 377-87.

18. Andrianov, V.K., Bulychev, A.A., and Kurella, G.A., 1970. Biophys. $\underline{15}$, 199-200.

19. Spanswick, R.M., 1972. Biochim. biophys. Acta, $\underline{288}$, 73-89.

20. Volkov, G.A., 1973. Biochim. biophys. Acta $\underline{314}$, 83-92.

21. Saito, K., and Senda, M., 1973. Pl. Cell Physiol. $\underline{14}$, 147-56.

22. Saito, K., and Senda, M., 1973. Pl. Cell Physiol. $\underline{14}$, 1045-52.

23. Hansen, U.P., 1971. Biophysik $\underline{7}$, 223-27.

24. Walker, N.A., 1962. A. Rep. Div. Pl. Ind. C.S.I.R.O., pp. 80.

25. Kitasato, H., 1968. J. gen. Physiol. $\underline{52}$, 60-87.

26. Lucas, W.J., 1975. J. exp. Bot. $\underline{26}$, 271-86.

27. Richards, J.L., and Hope, A.B., 1974. J. membrane Biol. $\underline{16}$, 121-44.

28. Lucas, W.J., 1976. J. exp. Bot. $\underline{27}$, 19-31.

29. Hope, A.B., 1965. Aust. J. biol. Sci. $\underline{18}$, 789-801.

30. Lucas, W.J., Ferrier, J.M., and Dainty, J., 1976. Plasmalemma Transport of OH^- in Chara corallina: Dynamics of Activation and Deactivation. J. membrane Biol. In the Press.

31. Lucas, W.J., and Dainty, J., 1976. Spatial Distribution of Functional OH^- Carriers Along a Characean Internodal Cell: Determined by the Effect of Cytochalasin B on $H^{14}CO_3^-$ Assimilation. J. membrane Biol. In the Press.

Regulation of Cell Membrane Activities in Plants
E. Marrè and O. Ciferri eds.
© *1977, Elsevier/North-Holland Biomedical Press, Amsterdam*

REGULATION OF PHOSPHATE TRANSPORT

IN <u>CHLORELLA PYRENOIDOSA</u> AND <u>CANDIDA TROPICALIS</u>

G. Ducet, F. Blasco and R. Jeanjean

Laboratoire de physiologie cellulaire,

associé au C.N.R.S. (ERA 070619 et RCP 285)

U.E.R. de Luminy, 70 Rte. L. Lachamp.,

13288 MARSEILLE CEDEX 2

INTRODUCTION

It is well known that phosphate uptake by cells depends on trans-port by a carrier. Indeed it has been found often that cells may contain two transport systems, differing in their affinity for phos-phate. However Borst-Pauwels[1] and Jeanjean[2] suggested that the two systems may be just apparent if transport is mediated by a multisite carrier.

In many cases the second system, that in general has a higher affinity for the substrate, appears only when cells suffer from phos-phate starvation. This has been demonstrated by Rosenberg et al[3] for <u>B. cereus</u> and in <u>Candida tropicalis</u> by Blasco[4] who reported that starvation induces a high affinity phosphate transport system which disappears quickly when the deficient culture medium is supplemented with phosphate. The protein nature of the phosphate carrier was demonstrated by Medveczky and Rosenberg[5, 6] who isolated a binding protein that stimulates phosphate uptake in permease mutants or in bacteria that, following osmotic shock, have lost the capacity for phosphate transport.

Induced phosphate transport is generally described as genetically controlled by derepression. The decrease in the rate of phosphate uptake that is observed when phosphate is supplied to deficient cells is related to an allosteric effect of phosphate on the carrier.

This paper shows that assuming a direct or indirect degradation of the carrier by phosphate, it is possible to explain at last quali-tatively our experimental data as the result of the balance between continuous synthesis and degradation of the carrier.

MATERIALS AND METHODS

Growth conditions for <u>Chlorella pyrenoidosa</u> (var. Lefèvre) and <u>Candida tropicalis</u> (strain 101) as well as the other experimental methods have been reported elsewhere[2, 4].

RESULTS

Chlorella and Candida behaved differently as regard to phosphate
uptake systems.

Jeanjean[2] found that Chlorella grown on a phosphate-containing
medium has always two transport systems, one with low affinity
(app. Km   3 $\cdot 10^{-4}$ M) and the other with high affinity
(app. Km $2.5 \cdot 10^{-6}$ M). In synchronous cultures, an increase in the
uptake rate was observed after illumination. This increase was then
followed by a decrease to a constant level if the rate of uptake is
related to unit of cell surface (and not to the volume which increases
until the division phase is attained)[7]. This constant uptake rate is
observed for the two systems. During phosphate starvation the rate of
phosphate uptake increases until a plateau is reached after about 5
hours although growth does not stop before approximatively 10 hours.
The increase rate is reported to affect the two systems, but the
effects of SH reagents and pH seem to indicate that during phosphate
starvation a new transport system, more active at alcaline pH, is
synthesized[8]. These results however are not completely convincing
since they may be explained also by changes in the membrane structure.
Indeed a change in the cell membrane, which acts as a solvent for the
carrier, may affect accessibility and mobility of the carrier.

Fig. 1 to 5 Effects of different conditions on phosphate uptake by
Candida tropicalis during phosphate starvation. V in mmoles of
PO_4 uptaken per liter of cell volume per min.

Fig. 1 1, Phosphate uptake as a function of the time of starvation. At
the time indicated by the arrow CH 10^{-3} M (2), PO_4 10^{-3} M (3) or
CH + PO_4 (4) were added.

Insert : decreased rate after adding PO_4, log scale.

Fig. 2 1 as in Fig. 1. At the time indicated by the arrow PO_4 was
added to give a final concentration of 2.10^{-3} M (2)
or 10^{-5} M (3).

Fig. 3 Experimental conditions as Fig. 1 except that MP 10^{-3} M was
used in place of CH.

Fig. 4 1 as in Fig. 1. At the time indicated by the arrows CH 10^{-4} M
(2 and 4) and CH 10^{-3} M (3 and 5) were added.

6 CH 10^{-3} M from time 0, then (arrow) centrifuged washed and transferred in a medium without PO_4 and CH.

Fig. 5 1 as in Fig. 1. At the time indicated by the arrows MP 10^{-3} M
and PO_4 10^{-3} M (2), MP 10^{-3} M (3 and 4) were added.

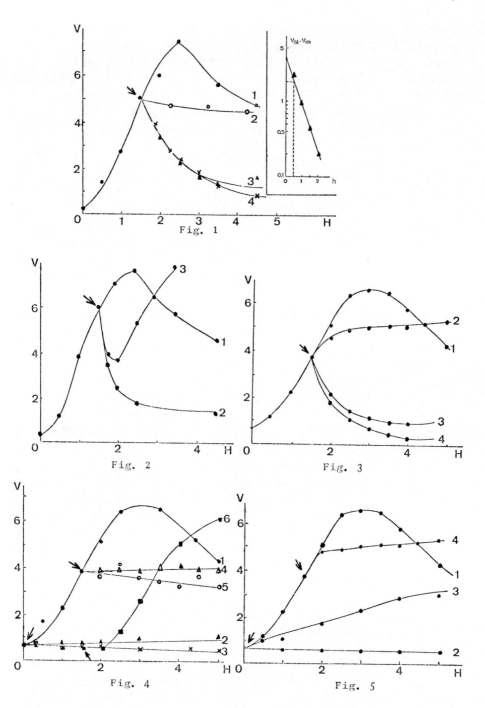

Fig. 1

Fig. 2

Fig. 3

Fig. 4

Fig. 5

Candida appears to possess only one transport system characterized by a low affinity (app. Km $1.2 \cdot 10^{-3}$ M) on phosphate-containing medium. However a second system with high affinity (app. Km $4 \cdot 10^{-6}$ M) becomes evident during phosphate starvation. The increased rate reaches a maximum which lasts a short time and thereafter a decrease in the rate is observed until a plateau is reached after about 5 hours. Candida growth stops only after 10 hours of phosphate starvation[9].

The phosphate transport systems of the two organisms have however certain common features. The increased uptake rate during phosphate starvation is dependant on protein synthesis as judged by the effect of relatively specific inhibitors such as cycloheximide (CH) and 6-methylpurine (MP). It has been demonstrated that in short term experiments CH does not act as an inhibitor of phosphate uptake. Rate increases show a lag period and representative curves are sigmoidal. Addition of phosphate induces a quick decrease in the rate of uptake, probably after a short lag period. If phosphate is added in limiting quantities, the decrease in the rate is followed by a new increase. Except for the lag period, increase and decrease seem to be exponential with a half-time smaller for decrease than for increase.

When protein synthesis inhibitors are added to phosphate starved cells, the rate of uptake appears to be blocked at the level attained before their addition or, depending on the concentration of the inhibitors, at a slightly lower or upper level. When cells starved for phosphate are treated with the inhibitors and phosphate, the decrease is more pronounced than that induced by phosphate alone. When the inhibitors are added to the phosphate-sufficient medium, the constant steady state rate decreases even at concentrations of inhibitors that have practically no effect on phosphate-starved cells.

After osmotic shock, uptake rates decrease but increase again if cells are resuspended in phosphate-deficient medium. This increase is sensitive to the action of the inhibitors.

Figures 1 to 5 report some observations obtained on Candida[4].

DISCUSSION

The above reported results may be summarized as follows :

1) The rate of phosphate uptake in non-starved cells is constant per unit of cell surface.

2) During phosphate starvation a) rate increase is sigmoidal b) the rate increases to a plateau directly for Chlorella or after reaching a maximum for Candida.

3) Addition of inhibitors of protein synthesis blocks the increase rate and even decreases it.

4) Addition of phosphate to the culture medium after phosphate starvation decreases the uptake rate.

These data suggest that the rate of uptake of phosphate results from a balance between a continuous synthesis and a continuous degradation of the carrier. Such a hypothesis assumes that the carrier be synthesized in the cytoplasm in a soluble form (Ss) which binds to the cell membrane in a transport form (St). Analytically only the transport form may be quantitatively estimated by measuring the transport rate. Degradation of the carrier results from catabolism and is also linked to phosphate concentration.

We have computed two possibilities 1) phosphate transport wears out the carrier. 2) phosphate accelerates catabolism of the soluble form. (This is analogous but not identical to the idea of carrier inactivation by phosphate). Corresponding schemes are as follows :

$$\xrightarrow{k_1} \; Ss \;\underset{\xleftarrow[\;k_4\;]{}}{\xrightarrow[\;]{k_3}}\; St \;\xrightarrow{k_5} \qquad\qquad (1)$$
$$\downarrow{\scriptstyle k_2}$$

$$\xrightarrow{k_1} \; Ss \;\underset{\xleftarrow[\;k_4\;]{}}{\xrightarrow[\;]{k_3}}\; St \qquad\qquad (2)$$
$$\swarrow{\scriptstyle k_2}\;\;\downarrow{\scriptstyle k'_2}$$

With k_1 for permanent synthesis, k_2 degradation of Ss by catabolism, k'_2 degradation of Ss in presence of phosphate in excess, k_3 and k_4 reversible binding of Ss to the cell membrane, k_5 degradation of St by phosphate transport.

Corresponding differential equations when phosphate is present are

System 1
$$\frac{dSs}{dt} = k_1 - (k_2+k_3)\,Ss + k_4\,St$$

$$\frac{dSt}{dt} = k_3\,Ss - (k_4+k_5)\,St$$

System 2
$$\frac{dSs}{dt} = k_1 - (k_2+ k'_2+k_3)\,Ss + k_4\,St$$

60

$$\frac{dSt}{dt} = k_3\ Ss - k_4\ St$$

At the beginning of starvation either k_5 or k'_2 becomes nul. For system 1 $\frac{dSt}{dt} = k_5\ St_P{}^*$ and for system 2 $\frac{dSt}{dt} = 0$. Only for system 2 is a lag period to be expected, with a sigmoidal increase of St with time to a plateau. At the beginning of starvation $St_P = \frac{k_1.k_3}{k_4\ (k'_2+k_2)}$, at the plateau $St_O = \frac{k_1.k_3}{k_4.k_2}$ and $\frac{St_O}{St_P} = \frac{k'_2+k_2}{k_2}$.

For this simple system, integration is easy :

$$St = St_{initial} + C_1\ exp.\lambda_1 t + C_2\ exp.\lambda_2 t$$

λ_1 and λ_2 are the roots of the quadratic $x^2 + (k_2+k_3+k_4)x + k_2.k_4 = 0$
C_1 and C_2 are given for initial conditions $C_1\lambda_1 + C_2\lambda_2 = 0$ and $C_1 + C_2 = St_{initial} - St_{final}$.

At the beginning of starvation, $St_{initial}$ is St_P and k'_2 is not used (case 1). When phosphate is added to starved cells, $St_{initial}$ is St_O and k'_2 acts (case 2). λ_1 and λ_2 are smaller in case 1 than in case 2 so that the decrease in rate after phosphate addition will be more rapid than the increase in the rate on phosphate starvation.

The addition of the inhibitors of protein synthesis will decrease k_1, say to k'_1 : λ_1 and λ_2 are not modified but C_1 and C_2 are. The shape of the curve will be the same but on a lower scale. Final values of St will be as the ratio of k_1 and k'_1. In the case of our results, the slow decrease observed when the inhibitors alone are employed means that inhibition is weak. This may also be deduced from the large concentrations necessary to block the rate increase, as compared to those which inhibit protein synthesis in cell free systems.

The proposed model may be employed for the results obtained in the case of Chlorella[2]. It may be more real to assume that there is a maximum number of sites (S_M) per unit of cell surface which can be occupied by the carrier. This leads to non linear differential equations :

$$\frac{dSs}{dt} = k_1 - (k_2+k'_2)\ Ss - k_3\ Ss\ (S_M-St) + k_4\ St$$

St_P and St_O are the steady state values of St when phosphate is present or not, respectively.

$$\frac{dSt}{dt} = k_3 \ Ss \ (S_M - St) - k_4 \ St$$

which can not be integrated exactly. Numerical computation with arbitrary values of the k's gives curves whose shape is not very different from those of the simple model.

On the other hand, these two models do not describe the maximum rate observed with Candida before reaching the plateau. Curves showing maxima can be obtained if there are more than two forms of carriers, i.e. soluble carrier is fixed on the cell membrane in an inactive form (St_1) which then gives rise to an active form (St_2). Degradation occurs from Ss only or from Ss and St_1. (Scheme 3).

This leads to differential equations, linear or non linear, of the same type as encountered in compartmental analysis and it is known that, depending on the k's there might be damped oscillations of St_2.

These models may represent our experimental results but, if possible, it will be necessary to compute the various k's before drawing any conclusion.

From these models it may be predicted that low concentrations of phosphate in the culture medium will give rise to a higher level of St (k'_2 is in some way related to phosphate concentration). The following table, from Jeanjean[2] shows that this prediction is confirmed.

Time of starvation hours	Relative rate of uptake in cells cultured in	
	medium containing PO_4 5.10^{-6} M	PO_4 5.10^{-4} M
1	140	100
2	200	100
5	200	100

CONCLUSION

These models correspond to a metabolic control regulated by the carrier level and may explain many experimental results in which an

increased rate of uptake is induced by starvation of inorganic or
organic substrates. In addition, in the case of phosphate, the
soluble carrier is supposed to become integrated in the cell membrane
but it may be integrated elsewhere in the cell, for example in the
mitochondrion in which there is a phosphate transport system mediated
by a carrier. If this is true, it will be possible to show (e. g. by
immunology) that the phosphate carrier of the cell membrane is identical to that present in the mitochondrion.

REFERENCES

1. Borst-Pauwels, G.W.F.H. (1974) J. Theor. Biol. $\underline{48}$ 183-195.

2. Jeanjean, R. (1974) Thèse Marseille-Luminy.

3. Rosenberg, H., Medveczky, N. and LaNauze, J.M. (1969) Biochim.
    Biophys. Acta. $\underline{193}$ 159-167.

4. Blasco, F. (1976) Thèse Marseille-Luminy.

5. Medveczky, N. and Rosenberg, H. (1970) Biochim. Biophys. Acta.
    $\underline{211}$ 158-168.

6. Medveczky, N. and Rosenberg, H. (1971) Biochim. Biophys. Acta.
    $\underline{241}$ 494-506.

7. Jeanjean, R. (1973) C. R. Acad. Sc. Paris, $\underline{277}$ 193-196.

8. Jeanjean, R. (1975) Biochimie. $\underline{57}$ 1229-1236.

9. Blasco,F., Ducet, G. and Azoulay, E. (1976) Biochimie. $\underline{58}$ 351-357.

This work was supported in part by D.G.R.S.T. contrat 650 220,1975

Regulation of Cell Membrane Activities in Plants
E. Marrè and O. Ciferri eds.
© *1977, Elsevier/North-Holland Biomedical Press, Amsterdam*

K^+-Na^+ SELECTIVITY IN ROOTS, LOCALIZATION OF SELECTIVE FLUXES AND THEIR REGULATION

Wolf Dietrich Jeschke
Botanisches Institut der Universität Würzburg
Federal Republic of Germany

SUMMARY

The K^+-Na^+ selectivity of roots and of root to shoot transport is discussed and attributed essentially to selective transport at the plasmalemma and at the tonoplast. Evidence for a K^+-Na^+ exchange system at the plasmalemma and its differential selectivity is given. The affinity of univalent cations to the external site of this system corresponds to the sequence $K^+ > Rb^+ > Cs^+ (> Na^+) \gg Li^+$; at the internal site only the affinity towards Na^+ was estimated. For a study of K^+-Na^+ selectivity at the tonoplast changes in the K^+ and Na^+ content in the apical 10 mm of barley root tips were determined. Within the zone of cell expansion substantial changes in the K^+ and Na^+ content were found when low-salt roots were incubated in a solution containing 0.2 mM K^+ and 1 mM Na^+. The results suggest that a Na^+-K^+ exchange across the tonoplast is possible under these conditions. The regulation of the vacuolar Na^+-K^+ exchange is discussed.

INTRODUCTION

A feature of many plant species is a low level of Na^+ in comparison with K^+, while Na^+ may exceed K^+ in the solution of many soils, let alone saline soils. Collander[1] has shown a wide spectrum of selectivities ranging from a strong preference for K^+ in Zea and Helianthus to a predominance of Na^+ in the shoots of the saltre-sistant species Plantago maritima and Atriplex hortensis. Except of some halophytes which excrete Na^+ in the form of NaCl the K^+-Na^+ selectivity is regulated primarily by selective uptake and transport of K^+ and Na^+ by the roots. Additionally differential retranslocation of ions within whole plants substantially can affect the selectivity. In the present paper the mechanisms of K^+-Na^+ selectivity and the localization of selective fluxes in root cells shall be discussed.

For a consideration of possible sites of selective monovalent ion transport in roots the scheme of Fig.6 will be used. The series of cells extending from the rhizodermis to the xylem vessels is represented by a single cell in this figure. Three sites of selective ion fluxes will be distinguished: 1) the plasmalemma of the cortical cells, which is in contact with the external solution; the fluxes at this membrane are the influx ϕ_{oc} and the efflux ϕ_{co}. 2) the tonoplast membrane surrounding the vacuoles of the root cells (fluxes ϕ_{cv} and ϕ_{vc}) and 3) the plasmalemma of the xylem parenchyma cells, which separates the symplasm from the xylem vessels (fluxes ϕ_{cx} and ϕ_{xc}).

1) SELECTIVITY AT THE PLASMALEMMA

In plants of barley and Sinapis Pitman[2,3] found substantially higher K^+/Na^+ ratios in the shoots compared to roots. This situation is seen in Table 1 in which the K^+-Na^+ selectivity $S_{K,Na}$ (see foot-note in Table 1) in roots and shoots of barley is shown for various external K^+ and Na^+ concentrations.

TABLE 1

K^+-Na^+ SELECTIVITY IN ROOTS AND SHOOTS OF BARLEY PLANTS[*]

External solution		Selectivity $S_{K,Na}$[**]				
		Roots				
$[K^+]$ mM	$[Na^+]$ mM	Whole tissue	Vacuole	Cytoplasm	Exudate[***]	Shoots
0.5	9.5	15	–	–	40	65
2.5	7.5	5.8	8.6	34.8	21	24
8	2	2.2	–	–	9	7.5

[*]from data of Pitman[2] and Pitman and Saddler[4]

[**]the selectivity is expressed by the ratio $S_{K,Na} = \dfrac{[K^+]_i \cdot [Na^+]_o}{[K^+]_o \cdot [Na^+]_i}$ which relates the internal concentrations of the ions to the concentration in the external solution.

[***]xylem exudate from detopped barley plants

The high selectivity in the shoots is related to $S_{K,Na}$ of the xylem exudate rather than to $S_{K,Na}$ in the roots (Table 1). The apparent paradoxon that roots supply shoots with K^+ and Na^+ at a higher $S_{K,Na}$ than is found in the root cells can be resolved if the supply of K^+ and Na^+ to the xylem sap occurs symplasmatically, pro-vided the ratio K^+/Na^+ in the root symplasm is high (Pitman[2]). The K^+-Na^+ selectivity then is set up at the plasmalemma of the root cells.

Flux measurements. This suggestion was confirmed by flux measure-ments (Pitman and Saddler[4]) showing that $S_{K,Na}$ in the cytoplasm of barley roots was 34.9 and thus higher than in the xylem exudate; the selectivity $S_{K,Na}$ in the root vacuoles was 8.6 or close to the low selectivity in the root tissue as a whole (Table 1). As the influxes ϕ_{oc} of K^+ and Na^+ were similar in these experiments, Pitman and

Saddler[4] concluded that the high K^+-Na^+ selectivity in the cytoplasm was due to a Na^+ extrusion pump at the plasmalemma of root cells and that this pump was linked to an active influx of K^+.

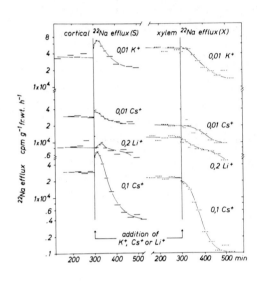

Fig.1. Effect of an addition of K^+, Cs^+ or Li^+ to the external solution on the cortical (S) or xylem (X) effluxes of ^{22}Na in barley roots. Roots were loaded for 22 h in 1 mM "Na-solution" containing in mequ 1^{-1}: NaH_2PO_4/Na_2HPO_4 (pH 5.8): 1; $Ca(NO_3)_2$: 6; $MgSO_4$: 1 and ^{22}Na at constant specific activity. They were washed out with unlabelled Na^+-solution for 5 h until a quasi-steady state was reached. At t = 300 min the elution was continued with 1 mM Na^+-solution + KCl, CsCl or LiCl at the concentrations (mM) indicated in the figure. The specific activity of ^{22}Na was different in the experiments, but the relative stimulations of the cortical efflux are directly comparable on the logarithmic ordinate scale (from Jeschke[19]).

A stimulation of the Na^+ efflux at the plasmalemma by K^+ in the external solution has been measured directly (Jeschke[5,6]). In these experiments Na^+-loaded barley roots were used, which had been equilibrated with 1 mM Na^+-solution (for the composition, see legend of Fig.1). When low concentrations of K^+ were added, there was a strong but transient increase in the tracer Na^+ efflux (Jeschke[6]) or a transient net Na^+ efflux (Jeschke and Stelter[7]) across the plasmalemma of the cortical cells. The K^+-dependent stimulation of the Na^+ efflux was followed by a severe decrease in the xylem transport of Na^+ and a slower increase in trans-root K^+ transport. As an example, the effect of 0.01 mM K^+ on the cortical and xylem Na^+ efflux in barley roots in the presence of 1 mM Na^+ is seen in Fig.1.

The K^+-dependent part of the stimulated Na^+ efflux ϕ_{co,K^+-dep} responds to the external K^+ concentration according to Michaelis-Menten kinetics and has the same Michaelis constant with respect to $[K^+]_o$ as system I K^+ uptake (Epstein et al.[8]) and as the initial net K^+ uptake in the same roots during K^+-Na^+ exchange (Jeschke[6,9]).

Besides K^+, Rb^+ and Cs^+ can stimulate the Na^+ efflux and inhibit the trans-root Na^+ transport selectively in barley (Fig.1, the effect of Rb^+ is not shown, it is similar to that of K^+). The relative affinities of various univalent cations to the K^+-Na^+ exchange system have been estimated by measuring the cation-dependent Na^+ efflux as a function of external concentrations. Again saturation curves were obtained, the K_m values suggest the following selectivity sequence:

$$K^+ > Rb^+ > Cs^+ (> Na^+) \gg Li^+ \qquad (1)$$

In this sequence Na^+ is included, its affinity follows from an inhibition of ϕ_{co,K^+-dep} by higher external Na^+ concentrations (Jeschke[9]) and from influx measurements (Rains and Epstein[10]) showing that Na^+ influx occurs at the same site as K^+ influx.

If there is an exchange of cytoplasmic Na^+ for external K^+ at the plasmalemma, Na^+ should have a finite affinity to the cytoplasmic site of the system which mediates this exchange. This affinity was estimated by using roots with varied and measurable cytoplasmic Na^+ contents $Q_{c(Na)}$ and inducing a K^+-stimulated Na^+ efflux by the addition of K^+ at constant $[K^+]_o$. The K^+-dependent Na^+ effluxes measured in these roots were not linearly related to $Q_{c(Na)}$, rather did they obey to a Michaelis Menten curve within the experimental errors. The apparent K_m corresponds to a $Q_{c(Na)}$ of $1.4 \, \mu mol \, g \, fr.wt.^{-1}$ or to a cytoplasmic Na^+ concentration of 28 mM (if the cytoplasm is assumed to occupy 5 % of the tissue volume). K^+ appears to have a lower affinity to the efflux site as Na^+ failed to stimulate the K^+ efflux in barley roots.

These data are in agreement with the presence of a K^+-Na^+ exchange system at the plasmalemma (1 in Fig.6). As discussed elsewhere (Jeschke[9]) the selectivity sequence (1) and the effect of inhibitors (Jeschke[11]) could indicate that the system is related to the membrane-bound ATPases of plant roots (Hodges[12]). Until now it cannot be decided from our data, whether K^+ influx and Na^+ efflux are linked in one transport system or whether proton extrusion is involved in K^+ influx (Pitman[13]) and proton influx mediates Na^+ efflux as suggested for bacteria (Harold and Papineau[14]). Certainly the stoichiometry between K^+ influx and Na^+ efflux is higher than 1 since net cation (K^+) uptake is normally present in addition to K^+-Na^+ exchange. The mechanism by which K^+ influx and Na^+ efflux are linked to each other require studies of the relation between proton fluxes and the K^+-Na^+ exchange.

The scheme in Fig.6 obviously permits two modes of K^+-Na^+ selectivity at the plasmalemma: selective uptake due to the affinity sequence at the external site (1) as suggested by Rains and Epstein[10] and selectivity by K^+-Na^+ exchange (Pitman and Saddler[4], Jeschke[6]). The relative contribution of these mechanisms will be referred to in the conclusion.

Electrochemical measurements. Active Na^+ extrusion as suggested by Pitman and Saddler[4] demands Na^+ to be at a lower electrochemical potential in the cytoplasm than in the external solution. In several species this is true for the intracellular or vacuolar Na^+ (Table 2) as judged by the Nernst equation. For other species (Table 2, d,e) in which the electropotential ψ_{vo} was remarkably low, Na^+ was suggested to be actively transported to the vacuole (Bowling and Ansari[15]; Shepherd and Bowling[16]) when $[Na^+]_o$ was low.

TABLE 2

COMPARISON OF OBSERVED K^+ AND Na^+ CONCENTRATIONS IN ROOTS WITH THE VALUES PREDICTED FROM THE NERNST EQUATION

Unless otherwise indicated the values of $[K^+]_i$ and $[Na^+]_i$ are concentrations in the whole tissue, i.e. essentially vacuolar concentrations. The index v refers to vacuolar, the index c to cytoplasmic concentrations. ψ_{vo} is the vacuolar electropotential

Plant, reference	Ext. conc. mM K^+	Ext. conc. mM Na^+	ψ_{vo} mV	$[K^+]_i$ mM pred.	$[K^+]_i$ mM obs.	$[Na^+]_i$ mM pred.	$[Na^+]_i$ mM obs.
Hordeum, a	0.22	9.9	-110	17_v	34_v	750_v	68_v
b	-	1	- 90	-	-	36_c	92_c
b	0.2	1	-100	-	-	53_c	15_c
Vicia, c	1	1	-112	80	83	80	22
Helianthus, d	7	0.25	- 28	21	111	0.8	19
	7	10	- 32	25	83	35	39
Potamogeton,e	0.055	0.89	- 33	0.2	84	3.3	23
Allium, f	1	1	- 32	3.5_c	100_c	3.5_c	9.4_c

a) Pitman and Saddler[4], b) Jeschke[9], c) Pallaghy and Scott[24], d) Bowling and Ansari[15], e) Shepherd and Bowling[16], f) Macklon[17]

Only few data are available for the cytoplasmic Na^+ concentration. In onion roots (ψ_{vo} low) Na^+ appeared to be actively transported to the cytoplasm (Macklon[17]) but the difference in electrochemical potential was significantly lower than for K^+ (Table 2, f) showing a preference for K^+ also in this species. In barley roots the driving force on Na^+ is from the external solution to the cytoplasm (ϕ_{co} active) when Na^+ and K^+ are present externally (Table 2b). On the other hand, ϕ_{co} is passive in Na^+-loaded roots ($[Na^+]_o$ = 1 mM, $[K^+]_o$ = 0, Table 2b).

The electrochemical data thus are in agreement with an active extrusion of Na^+ in Vicia and Hordeum, when K^+ is present and $[Na^+]_o$ is not very low. They show on the other hand, that the K^+-stimulated Na^+ efflux measured routinely with Na^+-loaded roots (cp. Fig.1) initially occurs in the direction of the passive driving force. But the low steady state Na^+ level which is reached at the end of the transient fluxes is below the electrochemical potential in the external solution and requires K^+-linked active Na^+ efflux.

2) SELECTIVE K^+ AND Na^+ TRANSPORT AT THE TONOPLAST

The low $S_{K,Na}$ in vacuoles of barley root cells compared to the cytoplasm (Table 1) could not readily be explained if the transport of K^+ and Na^+ across the tonoplast were governed by the selectivity $S_{K,Na}$ (high K^+/Na^+ ratio) in the cytoplasm and if K^+ and Na^+ had equal affinities to the transport site(s) at the tonoplast. For excised barley roots the discrepancy between $S_{K,Na}$ in the vacuole and the cytoplasm was shown to be due to a regulatory change in K^+-Na^+ selectivity during transition from low-salt to high-salt state (Pitman et al.[18]). Initially the roots showed little preference for K^+ and a K^+-Na^+ discrimination was detectable only after 5 h salt uptake from a solution containing 2.5 mM K^+ and 7.5 mM Na^+. It was suggested that during initial uptake the K^+/Na^+ ratio in the cytoplasm was low and that the low $S_{K,Na}$ in the vacuoles was due to a considerable Na^+ accumulation during early stages of salt uptake.

That K^+ is preferentially located in the cytoplasm and Na^+ in the vacuoles of barley roots also at low external concentrations follows from Fig.2. Here the ratio of vacuolar and cytoplasmic Na^+ contents was measured by compartmental analysis at 1 mM $[Na^+]_o$ and varied $[K^+]_o$. 0.05 mM $[K^+]_o$ was sufficient to change this ratio from 18 in the absence of K^+ to about 50, indicating that only 2 % of the cellular Na^+ is then located in the cytoplasm. Compartmental analysis showed a preferentially vacuolar localization of Na^+ as opposed to K^+ also in onion roots (Macklon[17]) or coleoptile cells (Pierce and Higinbotham[20]). This differential intracellular localization

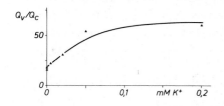

Fig.2. Dependence of the ratio vacuolar/cytoplasmic Na^+ content of barley roots (Q_v/Q_c) on the external K^+ concentration. Q_v and Q_c were measured by compartmental analysis (Jeschke[6]). Conditions and solution as in Fig.1, except that K^+ at varied concentration was continuously present through-out the experiments.

of K^+ and Na^+ may be deduced also from their distribution along the axis of roots which shows a pronounced accumulation of K^+ in the meristematic cells near the root tips (Brown and Cartwright[21]; Scott et al.[22]; Jeschke and Stelter[23] and literature cited therein).

Electrochemical evidence (Pierce and Higinbotham[20]; Macklon[17]) indicates active Na^+ transport across the tonoplast to the vacuole while K^+ transport between the cytoplasm and vacuoles of root cells often occurs close to electrochemical equilibrium (Pallaghy and Scott[24]; Macklon[17]; Davis and Higinbotham[25]). Jennings[26,27] drew attention to the significance of an inwardly directed Na^+ pump at the tonoplast for the K^+-Na^+ selectivity and to its significance in pro-ducing a low cytoplasmic Na^+ concentration in halophytes.

Conceivably an active transport of Na^+ across the tonoplast could be associated with a reverse K^+ transport for maintenance of electro-neutrality. This would in effect be a Na^+-K^+ exchange across the tonoplast and remove Na^+ from the cytoplasm. It may be asked, whether this exchange does in fact occur.

Indications for such an exchange can be derived from the results of Hooymans[28]. As opposed to Na^+, which appeared to be "irreversibly" accumulated in the vacuoles (Neirinckx and Bange[29]; Hooymans[28]), previously accumulated K^+ was transported from roots to shoots of barley plants. This transport was enhanced when K^+ in the external solution was replaced by 5 mM Na^+. Although these results were inter-preted to be due to an effect of Na^+ in the cytoplasm (Hooymans[28]), the amounts of Na^+ taken up (20 μmol g fr.wt.$^{-1}$) appear to be to large to be located in the cytoplasm. It is suggested therefore that the effect of Na^+ was due to a Na^+-K^+ exchange at the tonoplast.

This exchange could be demonstrated directly if Na^+ would increase the vacuolar efflux ϕ_{vc} of K^+. However, compartmental analysis hardly is suitable to show this increase as steady state conditions are re-quired and as separate and reliable determinations of ϕ_{vc} and ϕ_{cv} in roots are not possible with the present methods (Pitman[30]; Cram[31]).

70

Alternatively it will have to be shown that K^+ which was previously contained in the vacuoles can be replaced by Na^+. In order to investigate vacuolar K^+ and Na^+ exchanges ion profiles along the axis of single roots of barley and Atriplex were measured. These profiles can be used to determine the intracellular localization of ions, since the meristematic cells near the tip and the elongating or fully differentiated cells are distinguished by the proportions of cytoplasmic and vacuolar volumes in these cells (Jeschke and Stelter[23]).

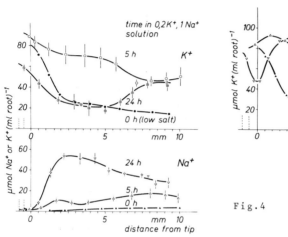

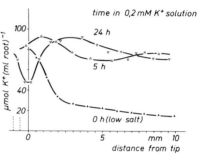

Fig.4

Fig.3

Fig.3 and 4. Longitudinal profiles of the K^+ and Na^+ content [/umol (ml root)$^{-1}$] in barley roots grown after germination for 4 days in 0.5 mM CaSO$_4$-solution (low-salt) and in roots, which were then excised and incubated for 5 or 24 h in solutions containing 0.2 mM K^+ and 1 mM Na^+ (Fig.3) or 0.2 mM K^+ (Fig.4). The solutions contained in mequ l^{-1}: NaH$_2$PO$_4$/Na$_2$HPO$_4$ (pH 5.8): 1; KCl: 0.2; Ca(NO$_3$)$_2$: 6; MgSO$_4$: 1, (Fig.3) or KH$_2$PO$_4$/K$_2$HPO$_4$ (pH 5.8): 0.2 plus Ca^{++} and Mg^{++} (Fig.4). The vertical bars denote the standard error of the mean (n = 3-4). The K^+ and Na^+ profiles were measured in the same roots. The profiles taken at 5 and 24 hrs were shifted along the abscissa according to root growth.

Longitudinal K^+ and Na^+ profiles of low-salt barley roots (Fig.3) show a pronounced - and presumably cytoplasmic - accumulation of K^+ near the tip as is known for roots grown in the presence of K^+ and Na^+ (Scott et al.[22]). The changes in ion profiles that occurred when excised low-salt roots were incubated in a solution containing 0.2 mM K^+ and 1 mM Na^+, are shown in Fig.3. During the first 5 h there was a substantial increase in the K^+ content, which was followed by a decrease between 5 and 24 h; Na^+, on the other hand, was taken up con-

present results appear to suggest the opposite changes. But this is not necessarily a contradiction. Firstly the events close to the root tips could not be detected in experiments with long excised roots. Secondly and more important, an increase in K^+ content and not a rapid influx of K^+ is observed in Fig.3. Together with the results of Scott et al.[22] K^+ influx may be expected to be low and comparable to Na^+ influx. The expanding cells then resemble the mature cells (Pitman et al.[18]) in having a low initial K^+-Na^+ selectivity. The crucial difference to the mature cells is the rapid increase in (vacuolar) K^+ content which is prerequisite to the Na^+-K^+ exchange seen in Figs. 3 or 5.

Moreover the models of Pitman et al.[18] and ours do not exclude each other. The additional assumption of a higher affinity of Na^+ for influx at the tonoplast agrees with the low initial K^+/Na^+ selectivity in excised roots.

Comparing the results of Pitman et al.[18] with those of Fig.3 and 5, it is evident that the Na^+-K^+ exchange seen in Fig.5 shows p o s s i b l e net fluxes which c a n occur in a certain physiological state. They indicate that Na^+ c a n be translocated into the vacuole in exchange for K^+ and thus reveal the affinities which govern the distribution of K^+ and Na^+ between cytoplasm and vacuoles.

As suggested by Jennings[27] preferential and active transport of Na^+ to the vacuoles could be of great importance in halophytes. This appears even more important in the model of Wyn Jones et al.[33] suggesting that K^+ and betaine or other suitable organic molecules act as the cytoplasmic osmotica while Na^+ salts are occluded in the vacuoles of halophytes. While flux data for Triglochin maritima (Jefferies[34]) indicate passive distribution of Na^+ between cytoplasm and vacuole, the suggestion of Wyn Jones et al.[33] is supported by results with Atriplex (Jeschke and Stelter[23]) which show cytoplasmic localization of K^+ and suggest an exchange of vacuolar K^+ for Na^+.

3) SELECTIVITY AT THE SYMPLASM - XYLEM BOUNDARY.

In barley the K^+-Na^+ selectivity set up at the plasmalemma (Pitman and Saddler[4]; Jeschke[6]) appears sufficient to explain the high K^+/Na^+ ratios in the xylem exudate and in the shoot (Table 1). But as ions are transported across a membrane before entering the xylem vessels, differential selectivity towards monovalent ions may be expected. This is true no matter whether the transfer occurs by passive movement across the plasmalemma of xylem parenchyma cells

(Bowling[35]), whether it includes a second active step (Läuchli, Spurr and Epstein[36]; Pitman[37]) or whether it involves uptake across the tonoplast of living vessel cells as envisaged by Higinbotham et al.[38]:

Some Na^+-K^+ discrimination at the symplasm - xylem boundary might be indicated by the higher $S_{K,Na}$ in the cytoplasm compared to the xylem exudate (Table 1). Evidence for a discrimination of K^+ in favour of Rb^+ in the transport from root to shoot was obtained by Marschner and Schimansky[39]. This was more pronounced in high-salt compared to low-salt roots. In the re-elution measurements (Fig.2) also the net uptake and transport of K^+, Rb^+ and Cs^+ were measured. Net K^+ uptake by Na^+-loaded roots was somewhat faster than uptake of Rb^+ and Cs^+. But there was a pronounced discrimination of K^+ in favour of Rb^+ and particularly of Cs^+ in transport across the roots. Although K^+, Rb^+ and Cs^+ transport can be compared only when their cytoplasmic concentrations are known, the results and those of Marschner and Schimansky[39] suggest that Rb^+ and particularly Cs^+ are less effectively translocated from the symplasm to the xylem vessels as indicated in Fig.6. Additionally inhibitory effects of Rb^+ and especially of Cs^+ in the absence of K^+ (Pirson[40]; El-Sheikh and Ulrich[41]) may contribute to the low transport of these ions.

Selective u p t a k e of Na^+ at the xylem-symplast boundary has been envisaged as a mechanism of selective K^+ and Na^+ transport. This re-absorption of Na^+ occurred in elder parts of roots (Jacoby[42]; Shone et al.[43]) or in stem tissue (Jacoby[42]; Rains[44]), it was evident when Na^+ was applied locally to maize roots grown in K^+ containing solutions (Marschner and Richter[45]). It is suggested that this re-absorption is due to the Na^+-selective system at the tonoplast (2 in Fig.6), which maintains $\left[Na^+\right]_c$ at a low and $\left[K^+\right]_c$ at a high value. When $\left[Na^+\right]$ in the xylem sap is high this can lead to a re-absorption of Na^+ which is then accumulated in vacuoles, possibly by an exchange for K^+. When roots as a whole are exposed to mineral nutrients as in the soil or in a culture solution, vacuolar accumulation of Na^+ probably occurs while Na^+ is moved towards the xylem vessels (Jennings[26]). Under these conditions re-absorption in the roots may be of minor significance compared to that in the stems.

CONCLUSIONS

K^+-Na^+ selectivity in roots and of root to shoot transport is suggested essentially to be due to selective transport at two sites, the plasmalemma (1) and the tonoplast (2 in Fig.6).

1) Selectivity is determined   p r i m a r i l y   at the plasmalemma. The exchange system (1 in Fig.6) links part of the K^+ influx
to Na^+ efflux theirby decreasing $[Na^+]_c$ and Na^+ transport across the
roots. The stoichiometry K^+/Na^+ is higher than 1: K^+ influx exceeds
even the transient Na^+ efflux when Na^+-loaded roots are exposed to

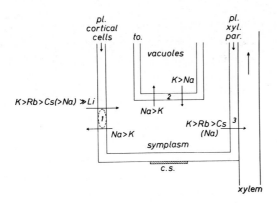

Fig.6. Diagram to show the localization of univalent cation-specific
transport sites within a root. The "cell" represents the series of
cells extending from the cortex to the xylem vessels within the
stele. pl. = plasmalemma; to. = tonoplast; xyl.par. = xylem parenchyma; c.s. = casparian strip.
    The arrows indicate the directions of   p o s s i b l e   net ion
fluxes involved in univalent cation transport. The indicated linkage
of K^+ influx and Na^+ efflux at the plasmalemma (1) does not imply a
stoichiometry 1/1. Cation influx at this site can also be linked to
anion influx or H^+ efflux. The arrows do not specify which transport
is active although Na^+ efflux at (1) and Na^+ influx at (2) appear to
involve active transport. The sequences as $Na^+ > K^+$ refer to
a f f i n i t i e s   and not to the magnitudes of fluxes.

external K^+; in the continuous presence of K^+ and normal concentrations of Na^+ (about 1 mM for barley) the K^+-dependent Na^+ efflux
accounts for a small fraction of the net K^+ influx only. The system
has the same high affinity towards K^+ as the kinetic system I of ion
uptake. However, in barley roots K^+-Na^+ exchange by this system is
effective up to concentrations of 10 mM Na^+ (Jeschke[9]).

    As seen in Fig.6, cation uptake by this system is K^+-Na^+ selec-
tive also as a consequence of the affinity sequence (1) at its external site. This type of uptake selectivity which was attributed to
carrier competition (Rains and Epstein[10]) may be significant under
conditions of increased uptake as during initial salt uptake by low-
salt roots (Rains and Epstein[10]; Pitman et al.[18]) or when increased
transpiration decreases the symplasmatic ion content (Pitman[46]). As

shown elsewhere (Jeschke[9]) carrier competition per se is not sufficient, however, to explain the finding that Na^+ influx is "virtually abolished" (Rains and Epstein[10]) in the presence of K^+. This can be accounted for by the combined effect of K^+ in competing effectively with Na^+ for influx and in effecting a K^+-linked Na^+ efflux.

2) Although K^+-Na^+ selectivity primarily is set up at the plasmalemma and the main function of the tonoplast appears to be salt u p t a k e   for generating turgor, the Na^+-K^+ selective properties of the tonoplast should not be neglected. Electrochemical measurements (Pierce and Higinbotham[20]; Macklon[17]) demand active tonoplast influx of Na^+ more than of K^+. The K^+/Na^+ ratio inside the tonoplast is much smaller than outside (Pitman and Saddler[4]; Macklon[17]; Pierce and Higinbotham[20]). This may be described as in Fig.6 by indicating that - equal Na^+ and K^+ activities on both sides provided - net influx of Na^+ across the tonoplast is favoured above that of K^+ while K^+ efflux when present is favoured over that of Na^+. As the activities of K^+ and Na^+ are not equal in the cytoplasm, the   m a g n i -
t u d e   of K^+ influx normally exceeds that of Na^+. The Na^+-selective properties of the tonoplast are revealed during the initial salt uptake by low-salt roots (Pitman et al.[18]). A substantial net Na^+-K^+ exchange across the tonoplast as suggested by Fig.6 apparently can be seen only under certain physiological conditions (Figs.3 and 5) or when the external medium is fully exchanged (Hooymans[28]). However, slow net Na^+-K^+ exchange across the tonoplast may be important during redistribution of K^+ from senescing tissues. The Na^+-K^+ selective properties of the tonoplast are suggested to be responsible also for the re-absorption or retention of Na^+ in roots (Jacoby[42]; Marschner and Richter[45]). In these cases, however, net increases in the vacuolar Na^+ content will have to be shown experimentally.

3) The relative contributions of selective K^+ and Na^+ transport at the plasmalemma and the tonoplast may vary considerably between different species. Especially in halophytes (Flowers[47]) may the K^+-Na^+ selectivity at the plasmalemma be smaller and Na^+ transport to the shoot be favoured.

ACKNOWLEDGEMENTS

This study was supported by the Deutsche Forschungsgemeinschaft. The skillful technical assistance of Miss H. Eschenbacher and G. Carl is gratefully acknowledged.

REFERENCES

1. Collander, R. (1941) Plant Physiol. 16, 691-720.

2. Pitman, M.G. (1965) Aust. J. biol. Sci. 18, 10-24.

3. Pitman, M.G. (1966) Aust. J. biol. Sci. 19, 257-269.

4. Pitman, M.G. and Saddler, H.D.W. (1967) Proc. nat. Acad. Sci. (U.S.) 57, 44-49.

5. Jeschke, W.D. (1970) Planta 94, 240-245.

6. Jeschke, W.D. (1973) in Ion transport in plants (W.P. Anderson, ed.) p. 285-296, London-New York: Academic Press.

7. Jeschke, W.D. and Stelter, W. (1973) Planta 114, 251-258.

8. Epstein, E., Rains, D.W., Elzam, O.E. (1963) Proc. nat. Acad. Sci. (U.S.) 49, 684-692.

9. Jeschke, W.D. (1976) in preparation.

10. Rains, D.W. and Epstein, E. (1967) Plant Physiol. 42, 314-318.

11. Jeschke, W.D. (1974) in Membrane transport in plants (U. Zimmermann and J. Dainty, eds.) p. 397-405, Berlin-Heidelberg-New York: Springer-Verlag.

12. Hodges, T.K. (1973) Advan. Agron. 25, 163-207.

13. Pitman, M.G. (1970) Plant Physiol. 45, 787-790.

14. Harold, F.M. and Papineau, D. (1972) J. Membrane Biol. 8, 45-62.

15. Bowling, D.J.F. and Ansari, A.Q. (1971) Planta 98, 323-329.

16. Shepherd, U.H. and Bowling, D.J.F. (1973) New Phytol. 72, 1075-1080.

17. Macklon, A.E.S. (1975) Planta 122, 109-130.

18. Pitman, M.G., Courtice, A.C., Lee, B. (1968) Aust. J. biol. Sci. 21, 871-881.

19. Jeschke, W.D. (1976) Transmembrane ionic exchanges in plants. Internat. Workshop Rouen - Paris.

20. Pierce, W.S. and Higinbotham, N. (1970) Plant Physiol. 46, 666-673.

21. Brown, R. and Cartwright, P.M. (1953) J. exp. Bot. 4, 197-221.

22. Scott, B.I.H., Gulline, H., Pallaghy, C.K. (1968) Aust. J. biol. Sci. 21, 185-200.

23. Jeschke, W.D. and Stelter, W. (1976) Planta 128, 107-112.

24. Pallaghy, C.K. and Scott, B.I.H. (1969) Aust. J. biol. Sci. 22, 585-600.

25. Davis, R.F. and Higinbotham, N. (1976) Plant Physiol. 57, 129-136.

26. Jennings, D.H. (1967) New Phytol. 66, 357-369.

27. Jennings, D.H. (1968) New Phytol. 67, 899-911.

28. Hooymans, J.J.M. (1974) Z. Pflanzenphysiol. 73, 234-242.

29. Neirinckx, L.J.A. and Bange, G.G.J. (1971) Acta Bot. Neerl. 20, 481-488.

30. Pitman, M.G. (1971) Aust. J. biol. Sci. 24, 407-421.

31. Cram, W.J. (1975) in Ion transport in plant cells and tissues (D.A. Baker and J.L. Hall, eds.) p. 161-191, Amsterdam - London: North-Holland.

32. Richter, Ch. and Marschner, H. (1973) Z. Pflanzenphysiol. 70, 211-221.

33. Wyn Jones, R.G., Storey, R., Pollard, A. (1976) Transmembrane ionic exchanges in plants. Internat. Workshop, Rouen-Paris.

34. Jefferies, R.L. (1973) in Ion transport in plants (W.P. Anderson, ed.) p. 297-321, London-New York: Academic Press.

35. Bowling, D.J.F. (1973) in Ion transport in plants (W.P. Anderson, ed.) p. 483-491, London-New York: Academic Press.

36. Läuchli, A., Spurr, A.R., Epstein, E. (1971) Plant Physiol. 48, 118-124.

37. Pitman, M.G. (1972) Aust. J. biol. Sci. 25, 243-257.

38. Higinbotham, N., Davis, R.F., Mertz, S.M., Shumway, L.K. (1973) in Ion transport in plants (W.P. Anderson, ed.) p. 493-506, London-New York: Academic Press.

39. Marschner, H. and Schimansky, Ch. (1971) Z. Pflanzenernähr. Bodenk. 128, 129-143.

40. Pirson, A. (1939) Planta 29, 231-261.

41. El-Sheikh, A.M., Ulrich, A., Broyer, T.C. (1967) Plant Physiol. 42, 1202-1208.

42. Jacoby, B. (1964) Plant Physiol. 39, 445-449.

43. Shone, M.G.T., Clarkson, D.T., Sanderson, J. (1969) Planta 86, 301-314.

44. Rains, D.W. (1969) Experientia 25, 215-216.

45. Marschner, H. and Richter, Ch. (1973) Z. Pflanzenernähr., Bodenk. 135, 1-15.

46. Pitman, M.G. (1965) Aust. J. biol. Sci. 18, 987-998.

47. Flowers, T.J. (1975) in Ion transport in plant cells and tissues (D.A. Baker and J.L. Hall, eds.) p. 309-334, Amsterdam - London: North-Holland.

Regulation of Cell Membrane Activities in Plants
E. Marrè and O. Ciferri eds.
© *1977, Elsevier/North-Holland Biomedical Press, Amsterdam*

SUGAR-PROTON COTRANSPORT SYSTEMS

W.Tanner, E.Komor, F.Fenzl and M.Decker
Fachbereich Biologie und Vorklinische Medizin
Botanik I, Universität Regensburg
8400 Regensburg, FRG

INTRODUCTION

In the past few years it has become evident that sugars and amino acids are actively transported by most biological systems together with cations. A ternary complex consisting of the substrate, the cation and a transport protein of the membrane is assumed to be formed; the membrane potential and the concentration difference of the cation are the driving forces for this active transport. Whereas in animals the cation playing such an important role is the sodium and the corresponding transport processes are called sodium cotransport [1-3] systems, the proton plays the analogous role in a number of bacteria [4,5], fungi [6-8], and algae [9,10]. So far this clear division into a sodium and a proton cotransporting world knows only few exceptions (e.g. ref 11).

In the following some well and some less well understood features of sugar-proton cotransport in Chlorella vulgaris shall be summarized and discussed in comparison with other sugar-proton cotransport systems.

MATERIALS AND METHODS

They have been published previously [9,10,12-21].

RESULTS AND DISCUSSION

1. How many protons enter the cell together with the sugar?
Hexose transport in autotrophically grown Chlorella vulgaris cells is an inducible process [12,13]. When glucose or 6-deoxyglucose is added to a suspension of induced cells, protons rapidly disappear in a transient manner [9] and after correcting for the buffer capacity of the suspension, a stoichiometry of 1 H^+ entering the cell per 1 sugar has been measured. For the sugar analogues 3-O-methylglucose and 1-deoxyglucose, on the other hand, significantly higher values have been obtained (Table 1), although these sugars are taken up by the same system as glucose and 6-deoxyglucose [14].

TABLE 1

H^+/Sugar Stoichiometries

Organism	Sugar	H^+/Sugar	Ref.
Chlorella vulgaris	glucose	1.0	
	6-deoxyglucose	1.0	
	3-0-methylglucose	1.6	10
	1-deoxyglucose	2.0	
E.coli	lactose	1.0	22
Saccharomyces carlsbergensis	maltose	2-3	6
Neurospora crassa	glucose	0.8-1.4	7

In Table 1 the proton:sugar stoichiometries known so far from other
organisms are included. Although it is conceivable,of course, that
different stoichiometries are realized in nature in different organisms, it has to be pointed out that there exists a number of reasons, why the apparent stoichiometry might be either too high or
too low. An example of the former case observed with Chlorella will
be discussed in 4 (1), an example of the latter one in 3. In addition, the measured value obviously also will be too low if the rate
of H^+ disappearance from the medium is a net rate determined by the
actual rate of H^+ entry counterbalanced partly by an increased
pumping out of protons.

   Proton cotransport with substrate can only work if the protons
running into the cell down their electrochemical gradient are pumped
out of the cell again. For this a proton translocating ATPase is
assumed to exist in all cells possessing proton cotransport systems.
Evidence for such an ATPase is available from bacterial [23] and
fungal cells [8,24]. In Chlorella a respiratory increase is observed,
when 6-deoxyglucose is added to induced cells. From the extra oxygen
taken up the requirement of 1 ATP per 1 sugar transported can be calculated [15]. A H^+-translocating ATPase in Chlorella should therefore
pump 1 H^+ per 1 ATP hydrolyzed.

   2. How much work can the proton motive force (pmf = Δp) do?
Since the pmf = $\Delta\Psi$-ZpH [25] it was necessary to try to determine the
pH difference between the medium and the inside of Chlorella as
well as the membrane potential across the plasmalemma. With these
data it would be possible to decide, whether the proton motive
force can completely account for the degree of sugar analogue accu-

mulation observed. The estimate of the inner pH with the weak acid
5,5-dimethyl-2,4-oxazolidine-dion (DMO) yielded an inside pH of
about 7 at an outside one of 6 (fig. 1).

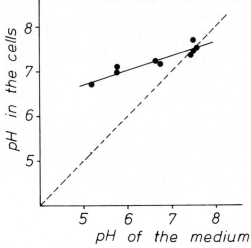

Fig. 1. Cellular pH in
relation to the extra-
cellular one. For de-
tails see ref 16.

At an outside pH of 7.5 and above no ΔpH exists any more. Thus the
ΔpH observed alone is neither able to explain, why Chlorella cells
accumulate 6-deoxyglucose 1600 fold [17] at pH 6.0 nor why a considerable accumulation is observed even beyond an outside pH of 8 [16].

To estimate the membrane potential the equilibrium distribution
of the lipophilic cation tetraphenylphosphonium (TPP^+) [3] can be used
as an good sensor for the membrane potential $\Delta\Psi$ of Chlorella [18]; a
value of 135 mV inside negative has been obtained. Although it is
realized that with this method one runs into the same problem of
measuring some ill-defined overall $\Delta\Psi$, as one measures an overall
ΔpH with the DMO method, but controls indicated, that the possible
contribution to the overall $\Delta\Psi$ by cell organells should have been
only about -20 mV.

From these values then a proton motive force of roughly 200 mV
(or 4,6 Kcal/mole) can be calculated, which indeed is sufficient to
explain a sugar accumulation of 1600 fold (Table 2). However, these
considerations are valid only, if sugar uptake in Chlorella proceeds
electrogenic, which means that a net charge has to be transported
during each cycle of carrier-sugar/carrier movement. That this is
the case in Neurospora has beautifully been demonstrated by Slayman
and Slayman [7]. As in Neurospora electrogenic sugar transport in
Chlorella would be expected to transiently depolarize the membrane
potential. Since not only the equilibrium distribution of TPP^+ but

TABLE 2

Free energy (ΔG) values of maximal accumulation of
6-deoxyglucose and of the proton motive potential
difference in Chlorella. For details see ref 18.

		ΔG (Kcal/mole)
Maximal 6-deoxyglucose accumulation	1600	4.4
Δ pH	1.1	1.5 ⎫ 4.6
	-135 mV	3.1 ⎭

also the rate of influx of TPP$^+$ into the cells is a function of $\Delta\Psi$,
it was possible to demonstrate electrogenic sugar transport by a
transient inhibition of TPP$^+$ influx into the cells with sugars.
Fig. 2 shows that only in induced cells the rate of TPP$^+$ influx is

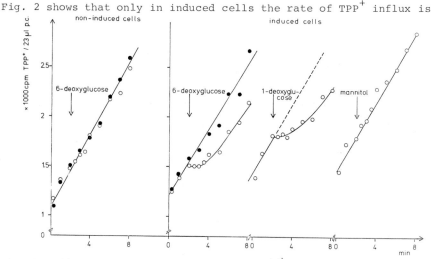

Fig. 2. Effect of sugars on the rate of TPP$^+$ uptake. For details see
ref 18.

inhibited for a short time with 6-deoxyglucose, whereas in non-induced cells no effect is seen. The same is true for induced cells, if
a substrate not transported like mannitol is added. When the rate of
TPP$^+$ entry is calibrated against $\Delta\Psi$ [18] the depolarizing effects of
the sugars transported can quantitatively be determined. Fig. 3 shows
that a maximum depolarization of 70 mV is caused by the sugars and
that after 1-3 min a new steady state potential about 20 mV less negative than the original one is reached.

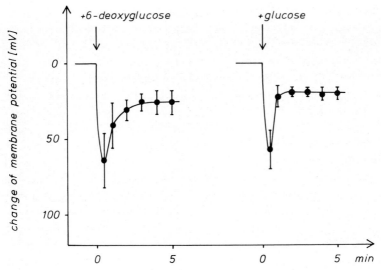

Fig. 3. Depolarization of $\Delta\Psi$ by the addition of transportable sugars. For details see ref 18.

   With the same method the pH dependence of $\Delta\Psi$ has been measured and in fig. 4 it can be seen that at an outside pH of 8 still a $\Delta\Psi$ of -120 mV exists, which can easily account for the sugar accumulation at this high pH, when no ΔpH can drive uptake any more [16].

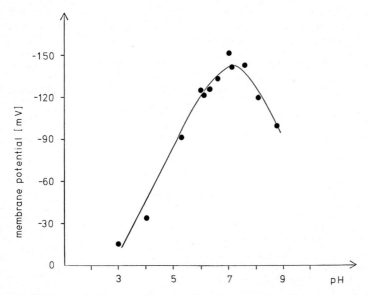

Fig. 4. $\Delta\Psi$ in relation to external pH. For details see ref 18.

3. How is the pmf doing the transport work?

For Na^+-cotransport systems it has been observed that the presence of Na^+ either influences the Vmax or the Km for entry of the substrate [2]. For proton cotransport systems such data are so far available only for Chlorella [16,19]. There it has been observed that a change of the proton concentration of the medium drastically changes the Km for 6-deoxyglucose uptake. Interestingly, depending on the pH either transport with a fairly high Km (5×10^{-2} M at pH $>$ 8.0) or transport with low Km (3×10^{-4} M at pH 6.0 and lower) is observed (Fig. 5). At intermediate pH values two-phasic curves reflecting both these systems are obtained. We do not think that these data indicate that two independent sugar transport systems exist and rather prefer the interpretation that a protonated and an unprotonated form of the same transport protein is responsible for the phenomenon. The reasons for this interpretation are the following ones:

(a) it depends on the pH of the medium whether 1 or 2 systems are observable; (b) when the high affinity system decreases with increasing pH, the low affinity system increases in a mirror like fashion [16]; (c) at a pH value of about 7 both systems are active to 50%, which corresponds to the Km value for protons of the high affinity system of 0.14 mM (= pH 6.85; pK of the acidic group !?).

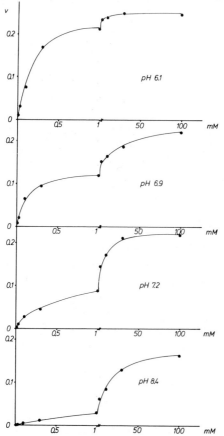

Fig. 5. Concentration dependence of 6-deoxyglucose uptake at various pH values. For details see ref 19.

For these reasons the model of fig. 6 has been postulated to describe sugar-proton cotransport in Chlorella. At low pH the high affinity system, the actual active transport is at work, whereas at high pH the low affinity system essentially responsible only for facilitated diffusion is operative. If this model is correct a decreasing H^+/sugar stoichiometry with increasing pH would be expected at high sugar concentration (the low affinity system has to be saturated for this experiment!). This has indeed been measured: whereas the stoichiometry for 6-deoxyglucose is 1 at pH 6.0 it is 0.25 at pH 8.0 indicating that at the latter pH three molecules out of four enter the cell without a proton [16].

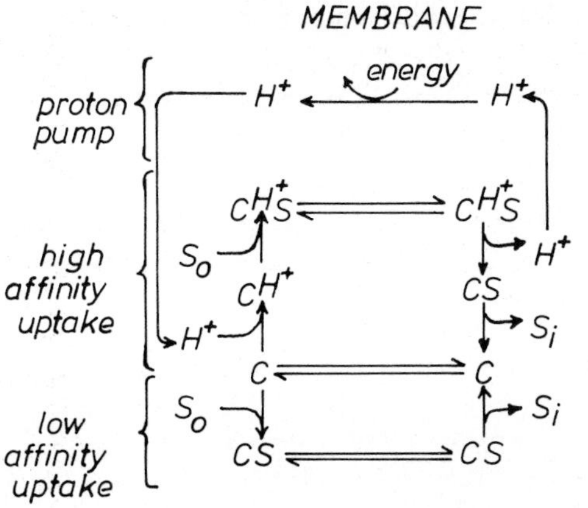

Fig. 6. Model of sugar-proton cotransport in Chlorella vulgaris.

Two points should be emphasized in relation to the model, however. The first one is raised by the question whether two-phasic kinetics are at all expected according to our transport model. To make the argument short here: two-phasic uptake kinetics according to the model can only be obtained if the translocational steps are not the rate limiting ones of the whole transport cycle. Generally, this is not assumed in almost all the transport literature [26]. The fact, however, that a transient depolarization of 70 mV (fig. 3) is not at all visible in the time-course of sugar uptake, indeed supports this assumption. For Neurospora the same conclusion has been reached (Slayman, personal communication).

Finally the question remains unresolved, in what way a $\Delta \Psi$ achieves transport work when no ΔpH is existent. One speculation on this subject has been that a lack of a ΔpH in the bulk phases on both sides of the membrane not necessarily excludes microscalar pH differences along the transport channel [27].

4. A few more problems:

(1) The problem of different H^+/sugar stoichiometries for different sugars, although they use the same transport system: As already mentioned above (RESULTS and DISCUSSION 1.) the amount of protons co-transported may differ by a factor more than 2 for different sugars, although these sugars all use the same transport system. A similar situation has been observed for Na^+ and various amino acids in animal cells [28]. In Chlorella the reaction order for protons for 1-deoxyglucose transport is 1 and thus the same as for 6-deoxyglucose transport, although their H^+/sugar stoichiometries differ considerably [20]. This means that the ternary complex ought to be CSH^+ in either case. The stoichiometry larger 1 for 1-deoxyglucose could however come about by the sequence described in model II or III of fig. 7. In model II some substrate S stays on the carrier and leaves the cell again but the carrier looses the H^+ inside the cell. In model III S triggers the entry of CH^+ without going along in each case. Both models would lead to an increased H^+/sugar ratio although the ternary complex would contain all partners in a ratio 1:1:1.

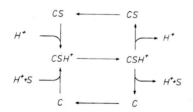

Model II:

Model III:

Fig. 7. Models for the uptake of sugars with a H^+/sugar stoichiometry larger than 1. For details see ref 20.

That model III is correct for Chlorella has been shown in an experiment based on the following idea: it is known that very little substrate leaves preloaded Chlorella cells when no substrate is outside the cells. Thus the addition of a substrate outside the cells increases the rate of efflux of preloaded cells by a factor of 80 (transstimulation of efflux) [29]. Obviously the carrier C or the carrier proton complex CH^+ does not "see" the inside except in the presence of S in the medium. This mechanisms is extremely sensible, since it prevents a futile cycle of CH^+/C transport, which would lead to an uncoupling phenomenon, i.e. a breakdown of the H^+ gradient and an increased H^+-pumping activity without useful transport work.

In the case of model III 1-deoxyglucose would be expected to start this futile cycle. CH^+ will now "see" the inside of the cell and transstimulate the efflux of an inner substrate to a greater extent than 1-deoxyglucose itself will enter the cell. In Table 3 it can be seen that in cells preloaded with 6-deoxyglucose 1 molecule leaves the cell for each molecule 6-deoxyglucose entering it. With 1-deoxyglucose in the outside more than two molecules leave the cell for each 1-deoxyglucose entering it.

TABLE 3

CARRIER-MEDIATED EFFLUX OF INTERNAL 6-DEOXYGLUCOSE COMPARED DURING STEADY-STATE CONDITIONS WITH CARRIER-MEDIATED INFLUX OF EXTERNAL 6-DEOXYGLUCOSE OR EXTERNAL 1-DEOXYGLUCOSE

The experiments have been performed as described in ref 20.

Conditions	Ratio	Number of experiments
6-Deoxyglucose outside:		
Efflux of 6-deoxyglucose/influx of 6-deoxyglucose	1.03±0.22	(8)
1-Deoxyglucose outside:		
Efflux of 6-deoxyglucose/influx of 1-deoxyglucose	2.22±0.61	(7)

Although this sort of playing with models seems to be very academic, the phenomenon described might very well have important biological consequences. Thus any modified substrate for ion-cotransport systems could be a potent uncoupling poison in small concentrations if it opens the ion gate without being transported itself. It could very well be that the toxin helminthosporoside, which interacts with α-galactoside transport sites in sugar cane [30] is a toxin just acting essentially in the way described here.

(2) The problem of purifying the inducible transport protein: As shown before [31] it is possible with the double labelling technique of Kolber and Stein [32] to obtain one membrane bound protein in induced cells that is absent in non-induced ones. The best purification of the specifically double labelled material has so far been obtained by purifying cell walls; after solubilisation of cell wall associated membranes the protein in question has been found to be 12 fold enriched as compared to the total cell extract. In fig. 8b the doubly labelled protein of the cell wall fraction is seen to be considerably more pure than a presumably plasmalemma enriched fraction (fig. 8a) prepared according to the method of Lembi and Morré [33].

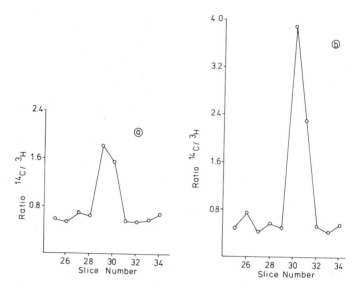

Fig. 8. Membrane proteins solubilized in SDS and separated on SDS polyacrylamide gels. Membranes were separated from a mixture of induced cells (in the presence of ^{14}C-Phenylalanine) and non-induced cells (incubated with ^{3}H-phenylalanine);
a) procedure of Lembi and Morré was followed (plasmalemma enriched fraction)
b) cell walls were purified on sucrose gradients and membranes associated with these walls were used.

The inducible transport protein was found to be completely absent from the soluble fraction. It has a molecular weight of 30 000 and the properties of an intrinsic membrane protein.

(3) The problem why in vitro vesicles don't work so far: Membrane vesicles prepared from Chlorella cells induced for sugar transport could so far not successfully be used to study transport in vitro. One reason for many failures in our laboratory most likely is the following one: it has been observed that during cell breakage rapidly free fatty acids are produced, which severely inhibit sugar-proton cotransport even of any intact cell not broken during homogenisation [21]. Fig. 9 shows that already 1.5 min after cell breakage at 0° the cell homogenate contains sufficient inhibitory fatty acids to completely stop 6-deoxyglucose uptake into intact cells (the lipid extract of the cell homogenate was added to intact cells). When lipids were extracted from unbroken cells this extract did not show any inhibitory effect (curve b in fig. 9). It has been observed [21] that in Chlorella the content of free fatty acids increases by factor 20 due to cell breakage and in S.cerevisae by a factor 14. The

main problem, therefore, to obtain actively transporting vesicles from eucaryotic cells will be for the future to prevent those lipases from working, which obviously are activated in some way during cell breakage. Before this is achieved it is unlikely that substrate proton cotransport from higher organisms will be measurable in vitro.

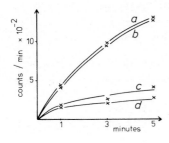

Fig. 9. Inhibition of 6-deoxyglucose uptake by lipid extracts of whole cells (b) and disrupted cells ((c) extracted 1.5 min after disruption and (d) extracted 30 min after disruption). For details see ref 21.

CONCLUSIONS

Whereas the cotransport of protons with sugars and amino acids thus is well established in bacteria and in a number of lower plants, the main lack in our knowledge relates to the active transport of organic solutes in higher plants. Recently we have obtained evidence, however, that higher plants also belong to the proton cotransporters: it has been shown [34] that active sucrose uptake by the cotyledons of Ricinus communis is associated with a concomittant disappearance of protons from the uptake medium. It seems likely, therefore, that in the near future active transport of organic solutes will be basically understood in animals and plants alike.

Acknowledgement: The experimental work reported herein has been supported by the Deutsche Forschungsgemeinschaft.

REFERENCES

1. Crane, R.K. (1962) Fed. Proc. 21, 891-895.
2. Schultz, S.G. and Curran, P.F. (1970) Physiol. Rev. 50, 637-717.
3. Heinz, E., Geck, P. and Pietrzyk, C. (1976) Ann. N. Y. Acad. Sci. 264, 428-441.
4. West, I.C. (1970) Biochem. Biophys. Res. Commun. 41, 655-661.
5. West I. and Mitchell, P. (1972) Bioenergetics 3, 445-462.
6. Seaston, A., Inkson, C. and Eddy, A.A. (1973) Biochem. J. 134, 1031-1043.
7. Slayman, C.L. and Slayman, C.W. (1974) Proc. Natl. Acad. Sci. U.S.A. 71, 1935-1939.
8. Misra, P.C. and Höfer, M. (1975) FEBS-Letters 52, 95-99.

9. Komor, E. (1973) FEBS-Letters 38, 16-18.

10. Komor, E. and Tanner, W. (1974) Eur. J. Biochem. 44, 219-223.

11. Stock, J. and Roseman, S. (1971) Biochem. Biophys. Res. Commun. 44, 132-138.

12. Tanner, W. (1969) Biochem. Biophys. Res. Commun. 36, 278-283.

13. Haaß, D. and Tanner, W. (1974) Plant Physiol. 53, 14-20.

14. Komor, E. and Tanner, W. (1971) Biochim. Biophys. Acta 241, 170-179.

15. Decker, M. and Tanner, W. (1972) Biochim. Biophys. Acta 266, 661-669.

16. Komor, E. and Tanner, W. (1974) J. Gen. Physiol. 64, 568-581.

17. Komor, E., Haaß, D., Komor, B. and Tanner, W. (1973) Eur. J. Biochem. 39, 193-200.

18. Komor, E. and Tanner, W. (1976) Eur. J. Biochem., in print.

19. Komor, E. and Tanner, W. (1975) Planta 123, 195-198.

20. Grüneberg, A. and Komor, E. (1976) Biochim. Biophys. Acta, in print.

21. Decker, M. and Tanner, W. (1976) FEBS-Letters 60, 346-348.

22. West, I.C. and Mitchell, P. (1973) Biochem. J. 132, 587-592.

23. West, I.C. and Mitchell, P. (1974) FEBS-Letters 40, 1-4.

24. Slayman, C.L., Long, W.S. and Lu, C.Y.-H. (1973) J. Membrane Biol. 14, 305-338.

25. Mitchell, P. (1967) Adv.Enzymol.Relat.Areas Mol.Biol. 29, 33-87.

26. Kotyk A. (1973) Biochim. Biophys. Acta 300, 183-210.

27. Mitchell, P. (1968) Chemiosmotic Coupling and Energy Transduction Glynn Res. Ltd. Bodmin. pp 88-102.

28. Wheeler, K.P. and Christensen, H.N. (1967) J. Biol. Chem. 242, 3782-3788.

29. Komor, E., Haaß, D. and Tanner, W. (1972) Biochim. Biophys. Acta 266, 649-660.

30. Strobel, G.A. (1974) Proc. Natl. Acad. Sci. U.S.A. 71, 4232-4236.

31. Tanner, W., Haaß, D., Decker, M., Loos, E., Komor, B. and Komor, E. (1974) in: Membrane transport in plants. Eds. U. Zimmermann, J. Dainty, Springer Verlag, Berlin, pp 202-208.

32. Kolber, A.R. and Stein, W.D. (1966) Nature 209, 691-694.

33. Lembi, C.A. and Morré, D.J. (1971) Planta 99, 37-45.

34. Komor, E., Rotter, M. and Tanner, W. (1976) Proc. Natl. Acad. Sci. U.S.A., submitted for publication.

Regulation of Cell Membrane Activities in Plants
E. Marrè and O. Ciferri eds.
© *1977, Elsevier/North-Holland Biomedical Press, Amsterdam*

A ROLE OF ELECTROGENIC PUMPS IN PRODUCING
TURGOR IN NITELLA

D. S. Fensom
Biology Department
Mount Allison University
Sackville, N.B. Canada

INTRODUCTION

From time to time electroosmosis has been proposed as being a process of some physiological importance in plant cells. Teorell[1], Bennett-Clark[2] and Fensom[3] independently suggested that it might operate across cell membranes to contribute to the turgor pressure of the vacuole. Electroosmotic water movements have been known to be possible in woody tissue for many years (see Spanner[4], or Stamm[5]), but the first unequivocal measurements across living plant membranes were published by Fensom and Dainty[6]. These were made on the cell wall and membranes in series of <u>Nitella</u> <u>translucens</u>.

Arising from these experiments on <u>Nitella</u>, Dainty et al.[7] published a discussion of theoretical aspects of electroosmosis in relation to plant physiology. It was possible to show that electroosmosis could, ideally, add to cell turgor additional pressure of up to 1 atm per mV. But there appeared to be two cogent reasons why such additional pressure could not, in nature, be realized. In those days (1960-63) evidence in support of electrogenic ion pumps was either unavailable or seemed unreliable, therefore no driving currents were thought to be available for electroosmotic water flow across the plasmalemma. The second objection was even more serious. It was shown from diffusional measurements and transcellular osmosis that flow of water could occur very readily across the plasma membrane about 10^5 times as easily as electroosmotic flow. This lead to the view that the water passages through the membrane were much more numerous than the electroosmotic pores, and that any appreciable internal pressure built up by an inwardly directed electroosmotic flow would be lost by backflow through the water passages.

The situation has changed. Both these objections are probably now invalid.

Not only are electrogenic pumps respectable as a part of the membrane complex, they have been recognized and their magnitude determined in a number of living cells. Recently the theory of electroosmotic contribution to water removal in certain animal cells has been re-examined by Hill[8]. Electroosmosis looks promising as the mechanism operating in renal tubules, Malpighian tubules, intestinal mucosa and other animal tissue.

In plants, electrogenic influx pumps have been postulated for Cl^-, HCO_3^- and other anions, and also efflux pumps for H^+ (Higinbotham et al.[9], Higinbotham[10,11], Higinbotham et al.[12], Spanswick[13,14,15], Poole[16,17], Vredenberg et al.[18], Hope et al.[19], Marrè et al.[20], Lucas[21]).

What about the other objection to pressure build up in the cell, the backflow of water?

In 1969 Barry et al.[22,23,24] argued that the electroosmotic water flow measurements made on plant membranes to date were only partly due to coupling of water with ions moving in charged pores. The rest and larger part of the water flow was due to "transport number" effect of ions at the membrane leading to an induced inward _osmotic_ water flow. They envisage the externally applied current across the cell wall and membrane as moving cations through wall and membrane more readily than the anions which are repelled due to the negative charge of the cell-membrane complex. Thus an asymmetrical geometry effect results with more ions inside the membrane (on the influx end) than outside. Hence a situation is created which will promote osmotic inflow. Therefore, they claim, the measured influx of water will really have two components, an electroosmotic flow in pores which is instantaneous in build-up when E is applied, and an osmotic flow which tends to increase with time. Their own experiments and the re-examination of the experiment of Fensom et al.[6], MacRobbie et al.[25], or Wanless et al.[26] support this suggestion. But it is this osmotic influx component which will itself restrict backflow of water through the membrane. If the ion accumulation inside the membrane is uniformly distributed, as it might be, the backflow could be appreciably reduced.

Fortunately it is possible to test this hypothesis on _Nitella_ in several ways. Electroosmotic measurements on _Nitella_ can be compared with transcellular osmotic flows so that the osmolarity required to balance the apparent electroosmotic flow can easily be calculated. Alternatively it is possible to use experiments of transcellular osmosis with associated potential differences (p.d.) measurements to show that flows and backflows actually can be produced in _Nitella_. These are always associated with a p.d. which could possibly, under an electroosmotic situation, balance the backflows. Finally, it is possible to propose a membrane model which will allow the situation to be theoretically treated so that the true electroosmotic component of flow may be compared with the electrically induced osmotic flow.

a) Comparison of electroosmotic measurements with balancing osmotic flows.

The flow of water per cm^2, J, induced by 0.25 μamp cm^{-2} (about 5 mV across each membrane complex, Williams et al.[27]) which is well below the action potential level, has been measured in _Nitella translucens_ (MacRobbie et al.[25]). A typical value of flow may be taken $J = 3.7 \times 10^{-8}$ cm^3 cm^{-2} sec^{-1}. This flow is apparently electroosmotic, but probably includes a considerable osmotic component. However, suppose it be assumed that it is all electroosmotically induced, that is cations and water are coupled in their flow through charged pores.

But $J = L_p \Delta P$, where L_p is the hydraulic conductivity of the membrane, close to 10^{-5} cm^3 cm^{-2} sec^{-1} atm^{-1} (Dainty et al.[28]), and ΔP is the pressure difference in atm. The back pressure required to balance this apparent electroosmotic flow would therefore be:

$$\Delta P = \frac{J}{L_p} = \frac{3.7 \times 10^{-8}}{10^{-5}} \text{ atm} = 3.7 \times 10^{-3} \text{ atm.}$$

This pressure can be induced by a concentration difference of 1.6×10^{-4} M ($= 1.6 \times 10^{-1}$ mM) across the membrane complex. Barry et al.[22,23] assert that somewhat less than half this measured flow was really electroosmotically driven (water coupled in charged pores), the rest being an osmotic effect due to preferential cation transport through the membrane. On this basis, an electro-osmotic flow equivalent to $< 8 \times 10^{-2}$ mM of concentration difference has accompanying it an osmotic flow induced by 8×10^{-2} mM. Thus backflow through the membrane under an applied p.d. (i.e. apparent electroosmosis) is reduced in this case, and will only become appreciable when the internal turgor has risen to balance inflow.

b) Comparison of transcellular osmotic flow and induced potential differences.

Tazawa et al.[29] have reported that a pressure flow through <u>Nitella</u> generated by transcellular osmosis will produce a p.d. across the cell (measured externally at each end) with the downstream side electro-positive. This was not interpreted as being a "streaming potential", but rather to be due to ion accumulations at the membrane wall surfaces. From these experiments an osmotic pressure gradient of 0.2 M sucrose ($P = 0.45$ joules cm^{-3}) produced a flow of 0.6 μl min^{-1} ($J = 10^{-5}$ cm^3 cm^{-2} sec^{-1}) and a potential difference of about 10 mV ($E = 10^{-2}$ volts) with the downstream side +. When the pressure gradient was removed, a backflow immediately set in.

These two examples a) and b) show that electrically driven flow across both ends of a <u>Nitella</u> cell induces an osmotic potential gradient to add to the flow, while an osmotic pressure driven flow produces an electrical potential gradient of polarity opposite to that which induces electroosmosis in the same direction. Alternatively one can say that electroosmotic flow itself tends to reduce pressure backflow, but pressure produced flow tends to increase electrically driven back-flow in a transcellular osmosis situation.

c) Electroosmosis in a membrane model.

The two previous sections have dealt with simultaneous measurements across both ends of <u>Nitella</u> cells in an osmometer. These conditions are highly artifi-cial as well as being complicated by the cell wall in series with several mem-branes, with one end of a cell acting differently (at least in terms of polariza-tion) to the other. I shall now propose a model of a membrane, Fig. 1, which can be analysed somewhat more simply. Of course it may be naive compared with Nature,

FIGURE I. Model of a cell membrane showing one pore.

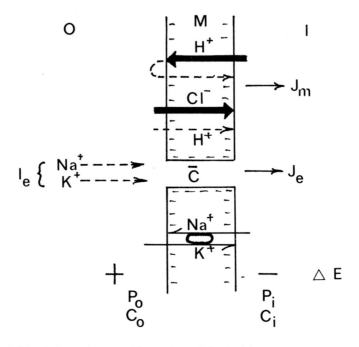

O is outside, M is membrane wall complex, I is inside.

$\bar{c}$ is concentration of cations in a pore.

J_e is the electroosmotic-water coupled water influx.

I_e is the electroosmotic-ion coupled ion influx.

J_m is the non-ion coupled water flux through the membrane.

P is the hydraulic pressure, P_i inside, P_o outside the membrane.

C is the concentration, C_i inside, C_o outside the membrane.

ΔE is the electrical potential difference across the membrane (inside negative).

Assume:

(1) Electrogenic pumps are unaltered by applied emfs, but can be altered by some membrane conditioners or plant growth regulating compounds.

(2) The electrogenic pumps are either (or both) HCO_3^- or OH^- influxes or H^+ efflux Cl^- influx <u>may</u> be electrogenic.

(3) The H^+ shunt is passive, $\therefore$ increases under greater p.d. applied.ΔE. A similar effect would be produced by OH^- acting passively in reverse direction.

(4) J_e (= osmotic flux) is passive, but depends on C_i-C_o, which is itself ion sensitive, i.e. as ΔE increased electrogenically, cations try to enter the cell and <u>do</u>, anions try to leave but cannot. $\therefore$ anions build up at inner surface.

(5) K^+, Na^+ influxes are coupled with water when they are "passive" fluxes (the active part of K^+ may be different). Na^+ efflux and some K^+ influx are shown as coupled, but not electrogenic in this model (MacRobbie,[37]).

but as an hypothesis for testing it may have value, so let us consider it.

In this model both the solution flux, J cm^3 cm^{-2} sec^{-1}, and the electric current I amps cm^{-2}, are divided into two parts, $J = J_m + J_e$ and $I = I_m + I_e$, where the subscript m is membrane flow not coupled with ions in the membrane itself, and e is electroosmotic flow, where the cations Na$^+$ and K$^+$ (but not H$^+$ or OH$^-$) move passively in the electric gradient (after Hill[8]). J_e and I_e are thought to be through pores with negatively charged walls and the coupling of H$_2$O with Na$^+$ and K$^+$ is assumed to be fairly tight in this ion-water channel.

The Onsager relations may be used (see Dainty et al.[7] for details)

$$J = J_m + J_e = L_p P + L_{PE} E \tag{1}$$

$$I = I_m + I_e = L_{EP} P + L_E E \tag{2}$$

P is the hydraulic pressure difference in joules cm^{-3} (1 joule cm^{-3} = 10 atm).
L_p is the hydraulic pressure permeability of the membrane-wall complex in cm sec^{-1} (joule cm^{-3})$^{-1}$.
E is the electrical potential difference in volts assisting the solution flux
  (i.e. downstream is — when the wall-membrane complex has a negative surface charge).
L_{PE} is the electro-kinetic cross coefficient, cm^3 cm^{-2} sec^{-1} volt^{-1} or amps cm^{-2} (Joule cm^{-3})$^{-1}$ = L_{EP}.
L_E is the conductance of the membrane, ohm^{-1} cm^{-2}.

Following Hill[8] it is convenient to define the coupling coefficient q as the fraction of the overall hydraulic conductance L_p which is electroosmotically active. When the volume flux through the electroosmotic channels, hydraulically driven is $J_e = q L_p P$, similarly the ion-water coupled electric current, electrically driven is $I_e = r L_E E$, where r is the fraction of the overall electric conductance L_E, which is electroosmotically active (∴ r is the coupling coefficient for ion-water in electroosmotic channels).

The hydraulic pressure-driven current will be carried by the cations Na$^+$ and the part of K$^+$ which is passive (the membrane being negatively charged). This is the current in the electroosmotic pathway I_e, and when $I_m = 0$, $I_e = J_e \bar{c} F = L_{PE} P$, where $\bar{c}$ is the concentration of ions in the pore in equivalent cm^{-3} and F is the faraday ($\approx 10^5$ coulombs). ∴ $I_e = q L_p \bar{c} F P$ and $L_{PE} = q L_p \bar{c} F$.

The rest of the current I has two components, one due to the pore conductance L_{E_e} and the other to membrane conductance L_{E_m} (which is assumed to be by ion flow uncoupled with water and is probably H$^+$ and OH$^-$).

The appropriate equations to describe the hydraulic and electric fluxes are therefore:

solution: $$J = (L_{P_e} + L_{P_m})P + L_{PE} E \tag{3}$$

ion current: $$I = L_{PE} P + (L_{E_e} + L_{E_m})E \tag{4}$$

If the ion-water coupled flow, hydraulically driven in electroosmotic channels
be q x total hydraulic flow, then

$$L_{P_e} = q \; L_p \quad \text{and} \quad L_{PE} = q \; L_p \; \bar{c} \; F \tag{5}$$

Similarly, if the ion-water coupled current, electrically driven in charged pore
channels be r x total electric current, then

$$L_{E_e} = r \; L_E \quad \text{and} \quad L_{EP} = r \; L_E \; \frac{1}{\bar{c} \; F} \tag{6}$$

but in a perfect electroosmotic system, for example in the charged pores under
consideration,

$$L_{PE}^2 = L_{P_e} \cdot L_{E_e} \quad (\text{c.f. Dainty et al.}[7], \text{ or Hill}[8])$$

as in the pores of our membrane complex

$$\therefore \; L_{PE} = \sqrt{(qL_p \cdot rL_E)} \quad \text{and} \quad L_E = \frac{L_{E_e}}{r} = \frac{q}{r} \; L_p (\bar{c} \; F)^2 \tag{7}$$

where L_p and L_E are conductivity coefficients for the whole membrane, as explained above.

It is fairly easy to marshall evidence to show that neither r nor q can be
1. less obvious, but now fairly well supported (see Discussion) is the contention
that neither r nor q is likely to be zero. We shall assume that this is the
case and therefore $L_{PE} \neq 0$.

d) Calculation of q, L_{PE} and $\bar{c}$ for the model.

In an electroosmometer, J/I may be measured when the hydrostatic pressure
difference is 0. But we must now assume that the osmotic pressure component induced by the applied current causes a flow of water through the membrane (uncoupled with ions) and some ion flow (uncoupled with water). In the latter case,
the uncoupled ion flow is probably mainly due to H^+ and OH^- (see Williams et al.[30]
Kitasato[31], Wanless et al.[26]) and seems to be about 80% of the total ion flow at
a current of 1 to 2 µamps cm^{-2}. i.e. r ≈ 0.2.

The true electroosmotic efficiency through the charged pores is J_e/I_e.

When P = 0,
$$\frac{J_e}{I_e} = \frac{L_{PE}}{L_{E_e}} = \frac{1}{\bar{c} \; F} \quad \text{if } L_{PE}^2 = L_p \cdot L_E \tag{8}$$

but P is not 0. When a current is applied it becomes P_{os}, i.e. the extra osmotic
pressure induced by the current and operating to add to water flow. None the less
as a first approximation we can assume that $J_e \approx q \; J$ and $I_e \approx r \; I$, where J and I
are the measured water fluxes and currents in an electroosmometer.

$$\therefore \; \frac{J_e}{I_e} \approx \frac{q \; J}{r \; I} \approx \frac{1}{\bar{c} \; F} \tag{9}$$

$$\therefore \; \bar{c} \approx \frac{r}{q} \cdot \frac{1}{(J/I)F} \tag{10}$$

taking r = 0.2 (Williams et al.[30]), q = 0.4 (Barry et al.[22,23], Wanless et al.[26]),
J/I = 8 x 10^{-2} cm^3 c^{-1} (MacRobbie et al.[25]) at .25 µamp,

$$\bar{c} \approx \frac{.2}{.4} \approx \frac{1}{8 \times 10^{-2} \times 10^5} \approx 6 \times 10^{-5} \text{ equivalents cm}^{-3}$$

But $L_{PE} = q \, L_P \, \bar{c} \, F$. Substituting for $\bar{c}$ from equation (10) and taking L_P as $10^{-4} \text{ cm}^3 \text{ cm}^{-2} \text{ sec}^{-1} \text{ (joule cm}^{-3})^{-1}$

$$L_{PE} \approx \frac{r \, L_P}{(J/I)} \approx \frac{.2 \times 10^{-4}}{8 \times 10^{-2}} \approx 2.5 \times 10^{-4} \text{ xm}^3 \text{ cm}^{-2} \text{ sec}^{-1} \text{ volt}^{-1}.$$

For these calculations of approximation I have assumed that q is about 0.4, that is 40% of the total solution flow induced electroosmotically. However, it is possible to analyze the situation more thorougly.

Since we cannot measure (J_e/I_e) with certainty, we must use (J/I), (when there is no hydrostatic pressure difference). In this case, the flow J is partly driven by the hydraulic pressure difference P_{os}.

$$\frac{J}{I} = \frac{L_P \, P_{os} + L_{PE} \, E}{L_{PE} \, P_{os} + L_E \, E} \tag{11}$$

substituting for L_{PE}, L_E, and $\bar{c}$

$$\frac{J}{I} = \frac{P_{os} + r \cdot \dfrac{E}{(J/I)}}{P_{os} \cdot \dfrac{r}{(J/I)} + \dfrac{r}{q} \cdot \dfrac{1}{(J/I)^2} E}$$

$$\therefore \; P_{os} \cdot r + \frac{r}{q} \cdot \frac{1}{(J/I)} \cdot E = P_{os} + r \cdot \frac{1}{(J/I)} \cdot E$$

$$\therefore \; \frac{r}{(J/I)} \cdot E \left(\frac{1}{q} - 1\right) = P_{os} (1 - r)$$

$$\therefore \; \frac{P_{os}}{E} = \frac{r}{(J/I)} \cdot \frac{\left(\frac{1}{q} - 1\right)}{(1 - r)} = \frac{r}{q} \cdot \frac{(1 - q)}{(1 - r)} \cdot \frac{1}{(J/I)}$$

i.e. $\dfrac{P_{os}}{E} = \dfrac{r}{q} \dfrac{(1 - q)}{(1 - r)} \dfrac{I}{J}$ or

$$\frac{\text{osmotic power}}{\text{electrical power}} = \frac{P_{os} \cdot J}{EI} = \frac{r}{q} \frac{(1 - q)}{(1 - r)} \tag{12}$$

In an electroosmotic situation where osmotic water flows (or pressures) are driven entirely by electrogenic pumps, the osmotic power cannot exceed the electrical power (by the first law of thermodynamics).

$$\therefore \; \frac{r}{q} \frac{(1 - q)}{(1 - r)} \text{ cannot exceed 1} \quad \therefore \; \frac{r}{(1 - r)} \leqslant \frac{q}{(1 - q)} \text{ and } r \leqslant q \tag{13}$$

Maximum turgor (P_{os}) will occur when q equals r at any measured electroosmotic efficiency (assuming that r and q can be neither 0 nor 1),

$$\therefore \; \frac{P_{os}}{E} \approx 25 \text{ (joules cm}^{-3}) \text{ Volt}^{-1} \text{ when } \frac{J}{I} \text{ is } 4 \times 10^{-2} \text{ cm}^3 \text{ c}^{-1} \text{ (200 H}_2\text{O/F)}$$

$$\approx 0.25 \text{ atm per mV of electrogenically added p.d.}$$

Conversely when E is generated by osmotic driving forces alone $q \lessgtr r$.

These calculations are based on a kind of Smidt model situation (see Dainty et al.[7]). A "friction model" would give similar simple solutions only under ideal conditions of no ion-water slippage.

e) "Basic flow", electroosmosis and turgor.

A small transcellular water flow has been found to occur in <u>Nitella</u> cells when placed in an electroosmometer (Fensom et al.[32]). This increases with light intensity and is influenced by membrane conditioners, by pH, and probably by HCO_3^- concentrations. The mere presence of an asymmetric water flow at constant pressure suggests that under conditions where turgor balances flow so that $J = 0$, the water flow into the cell would appear as added turgor. Basic flow may be represented as follows:

$$J = L_p \Psi + q \, L_p \, \bar{c} \, F \cdot E \tag{14}$$

where Ψ is the difference in water potential across the wall-membrane complexes in series with E. On the other hand, electroosmosis alone would give

$$\frac{J}{I}_{(P=0)} = \frac{r \, (1 - q)}{q \, (1 - r)} \cdot \frac{E}{P_{os}} \tag{15}$$

where r (for electrical power only) $\lessgtr q \lessgtr 0.2$.

Turgor, ΔP will result when $J = 0$, in which case

$$\Delta P_{turgor} = \Delta \pi + q \, \bar{c} \, F.E = \Delta \pi + r \, \frac{E}{(J/I_{(P=0)})} \tag{16}$$

in other words, additional turgor will result (when $J = 0$) if an increase is induced across the membrane complexes of

$\Delta \pi$, the osmotic part of water potential,

E , the biopotential across the membrane electrogenically driven in excess of Nernst potentials,

$\frac{1}{(J/I_{(P=0)})}$, the reciprocal of the measured electroosmotic efficiency, and

r , the degree or efficiency of the ion-water coupling in the electroosmotic pores. When $r = .2$, $E = 10^{-3}$ Volts and $J/I_{(P=0)} = 4 \times 10^{-2}$ cm^3 c^{-1}, the added turgor from electrical driving force alone would be 0.05 atm mV^{-1}.

N.B.   It can also be shown that
$$\Delta P_{turgor} = \Delta \pi + q \, \frac{(1 - r)}{(1 - q)} \, P_{os}$$

DISCUSSION

Electrogenic pumps, operating on the plasmalemma and tonoplast in plants, would seem to activate more processes than their own specific transfer of ions alone. They will alter the membrane potential and produce local membrane currents. The membrane potential will induce the approach or repulsion of ions at the membranes regardless of unstirred layers. If there are pores large enough to admit ions, electro-osmosis will occur. That the membrane is itself charged and that there are at least some pores large enough to admit ions (Armstrong[33], Fensom et al.[34]) seems very likely. Therefore it is probable that electroosmosis will be largely uni-directional - in plants mainly cation dependent and inwardly directed, at least as far as the net electroosmotic influx is concerned. But in addition, the attrac-tion of ions to each side of the membrane turns out to be asymmetrical due to the effects of charge and geometry of the cell wall on the outside (itself negatively charged), the membrane charge (usually negative) and the transport number effect (i.e. the greater permeation of cations - particularly H^+ - through the membrane compared with anions). This, in turn, means that the concentration of ions on both sides of the membrane will alter in favour of an inwardly directed osmotic effect if the inside becomes more electro-negative. This osmotic effect will not only restrict the backflow of water from electroosmotic components, but will add an osmotic pressure component of its own directed inwards to increase turgor.

Of course electrogenic pumps may reflect other cellular activities than membrane potential and current flow. They probably also reflect a direct transport of solute to the cytoplasm and a direct transport of organic solutes to sites very close to the plasmalemma or tonoplast. Therefore the accumulation of ions close to membranes is probably also accompanied by accumulation of organic solutes, both of which will act directly to alter the local osmotic potential. This would seem to be the explanation of "basic flow" across a Nitella cell (Fensom et al.[32]). It is not intended to suggest that such flows are electroosmotically induced alone. This also means that the limitations in magnitude of r relative to q cannot in real situations impose restriction: since both electrical and osmotic driving forces must occur, r may range from r < q (total electrical power) to r > q (total osmotic power).

The keys to this system lie in (a) the control of the electrogenic pumps by plant growth regulating compounds acting directly on the pump sites or by feedback from changes in metabolism, (b) the particular nature of the membrane itself, its pore size distribution, the relative permeation of H^+, H_2O and ions, and its ion pumps distribution. That electrogenic pumps can be altered by growth regulating compounds looks very likely and has been argued elsewhere (see Wanless et al.[26], Fensom et al.[32], Gillet et al.[33]). That H_2O permeation is relatively independent from passive ion permeation and also from H^+ permeation has been discussed else-where [26,32] and is reasonably certain. Hence membrane control of turgor would appear to be linked with electrogenic ion pump activity.

If increased pressure inside a cell (as opposed to decreased cell wall rigidity outside) can be induced by electrogenic pump activity, the question arises as to whether all the increased pressure induced inside a cell (by increased electrogenic pump activity) can be explained without resorting to water-ion coupling through electroosmosis at all. This would happen if there really were no charged pores in membranes large enough to admit monovalent cations like Na^+ or K^+. The Na^+ and K^+ would presumeably then be transferred across the membrane by some diffusional pathway (since the influx is largely passive), such as an organic-ion complex exchanging device. The water would then all move in separate channels under osmotic gradients. But the preceding equations show that this is impossible (if $q = 0$, P_{os}/E becomes $\propto$ or indeterminate), just as an analysis of the time course of J shows that a part of flow arises instantaneously with an applied p.d. in an electroosmometer[26].

Influxes can be increased either by increasing the external concentrations of ions or by applying an electric current (Williams et al.[30]), and are probably water coupled when moving passively (Wanless et al.[26]). On the other hand, electroosmotic efficiencies are clearly independent of "basic flow" and of transcellular osmotic permeabilities in a number of instances (Fensom et al.[32], Gillet et al.[33], Fensom, private communication), therefore "basic flows" are not themselves dependent on Na^+ or K^+ movements, nor changes in L_p caused by low pH, but would seem to be tied in with electrogenic pumps. According to the membrane model presented here, H^+ electrogenic efflux activity (or OH^- influxes) can be shunted back through the membrane uncoupled with water, i.e. not through the usual electroosmotic channels of charged pores large enough to carry water coupled with Na^+ and K^+. It is not intended to suggest here that electrogenic pumps are only H^+ efflux in type. All electrogenic pumps would potentially alter the p.d. across the membrane. But only a pump which shunted across the membrane without being coupled with water would act to build up P_{os} without changing $L_{PE}.E$ very much.

The original objections to an electrically driven contribution to turgor pressure turn out to have altered. Presumeably the H^+ ions may move readily through the membrane under electrogenic or applied p.d.'s without water coupling. This allows the internal turgor to build up from the combination of electroosmotic and electrically induced osmotic flows.

If the current model is valid, the maximum turgor pressures induced electrogenically will occur when $r = q$, that is when the proportion of current carried through electroosmotic pores to the total current is about equal to the proportion of water coupled electroosmotically compared with total water flow. Precise measurements of r and q have not yet been made. We know that with small applied currents somewhere between 18 and 35% of the current is carried by Na^+ and K^+. But some K^+ influx will be active, therefore r will be < 18 - 35% of the additional current. At a maximum it might perhaps be as high as 20%. If both r and q are around 0.2, and the electroosmotic efficiency is 4×10^{-2} cm^{-3} c^{-1} (= 200 H$_2$0/F), the maximum <u>added</u> turgor pressure could be up to 0.25 atm per mV of electrogenic p.d. or 0.25 atm m^{-1} induced by P$_{OS}$ if backflow did not occur.

Thus a cell which produced 40 mV of p.d. above its resting (or Nernst or Goldman) potentials, could add an additional 2 - 10 atm to its turgor. However, this assumes a reasonable uniformity of electrogenic activity along the cell membrane and the absence of backflow of water. In <u>Nitella</u> it is probable that the H$^+$ or HCO$_3^-$ electrical activity may occur in bands along the cell (Lucas et al.[35], Lucas[36]), therefore the model used in this paper is clearly oversimplified compared with Nature in this case. There may none-the-less be other cells where electrogenic uniformity occurs, particularly in young cells of multicellular tissue or guard cells of stomata, and if so, these arguments could then have more than theoretical pertinence.

REFERENCES

1. Teorell, T. (1958) Z. Phys. Chemie 15: 385-398.
2. Bennet-Clark, T.A. (1959) Plant Physiology II (Ed. F.C. Steward), Academic Press, New York.
3. Fensom, D.S. (1959) Can. J. Bot. 37: 1003-1026.
4. Spanner, D.C. (1975) Encyclopedia of Plant Physiology, Vol. I (Ed. M. Zimmermann and J. Milburn). Springer-Verlag (Berlin).
5. Stamm, A.J. (1926) Coll. Symp. Monograph 4: 246-257.
6. Fensom, D.S. and Dainty J. (1963) Can. J. Bot. 41: 685-691.
7. Dainty, J., Croghan, P.C. and Fensom, D.S. (1963). Can. J. Bot. 41: 953-966.
8. Hill, A.E. (1975) Proc. R. Soc. London. B. 190: 115-134.
9. Higinbotham, N., Graves, J.S. and Davis, R.F. (1970) J. Membrane Biol. 3: 210-222.
10. Higinbotham, N. (1970) Am. Zoologist 10: 393-403.
11. Higinbotham, N. (1974) Plant Physiol. 54: 454-462.
12. Higinbotham, N. and Anderson, W.P. (1974) Can. J. Bot. 52: 1011-1021.
13. Spanswick, R.M. (1972) Biochim. Biophys. Acta, 288: 73-89.
14. Spanswick, R.M. (1974) Can. J. Bot. 52: 1029-1034.
15. Spanswick, R.M. (1974) Biochim. Biophys. Acta, 332: 387-398.
16. Poole, R.J. (1973) Ion Transport in Plants (Ed. W.P. Anderson) Academic Press, New York.
17. Poole, R.J. (1974) Can. J. Bot. 52: 1023-1028.
18. Vredenberg, W.J. and Tonk, W.J. (1973) Biochim. Biophys. Acta, 298: 354-368.
19. Hope, A.B., Lüttge, U. and Ball, E. (1972) Z. für Pflanzenphysiol. 68: 73-81.
20. Marrè, E., Lado, P., Rasi-Caldogno, R., Colombo, R. and De Michelis, M.I. (1974) Plant Science Letters 3: 365-379.

21. Lucas, W.J. (1975) J. Exp. Bot. 26: 331-346.

22. Barry, P.H. and Hope, A.B. (1969) Biophys. J. 9: 700-728.

23. Barry, P.H. and Hope, A.B. (1969) Biophys. J. 9: 729-757.

24. Barry, P.H. and Hope, A.B. (1969) Biochim. Biophys. Acta, 193: 124-128.

25. MacRobbie, E.A.C. and Fensom, D.S. (1969) J. Exp. Bot. 20: 466-484.

26. Wanless, I.R., Bryniak, N. and Fensom, D.S. (1973) Can. J. Bot. 51: 1055-1070.

27. Williams, E.J., Johnston, R.J. and Dainty, J. (1964) J. Exp. Bot. 15: 1-44.

28. Dainty, J. and Hope, A.B. (1959) R.Br. Australian J. Biol. Sci. 12: 136-145.

29. Tazawa, M. and Nishizaki, Y. (1956) Jap. J. Bot. 15: 227-238.

30. Williams, E.J., Munro, C. and Fensom, D.S. (1972) Can. J. Bot. 50: 2255-2263.

31. Kitasato, H. (1968) J. Gen. Physiol. 52: 60-87.

32. Fensom, D.S., Barclay, S.J., Law, S. and Thompson, R.G. (1973) Can. J. Bot. 51: 1045-1053.

33. Gillet, C., Fensom, D.S. et Lefebvre, J. (1971) Experientia (Basel) 27: 853-854.

34. Fensom, D.S. and Wanless, I.R. (1967) J. Exp. Bot. 18: 563-577.

35. Lucas, W.J. and Smith, F.A. (1973) J. Exp. Bot. 24: 1-14.

36. Lucas, W.J. (1975) J. Exp. Bot. 26: 271-286.

37. MacRobbie, E.A.C. (1962) J. Gen. Physiol. 45: 861-878.

Regulation of Cell Membrane Activities in Plants
E. Marrè and O. Ciferri eds.
© *1977, Elsevier/North-Holland Biomedical Press, Amsterdam*

EFFECT OF ENVIRONMENTAL STRESS ON TRANSPORT OF IONS ACROSS MEMBRANES

J. Levitt
Department of Horticultural Science & Landscape Architecture
University of Minnesota
St. Paul, MN   55108

I. THE WATER STRESS

The water stress has been shown to affect ion transport in both the absorbing organs (the roots) and the photosynthesizing and transpiring organs (the leaves). At least some of the factors involved in these two effects are so different that the two kinds of organs must be treated separately.

a. The roots.

There can be no appreciable ion transport from air-dry soil into the roots, since there is no aqueous diffusion path between the two. If, however, some free water connects the soil to the roots, ion transport is possible though the rate must be influenced by the water stress. It is, therefore, not surprising that K uptake by young maize roots declined sharply with increasing soil water tensions (Mengel and Braunschweig 1972). Nutrient uptake by three grassland winter annuals similarly decreased at increasing soil moisture stress (Gerakis et al. 1975). Such decreases may logically be related to the decrease in availability of the aqueous diffusion path. In the case of the maize roots there was, indeed, a linear relation between the K diffusion rates and total K uptake. The depression in yield at the higher soil water tensions was in fact due to a reduced availability of K caused by restricted diffusion. Higher applications of K counterbalanced this effect to some extent. A similar avoidance of the adverse effects of drought on pearl millet was reported due to an adequate level of fertilization, especially with N (Lahiri et al. 1973).

On the other hand, a decrease in soil water content may, in certain cases, be expected to increase ion influx into the roots because of the increase in concentration   Thus, it was shown by the split-root technique that the uptake of ^{137}Cs by sorghum was controlled by two soil factors - ion diffusion path (i.e., availability of diffusion paths) and ion concentration (Shalhevet 1973). It was the diffusion path that controlled the uptake as soil moisture content decreased, but when the moisture was allowed to fluctuate in a drying and wetting cycle, uptake was controlled by concentration, and the drier the soil the higher the uptake. This effect of concentration has been ascribed to an accumulation of an ion species at the root surface, when transpiration is comparatively high (Dunham and Nye 1974).

When grown in nutrient solutions, the roots may show a different rate of ion transport from the rate in soils. When the matric potential at the onion root surface fell to about -25 bars, the root absorbing power was only 20% of that from a normal nutrient solution (Dunham and Nye 1974). Osmotic stress was also found to decrease ion transport. Mannitol and PEG solutions reduced Rb absorption by corn roots to 20% or less of the controls (Smith et al. 1973, Parrondo et al. 1975), and this was true whether the stress was applied before or during the uptake. Therefore, the effect in solution culture had no relation to the diffusion path or concentration of the ion (two important factors in soils), but it could be explained as a net change due to increased ion efflux. This explanation was only partially supported by a stress imposed following a period of Rb accumulation, which caused roots to lose slightly more Rb than the controls during the first 30 min., after which the rates of change were insignificant in both. Rate of recovery was about the same as from a water stress produced by desiccation (Parrondo et al. 1975). The osmotic effect is very rapid. When the water potential was lowered to -5 to -13 bars by addition of PEG 400 or 6000, both water and ion (K and phosphate) uptake by young wheat roots immediately decreased (Erlandsson 1975). NaCl had essentially the same effect as PEG. Erlandsson suggested an effect on the active ion transport mechanism. Transport from barley roots to the shoot was inhibited following a brief wilting which was insufficient to affect the rate of uptake of ^{86}Rb and L-leucine.

The water stress may not affect the transport of all ions equally. In wheat and yellow peas, the Ca:K and Ca:P ratios increase with decreasing soil moisture content (Mattson 1973). These results were explained by applying the theory of membrane equilibria.

b. The leaves.

1) Stomatal closure and ion transport.

It has long been known that water stress leads to stomatal closure. It has also been amply shown that stomatal movement is normally associated with ion transport (Raschke 1975a), specifically a transport of K ions into or out of the guard cells. It is, therefore, not surprising that stomatal closing induced by a water stress is accompanied by an efflux of K ions and a consequent osmotic loss of water and therefore of guard cell turgor. It has also been shown that abscisic acid (ABA) quickly accumulates in plants as soon as their net water content begins to decrease appreciably (Hsiao 1973). In pea seedlings, for instance, a loss of about 5% of their water content leads to about a 20X increase in ABA level of the shoots and a 7-10X increase in bound ABA (Doerffling et al. 1974). In Vitis vinifera, ABA-like inhibitors doubled in concentration within 15 mins. of excision, when the water potential had dropped to -15 bars (Loveys and Kriedemann 1973). In the case of two Ambrosia species, the ABA concentration began to increase in a cortical water potential range of -10 to -12 bars (Zabadal 1974). The increase in ABA was so abrupt, that a reduction in water potential of only 1 bar around the critical point might cause a significant increase. Phaseic acid may also increase (Loveys and Kriedemann 1974). Artificial treatment with ABA also leads to stomatal closure in the absence of water stress, and there is an accompanying efflux of K ions (Jones and Mansfield 1970). These three observations point to the following sequence of events. The water stress induces a direct strain - a small decrease in leaf water content, insufficient by itself to cause stomatal closure. This results in ABA accumulation which induces K ion efflux, accompanied by a marked loss of guard cell water and turgor, resulting in stomatal closure. This hypothesis in its turn leads to three further questions. a) How can ABA induce the K ion transport out of the cell? b) How does the drop in leaf water content induce ABA formation? c) Why does ABA specifically affect the guard cells and not the other leaf cells?

a) It must first be pointed out that ABA does alter ion transport in other cells as well as inducing K ion efflux from guard cells. It inhibits K ion uptake by maize roots (Shaner et al. 1975), and by Avena coleoptiles (Reed and Bonner 1974). It not only inhibits K ion uptake by sunflower hypocotyls, but it also antagonizes the IAA- and GA-enhanced uptake (Doerffling et al. 1973). Since ABA is, in general, an inhibitor, and since there is no evidence of a direct enhancement of K^+ efflux, the tentative conclusion is that ABA induces net efflux by inhibiting influx. The basic question then is, how can ABA produce this probable inhibition of K ion uptake and the resulting net efflux from guard cells. There is some evidence pertinent to this question from the above work with non-stomatal cells. When ABA inhibited ion accumulation by maize roots, it did not affect the ATP content of the tissues, nor the activity of the microsomal, K stimulated ATPases (Shaner et al. 1975). Nevertheless, in the case of Lemna, treatment with ABA caused a 50% reduction in phosphorylation with labeled ATP (Chapman et al. 1975).

The ABA-induced net efflux of K ions from plant cells in general, and guard cells in particular, should presumably be explainable by the mechanism of its action as a growth regulator. Although this action has not been fully explained, there is considerable indirect evidence pointing to an inhibition of H ion transport. (1) Recent evidence from many laboratories has shown that IAA stimulates cell enlargement by inducing H ion transport from the protoplast surface to the cell wall (Cleland 1971, Jacobs and Ray 1975). At least in the case of root growth, the evidence points to an ABA antagonism of this IAA-induced cell enlargement (Audus 1975) and, therefore presumably of the IAA-induced H ion transport. ABA also inhibited IAA-induced cambial activity in Abies balsamea (Little 1975). (2) This concept would also explain the above-mentioned reduction in phosphorylation by ABA, since Mitchell's chemiosmotic theory is now accepted

(see Racker 1975) and requires a H ion transport for phosphorylation.   (3)
According to our recent theory (Levitt 1974), the photosynthesis-induced H ion
transport in guard cell chloroplasts is the primary ion transport which leads
secondarily to K ion transport followed by stomatal opening. If this concept is
correct, the ABA effect would probably be only indirectly on K ion transport, due
to a direct effect on proton transport.   (4) Raschke (1975b) has shown that CO_2
and ABA are simultaneously required for stomatal closure, and concluded from this
that ABA inhibits the explusion of H ions from guard cells. Since Raschke and
Humble (1973) showed that the H ion efflux from guard cells is equal to the K ion
influx, this would agree with our explanation of the ABA-induced inhibition of
net K ion influx as a secondary effect, accompanying the primary inhibition of H
ion efflux.

In apparent opposition to this concept, IAA has been reported to produce the
same stomatal closure as ABA, in at least 4 different laboratories (Browning
1974) instead of antagonizing ABA as it does in roots (see above). Of course, it
has long been known that IAA can both stimulate and inhibit growth, depending on
its concentration. In fact, a marked increase in stomatal opening on application
of IAA has also been reported (Tal et al. 1974). Another possible explanation
of this apparent discrepancy is the effect of the location of the IAA on H ion
transport in the cell. ABA has been characterized from extracts of pea
chloroplasts (Railton et al. 1974). Therefore, it would be able to act directly
on the photosynthesis-induced H ion transport in the chloroplasts of the guard
cells. This may explain why IAA does not normally stimulate stomatal opening,
because it does not normally occur in the chloroplasts.

In agreement with this suggestion, recent evidence with labeled auxins
indicates that IAA stimulates proton transport and, therefore, cell enlargment by
attaching to a membrane. This membrane is probably the ER, according to Normand
et al. (1975), probably the plasma membrane or Golgi membrane, according to Venis
and Batt (1975). It would, therefore, be impossible for such membrane-bound IAA
to affect the action of ABA in the chloroplasts.

In further support of this concept, fusicoccin, a toxin stimulating cell
enlargment and inducing proton extrusion in plants tissues, is also able to
remove the inhibitory effect of ABA (Lado et al 1975). Since fusicoccin and IAA
both induce acidification and stimulation of K ion uptake, when they stimulate
cell enlargement, it is only reasonable to propose that their mechanism of action
is the same. Yet fusicoccin also stimulates stomatal opening, in opposition to
the above results with IAA (Squire and Mansfield 1974). Perhaps fusicoccin
accumulates in the cell loci favoring both effects, whereas IAA accumulates only
in the locus favoring cell enlargement.

b) It has been shown by Milborrow and Robinson (1973), that chloroplasts of
the cotyledons of avocado can synthesize ABA from mevalonic acid (MVA). It is,

therefore, conceivable that the small net water loss induced by a water stress may lead to the formation of MVA as a breakdown product of carotenoids in the chloroplasts, and the MVA would then be converted to ABA. The ABA would, thus, be present in the precise locus for reversing the photosynthetic H ion transport, and would therefore lead to a net K ion efflux and stomatal closure. The efflux would occur into the subsidiary cells, which lack chloroplasts and therefore would not be able to synthesize ABA. In favor of this concept, the small gravitational pressure of amyloplasts has been proposed to lead to ABA synthesis in these plastids (Audus 1975). A similar pressure change due to the decrease in epidermal turgor pressure, accompanying the drop in leaf water content on exposure to a water stress, could conceivably lead to ABA synthesis in the guard cell chloroplasts. Perhaps the pressure change may operate via a rupture of lysosomes, with a release of enzymes capable of breaking down the carotenoids to MVA.

c) The final question - why does ABA specifically affect the guard cells and not the other leaf cells - is answered above. For if explanation b) is correct, the ABA would accumulate only in green cells. This would include the guard cells but not the other epidermal cells since these have no chloroplasts. Therefore, the K ion efflux from the guard cells into the subsidiary cells would be unimpeded. Nearly all the remaining green cells of the leaf would be unable to induce this efflux since each green cell is surrounded by others tending to induce the same efflux in the reverse direction. Another factor is the apparent existence in the guard cells but not in the mesophyll cells of the complete enzyme system (including PEP carboxylase and malic enzyme) at a sufficient level of activity to support the K ion efflux (Levitt 1974).

On the basis of the above concept, the effect of water stress on ion transport and associated processes may be explained as in Table 1.

It should be emphasized that the guard cells are assumed to be in the steady state at all times, in the above analysis: (1) Water content of the guard cells can remain constant for a period of time only if water loss exactly balances water absorption by the guard cells. (2) K ion content of the guard cells can remain constant for a period of time, only so long as K^+ influx equals efflux. Opening and closing always involves a change in these two steady states.

2) Stomatal closure in the absence of ion transport.

It does not follow that water stress always leads to stomatal closure via ion transport. We must first distinguish clearly between hydropassive closing, when the loss of guard cell turgor is due solely to evaporative (gaseous) loss of water, and hydroactive closing, when active ion transport leads to osmotic (liquid) loss of water and, therefore, of cell turgor.

a) Hydropassive closing occurs when a leaf undergoes a marked drop in water content, due to an excess of transpiration over water absorption. If the guard cells show this same excess to a sufficient degree, they lose turgor and the stomata close. A sudden, net loss, however, is more likely to occur under artificial conditions (e.g., shoot excision) than under natural conditions. Recent evidence, in fact, indicates that under natural conditions a stress-induced stomatal closure may occur before the leaf can undergo a detectable decrease in water content. Lange and his co-workers (Schulze et al. 1972, 1973, 1974) have shown this kind of closure in the case of the stomata of desert plants, when the vapor pressure difference between the evaporating sites in a leaf and the surrounding air becomes sufficiently large. In the desert plant Prunus armeniaca, this mechanism is so effective that the total daily transpiration is less on a dry day (stomata closed) than on a moist day (stomata open). In other words, the stomata can anticipate a desiccation strain (a net loss of water) when exposed to an atmospheric water stress, and can close in an "attempt" to prevent it.

This response to a stress before experiencing a strain, seems to imply intelligence on the part of the plant, and appears unexplainable at first sight. It must be remembered however, that both stomatal opening and closing depend on the balance between the turgor of the guard cells (P_g) and that of the surrounding epidermal cells (P_e). When $P_g - P_e$ is large enough the stomata open, the degree of opening being proportional to $P_g - P_e$; otherwise they close. Hydropassive movement can, therefore, occur in two ways. (1) If the stomata are open and the net evaporative loss of water is greater from the other epidermal cells than from the guard cells, P_e will decrease more than P_g and the stomata will initially open wider because $P_g - P_e$ initially increases. This phenomenon has long been known. and is one explanation for the paradoxical "Iwanoff effect" (the sudden increase in transpiration immediately after excision of a shoot). It happens, for instance, 2 minutes after detachment of a needle from Taxus (Rottenberg and Koeppner 1972). Of course, if this net evaporative loss continues beyond the 2 minutes, the guard cells will lose more and more turgor and will soon close (Rottenberg and Koeppner 1972).

(2) If, on the other hand, the net evaporative loss is greater from the guard cells than from the other epidermal cells, then P_e becomes greater than P_g and the stomata must close. This phenomenon has not yet been demonstrated, and probably has rarely, if ever been proposed. It is my contention, however, that the "Lange effect" (the above described closing of stomata due to a large relative humidity gradient) is due to this second kind of hydropassive movement, and that it occurs in xerophytes but not in mesophytes. This difference can be logically explained by a paradoxical difference between the two. A search of the older literature (e.g., Sachs 1882) reveals that mesophytes have surface stomata, and probably because of this location the guard cells are at least as well

cutinized as the other epidermal cells. Xerophytes on the other hand, commonly have sunken stomata and the guard cells appear free of cutin; but the other epidermal cells are more heavily cutinized than in the case of mesophytes. Because of the more rapid diffusion of vapor than liquid, and of the larger surface area of the guard cells in contact with air than with the adjacent subsidiary cells, when the humidity gradient increases to a critical point, the guard cells of the xerophyte may be expected to show a net loss of water and of turgor before the other heavily cutinized epidermal cells, and the stomata will therefore close before there is a measurable net loss in leaf water content.

Once the stomata are closed, however, they will soon regain their turgor by osmotic absorption of water from the surrounding, turgid cells, which have not suffered an appreciable net loss. This may lead to a rhythmic opening and closing of the stomata. If, however, the guard cells of the xerophytes behave biochemically as do those of the mesophytes, the initial hydropassive closure should soon be followed by a synthesis of ABA within the guard cells but not in any of the other green cells, and this would lead to the ion efflux, preventing a second reopening. Which of these two alternatives, does occur, does not seem to have been determined.

As mentioned above, mesophytes do not possess the extreme adaptations of xerophytes, and therefore they cannot anticipate a decrease in leaf water content as the xerophytes do. The "Lange effect", therefore, does not occur. This explains the inability of Raschke and Kuhl (1969) to observe it in maize leaves. In further agreement with this concept, Hall and Kaufmann (1975) corroborated the "Lange effect" for Sesamum indicum and demonstrated that low CO_2 concentration decreased but did not eliminate the effect. This result is to be expected, since the CO_2 concentration, though an important factor in photoactive stomatal opening, has no direct effect on the water content and therefore on the turgor of the guard cells and on hydropassive closing. A more difficult exception to explain, at least until the guard cells are investigated in detail, is the apparent ability of four arctic and alpine species to show the Lange effect (Johnson and Caldwell 1976).

The effects of water stress on ion transport are tabulated in Table 1.

Table 1. Effects of water stress on ion transport.

Water Stress

Roots

In soil
- Decreased in diffusion paths → Decreased ion influx
- Increased concentration → Increased ion influx

In water culture
- Increased osmotic stress → Decreased ion influx

Leaves

Mesophytes
- Transpiration ↗ water absorption
- Decreased leaf water content
- ABA synthesis in green cells
- Reversal of Δ pH in chloroplasts
- K^+ ion efflux from green guard cells to non-green subsidiary cells
- Hydroactive stomatal closure

Some xerophytes
- Only guard cell transpiration > water absorption
- Pe ↗ Pg
- Guard cells pushed shut
- Hydropassive stomatal closure
- Secondary ABA synthesis only in guard cells

II. THE CHILLING LOW TEMPERATURE STRESS

As early as the beginning of this century, it was suggested that chilling injures sensitive plants by increasing their cell permeability, resulting in a leakage of cell solutes (see Levitt 1972). The leakage was later confirmed by several investigators (Lieberman et al. 1958, Lewis and Workman 1964, Katz and Reinhold 1965, Minchin and Simon 1973). Lyons et al. (1964) explained the premeability concept by the relative inflexibility of the membranes due to a lower content of unsaturated fatty acids in the chilling sensitive plants. But the inflexibility by itself would have the opposite effect -- a decrease in permeability. It would have to be followed by cracks in the membrane, produced by mechanical stresses, in order to increase the permeability. It may, therefore, be more reasonable to suggest a damage to ion pumps rather than to the passive permeability of the cell. This concept is supported by Lieberman et al. (1958) who found that most of the leakage was due to K ions. Furthermore, the membrane-associated enzymes show a rapid drop in activity at and below the phase change temperature as indicated by a sharp downward bend in the Arrhenius plot (Lyons 1973, Sechi et al. 1973). Several other lines of evidence favor the ion pump interpretation. The existence of a respiratory upset at chilling temperatures (Eaks and Morris 1956) might indicate a lack of energy for the active ion transport. A decrease in ATP concentration of cotton seedlings at chilling temperatures (Stewart and Guinn 1969) would also slow down the active ion transport. Finally it has been shown that the ATPase of ion pumps is inactivated below the phase transition temperature of the lipid with which it is associated (Sechi et al. 1973, Lee 1975). This has been corroborated for membrane ATPase of bean roots (Kuiper 1972), from cell wall, mitochondrial, and microsome fractions. Extraction of lipids from the ATPase decreased its activity, and this was partially restored by addition of lipid. All these factors would decrease, if not completely inhibit, the active ion uptake. Amar and Reinhold (1973), indeed, found a 3.5% loss of the leaf proteins when chilling was combined with osmotic shock, and suggested that this small amount of protein may be closely involved in the ion transport mechanism.

III. THE FREEZING LOW TEMPERATURE STRESS (OR CRYOSTRESS)

For well over half a century, suggestions have been made, from time to time, that freezing injures the plant by way of membrane damage (see Levitt and Dear 1970). More direct evidence of ATPase damage has been produced in the case of chloroplast membranes (Heber 1970, Garber and Steponkus 1976). Recent results in our laboratory (Palta 1976) now indicate that the initial freezing damage may involve the ion transport system of the cell.

In point of fact, evidence in support of this concept has long been available but ignored. It has long been known that freezing injury can be measured by the amount of ion efflux from the thawed tissues. This relation has been used since Dexter adopted it 40 years ago, as a method of measuring freezing injury (see Levitt 1972). The frozen and thawed tissue is immersed in distilled water under standard conditions and the injury is taken as proportional to the increase in conductivity of the water. The possible theoretical significance of this method has been completely overlooked, mainly because of the tacit (but erroneous) assumption that the ion efflux is essentially only from freeze-killed cells and therefore is simply the result, rather than a possible cause, of the injury. The conductivity method was, therefore, accepted as a measure of the percent of the cells killed.

a. Evidence of damage to ion transport system.

Palta (1976) has shown that this concept is erroneous, at least in the case of onion bulbs, for the following reasons:

1) The injury measured by the conductivity method is not paralleled by cell death. When injury is accompanied by an increase in conductivity, at least up to 50% of the maximum increase, all the cells are alive on thawing (when the conductivity is measured) as shown by cytoplasmic streaming, plasmolysis, and vital staining.

2) The increase in conductivity, when sufficiently large, is accompanied by a measurable decrease in cell sap concentration of all the cells, which at least approximately accounts for the increase in conductivity of the immersion water.

3) The increase in conductivity occurs in the absence of any detectable increase in cell permeability to tritiated water, and presumably to other polar molecules or ions.

4) The major component of the effluxed ions is K^+.

5) The increase in conductivity measured immediately after thawing may increase further with time after thawing, leading to eventual death, or may decrease, leading to repair and full recovery of the cells.

b. Possible sequence of events leading to damage to ion transport system.

The only logical explanation of these results is that extracellular freezing injures the cell by damaging its ion transport system. The probable sequence of events is as follows:

1) Ice forms first in the extracellular space and continues to grow at the expense of water moving out of the cells. Air is simultaneously expelled from the tissues due to the expansion of water on freezing.

2) During thawing, the intercellular spaces become infiltrated with thaw water. Two opposing processes are immediately initiated -- a) osmotic reabsorption of intercellular thaw water, and b) leaching of cell solutes by

intercellular thaw water. If there is no injury, a) will go to completion
quickly, the cells becoming turgid and air being pulled back into the
intercellular spaces. If there is some injury, b) will occur before a) is
complete and osmotic balance will be achieved before the cells can regain turgor.
The tissues will, therefore, remain flaccid and infiltrated, as long as the
leached solutes remain in the intercellular water outside the cells. Since the
reabsorption of these solutes is an active, metabolically dependent process, it
may conceivably be slowed down or halted by a combination of at least three
factors: (1) the low temperature, (2) the O_2 deficiency due to infiltration of
the intercellular spaces, and (3) the freeze-induced damage to the ion-pump of
the cell membrane.

The low temperature, by itself, cannot be a direct cause of the injury, since
it is ineffective in the absence of freezing. The O_2 deficiency is not likely to
be a factor in the frozen tissues, because of the low temperature and the freeze-d
dehydration, both of which must synergistically decrease the rate of respiration
and therefore of O_2 use. Even after thawing, the temperature is still low enough
to decrease the O_2 requirement, and the solubility of O_2 in water is increased.
Recent evidence, in fact, opposes a role for the O_2 deficiency. Tissues of
wheat, rye, bean and Lemna minor when vacuum-infiltrated were practically unable
to fix CO_2, due to the decrease supply of CO_2 to the chloroplasts; but their
respiration was unaffected (Macdonald 1975). The low temperature and the O_2
deficiency therefore can only be secondary factors at best, and the primary cause
of the continued infiltration of the intercellular spaces after thawing must be
the injury to the ion pump during the freezing. What, then, is the nature of
this injury?

c. Relation between damage to ion-transport system and membrane damage.

The above described effects of extracellular freezing on ion and water efflux
and influx must be intimately related to membrane properties, since both ion
transport and osmotic flow of water are controlled by the cell membrane. There
are, of course, many membranes in the cell -- the plasmalemma, the tonoplast, the
endoplasmic reticulum, the Golgi apparatus, the plastid thylakoids, etc. Which
of these is most likely to be injured by freezing? The plasma membrane
(plasmalemma) possesses properties that should make it more sensitive to
temperature stress than other cell membranes. In the case of the roots of Avena
sativa, it contained less protein and more lipid than the total membrane and the
mitochondrial fraction (Keenan et al. 1973). Although the phospholipids
accounted for a similar percent of the total lipids in all membranes, the plasma
membrane had a lower level of unsaturated fatty acids and a higher level of
palmitic acid than the other membranes. Further evidence pointing to the plasma

membrane as the seat of injury is the above-mentioned efflux of ions from the cell, which must occur through the plasmalemma since it is the boundary membrane separating the cell protoplasm from the external medium, or the intercellular spaces.

The final question is the actual biochemical or biophysical event during freezing which damages the ion-transport system and the membrane in which it resides. Both consist of lipid and protein, and it is unknown as yet whether the event involves a change in the lipids, or the proteins, or both.

In the case of freeze preservation, injury may be due to intracellular freezing, and therefore is perhaps not comparable to the extracellular freezing injury of higher plants. Nevertheless, the ion transport system again seems involved. Freeze-dried E. coli cells show damage to the K ion transport system, which may be partially repaired by incubation in nutrient medium (Israeli et al. 1974).

There is some evidence that hardening or acclimation of plants involves changes that tend to prevent damage to the ion transport system. Hardening of winter wheat increased the concentrations of AMP, ADP, and ATP in the leaves (Aksenova 1974). Similarly, both the degree of hardening and the content of ATP were higher in rye than wheat. Cold acclimated plants showed higher rates of phosphate absorption at a given temperature than did warm-acclimated plants (Chapin 1974, 1975).

IV. THE SALT STRESS

Since this environmental stress injures a plant by raising its cell content of ions above the toxic limit or the osmotic limit of the cell, ion transport must be of fundamental importance. There are several possible kinds of ion effects. Sensitive cells may actively absorb an excess of Na salt, and conversely may be prevented from absorbing an adequate amount of K^+ (see Levitt 1972). Resistant cells may actively excrete an excess of Na salt.

It is, thus apparent that one of the basic effects of at least several environmental stresses it to alter the ion transport system. An increased ion efflux occurs from the guard cells of the stomata without damage, in response to a water stress. This is an adaptive response. The two low temperature stresses induce a net injurious efflux by damaging the influx system, shifting the steady-state balance between the two. This decreased ion transport into the cell may be a basic mechanism of injury by several, if not most environmental stresses.

REFERENCES

1. Mengel, K. and Von Braunschweig, L. C. (1972) The effect of soil
    moisture upon the availability of potassium and its influence on the
    growth of young maize plants (Zea mays L.). Soil Sci. 114:142-148.

2. Gerakis, P. A., Guerrero, F. P., and Williams, W. A. (1975) Growth, water
    relations and nutrition of three grassland annuals as affected by drought.
    J. Appl. Ecol. 12:125-136.

3. Lahiri, A. M., Singh, Sudama, and Kackar, N. L. (1973) Studies on plant-
    water relationships: VI. Influence of nitrogen level on the performance
    and nitrogen content of plants under drought. Proc. Indian Nat'l Sci.
    Acad. Part. B. Biol. Sci. 39:77-90.

4. Shalhevet, Joseph (1973) Effect of mineral type and soil moisture content
    on plant uptake of ^{137}Cs. Radiat. Bot. 13:165-171.

5. Durham, R. J. and Nye, P. H. (1974) The influence of soil water content
    on the uptake of ions by roots: II. Chloride uptake and concentration
    gradients in soil. J. Appl. Ecol. 11:581-595.

6. Smith, Richard, C., St. John, Beverly H., and Parrondo, Rolando (1973)
    Influence of mannitol on absorption and retention of rubidium by excised
    corn roots. Am. J. Bot. 60:839-845.

7. Parrondo, Rolando T., Smith, Richard C., and Lazurick, Kenneth (1975)
    Rubidium absorption by corn root tissue after a brief period of water
    stress and during recovery. Physiol. Plant. 35:34-38.

8. Erlandsson, G. 1975. Rapid effects on ion and water uptake by changes
    of water potential in young wheat plants. Physiol. Plant. 35:256-262.

9. Mattson, Sante. (1973) Ionic relationships of soil and plant: IV. Ion
    uptake in relation to membrane activity and moisture. Acta Agric. Scand.
    23:11-16.

10. Raschke, K. (1975a) Stomatal action. Ann. Rev. Plant Physiol. 26:309-340.

11. Hsiao, T. C. (1973) Plant responses to water-stress. Ann. Rev. Plant
    Physiol. 24:519-570.

12. Doerffling, K., Sonka, Baerbel, and Tietz, D. (1974) Variation and
    metabolism of abscisic acid in pea seedlings during and after water
    stress. Planta 121:57-66.

13. Loveys, B. R. and Kriedemann, P. E. (1973) Rapid changes in abscisic acid-
    like inhibitors following alterations in vine leaf water potential.
    Physiol. Plant. 28:476-479.

14. Zabadal, Thomas J. (1974) A water potential threshold for the increase of
    abscisic acid in leaves. Plant Physiol. 53:125-127.

15. Loveys, B. R. and Kriedemann, P. E. (1974) Internal control of stomatal physiology and photosynthesis: I. Stomatal regulation and associated changes in endogenous levels of abscisic and phaseic acids. Aust. J. Plant Physiol. 1:407-415.

16. Jones, R. J. and Mansfield, T. A. (1970) Suppression of stomatal opening in leaves treated with abscisic acid. J. Exp. Bot. 21:714-719.

17. Shaner, Dale L., Mertz, Stuart M. Jr., and Arntzen, Charles J. (1975) Inhibition of ion accumulation in maize roots by abscisic acid. Planta 122:79-90.

18. Reed, Nu-May R. and Bonner, Bruce A. (1974) The effect of abscisic acid on the uptake of potassium and chloride into Avena coleoptile sections. Planta 116:173-185

19. Doerffling, K., Menzer, U., and Gerlach-Luessow, A. (1973) Antagonistic effects of abscisic acid, indoleacetic acid, and gibberellic acid on transport of potassium and phosphorus in sunflower epicotyls. Mitt. Staatsinst. Allg. Bot. Hamburg 14:19-23.

20. Chapman, K. S. R., Trewavas, A., and van Loon, L. C. (1975) Regulation of the phosphorylation of chromatin-associated proteins in Lemna and Hordeum. Plant Physiol. 55:293-296.

21. Cleland, R. (1971) Cell wall extension. Ann. Rev. Plant Physiol. 22:197-227.

22. Jacobs, M. and Ray, P. M. (1975) Promotions of xyloglucan metabolism by acid pH. Plant Physiol. 56:373-376.

23. Audus, L. J. (1975) Geotropism in roots. pp. 327-363. In: J. G. Torrey and D. T. Clarkson (Eds.) The Development and Function of Roots. Acad. Press. London.

24. Racker, E. (1975) Reconstitution, mechanism of action, and control of ion pumps. Tenth Hopkins Memorial Lecture. Biochem. Soc. Trans. 3:785-802.

25. Levitt, J. (1974) The mechanism of stomatal action - once more. Protoplasma 82:1-17.

26. Raschke, Klaus. (1975b) Simultaneous requirement of carbon dioxide and abscisic acid for stomatal closing in Xanthium strumarium L. Planta 125:243-259.

27. Raschke, K. and Humble, G. D. (1973) No uptake of anions required by opening stomata of Vicia faba: Guard cells release hydrogen ions. Planta 115:47-57.

28. Browning, G. (1974) 2-Chloroethanephosphonic acid reduces transpiration and stomatal opening in Coffea arabica L. Planta 121:175-179.

29. Tal, M., Imber, D., and Gardi, I. (1974) Abnormal stomatal behavior and hormonal imbalance in flacca, a wilty mutant of tomato. J. Exp. Bot. 28:51-60.

30. Railton, I. D., Reid, D. M., Gaskin, P., and MacMillan, J. (1974) Characterization of abscisic acid in chloroplasts of _Pisum sativum_ L. cv. Alaska by combined gas-chromatography-mass spectrometry. Planta 117:179-182.

31. Normand, G., Hartmann, M. A., Schuber, F., et Benveniste, P. (1975) Characterization de membranes de coleoptiles de Mais fixant l'auxine et l'acide N-naphthyl phtalamique. Physiol. Veg. 13:743-761.

32. Venis, M. A. and Batt, S. (1975) Membrane-bound receptors for plant hormones. Biochem. Soc. Trans. 3:1148-1151.

33. Lado, Piera, Rasi-Caldogno, Franca, and Colombo, Roberta (1975) Acidification of the medium associated with normal and fusicoccin-induced seed germination. Physiol. Plant. 34:359-364.

34. Squire, G. R., and Mansfield, T. A. (1974) The action of fusicoccin on stomatal guard cells and subsidiary cells. New Phytol. 73:433-440.

35. Milborrow, B. V. and Robinson, D. R. (1973) Factors affecting the biosynthesis of abscisic acid. J. Exp. Bot. 24:537-548.

36. Schulze, E. D., Lange, O. L., Buschbom, U., Kappen, L., and Evenari, M. (1972) Stomatal responses to changes in humidity in plants growing in the desert. Planta 108:259-270.

37. Schulze, E. D., Lange, O. L., Kappen, L., Buschbom, U., and Evenari, M. (1973) Stomatal responses to changes in temperature at increasing water stress. Planta 110:29-42.

38. Schulze, E. D., Lange, O. L., Evenari, M., Kappen, L., and Buschbom, U. (1974) The role of air humidity and leaf temperature in controlling stomatal resistance of _Prunus armeniaca_ L. under desert conditions. I. A simulation of the daily course of stomatal resistance. Oecologia 17:159-170.

39. Rottenburg, Wolfgang and Koeppner, Thusnelda (1972) Light and water supply as factors influencing the degree of stomatal opening in conifers. Ber. Dtsch. Bot. Ges. 85(7/9):353-362.

40. Sachs, J. (1882) Vorlesungen uber Pflanzenphysiologie. Verlag Wilhelm Engelmann. Leipzig.

41. Raschke, K. and Kuhl, U. (1969) Stomatal responses to changes in atmospheric humidity and water supply: experiments with leaf sections of _Zea mays_ L. in CO_2-free air. Planta 87:36-48.

42. Hall, Anthony E. and Kaufmann, Merrill R. (1975) Stomatal response to environment with _Sesamum indicum_ L. Plant Physiol. 55:455-459.

43. Johnson, D. A. and Caldwell, M. M. (1976) Water potential components, stomatal function, and liquid phase water transport resistances of four arctic and alpine species in relation to moisture stress. Physiol. Plant. 36:271-278.

118

44. Levitt, J. (1972) Responses of Plants to Environmental Stresses. Acad. Press. New York.

45. Lieberman, M., Craft, C. C., Audia, W. V. and Wilcox, M. S. (1958) Biochemical studies of chilling injury in sweet potatoes. Plant Physiol. 33:307-311.

46. Lewis, T. L. and Workman, M. (1964) The effect of low temperature on phosphate esterification and cell membrane permeability in tomato fruit and cabbage leaf tissue. Aust. J. Biol. Sci. 17:147-152.

47. Katz, S. and Reinhold, L. (1965) Changes in the electrical conductivity of Coleus tissue as a response to chilling temperatures. Isr. J. Bot. 13:105-114.

48. Minchin, Ann and Simon, E. W. (1973) Chilling injury in cucumber leaves in relation to temperature. J. Exp. Bot. 24:1231-1235.

49. Lyons, J. M., Wheaton, T. A., Pratt, H. K. (1964) Relationship between the physical nature of mitochondrial membranes and chilling sensitivity in plants. Plant Physiol. 39:262-268.

50. Lyons, J. M. 1973. Chilling injury in plants. Ann. Rev. Plant Physiol. 24:445-466.

51. Sechi, A. M., Bertoli, E., Landis, L., Parenti-Castelli, G., Lenaz, G., and Curatola, G. (1973) Temperature dependence of mitochondrial activities and its relation to the physical state of the lipids in the membrane. Acta Vitaminol. Enzymol. 27:177-190.

52. Eaks, I. L., and Morris, L. L. (1956) Respiration of cucumber fruits associated with physiological injury at chilling temperatures. Plant Physiol. 31:308-314.

53. Stewart, James McD., and Guinn, G. (1969) Chilling injury and changes in adenosine triphosphate of cotton seedlings. Plant Physiol. 44:605-608.

54. Lee, A. G. (1975) Fluorescence studies of chlorophyll a incorporated into lipid mixtures, and the interpretation of "phase" diagrams. Biochim. Biophys. Acta 413:11-23.

55. Kuiper, P. J. C. (1972) Temperature response of adenosine triphosphatase of bean roots as related to growth temperature and to lipid requirements of the adenosine triphosphatase. Physiol. Plant. 26:200-205.

56. Amar, L. and Reinhold, L. (1973) Loss of membrane transport ability in leaf cells and release of proteins as a result of osmotic shock. Plant Physiol. 51:620-625.

57. Levitt, J. and Dear, J. (1970) The role of membrane proteins in freezing injury and resistance. pp. 149-174. In: G. E. W. Wolstenholme and M. O'Connor (Eds.) The Frozen Cell. J. and A. Churchill, London.

119

58. Heber, U. (1970) Proteins capable of protecting chloroplast membranes against freezing. pp. 175-188. In: G. E. W. Wolstenholme and M. O'Connor (Eds.) The Frozen Cell. J. and A. Churchill, London.

59. Garber, M. P. and Steponkus, P. L. (1976) Alterations in chloroplast thylakoids during an *in vitro* freeze-thaw cycle. Plant Physiol. 57:673-680.

60. Palta, J. (1976) Effects of low temperature and water stress on the protoplasmic properties of *Allium cepa* cells. Ph.D. Thesis, Univ. of Minnesota.

61. MacDonald, I. R. (1975) Effect of vacuum infiltration on photosynthetic gas exchange in leaf tissue. Plant Physiol. 56:109-112.

62. Keenan, T. W., Leonard, R. T., and Hodges, T. K. (1973) Lipid composition of plasma membranes from Avena sativa roots. Cytobios 7:103-112.

63. Israeli, E., Giberman E., and Kohn, A. (1974) Membrane malfunctions in freeze-dried *Escherichia coli*. Cryobiology 11:473-477.

64. Aksenova, O. F. (1974) Adenosine phosphate system of winter cereals and effect of hardening on it. Biol. Nauki 17:81-84.

65. Chapin, F. Stuart III. (1974) Phosphate absorption capacity and acclimation potential in plants along a latitudinal gradient. Science 183:521-523.

66. Chapin, F. Stuart III. (1975) Morphological and physiological mechanisms of temperature compensation in phosphate absorption along a latitudinal gradient. Ecology 55:1180-1198.

Regulation of Cell Membrane Activities in Plants
E. Marrè and O. Ciferri eds.
© *1977, Elsevier/North-Holland Biomedical Press, Amsterdam*

A HYPOTHESIS ON CYTOPLASMIC OSMOREGULATION

R. Gareth Wyn Jones, R. Storey[*], R.A. Leigh[**], N. Ahmad
and A. Pollard.

Department of Biochemistry and Soil Science
University College of North Wales, Bangor,
Gwynedd, Wales.

INTRODUCTION

In a previous communication to the Workshop on 'Transmembrane Ionic Exchanges in Plants'[1], we presented a summary of analytical data on the accumulation of betaine in a number of plants and on the responses of betaine, proline, inorganic ions and other parameters to salt and osmotic stress in a number of glycophytic and halophytic plant species. Additional evidence on the activity of cytosolic malic dehydrogenase in the presence of various electrolytes and non-electrolytes and on K^+ and Na^+ concentrations and compartmentation in plant cells, particularly from Jeschke and Stelter[2], led to our proposing a model for cytoplasmic ionic and osmotic regulation in plants, especially those capable of betaine accumulation. Further consideration of the comparative biochemistry and physiology of osmo-regulation in lower plants and animals, especially marine invertebrates, and current physical chemical studies on protein-electrolyte-water interactions suggested that certain generalisations could be made about cytoplasmic osmo-regulation in most, if not all, eucaryotic cells.

This present paper will, in part, reiterate the hypothesis and some of the data which led to its formulation. The model suggests that the cytoplasmic ionic strength with K^+ the dominant cation does not normally exceed 200-250 mM and that, where physiological conditions require a lower osmotic potential, a molecule such as betaine is accumulated in the cytoplasm so as to cause, in conjunction with the ionic components, a minimum perturbation of protein-water interactions.

The levels of betaine accumulated in the three halophytic plant species which have been studied in some detail, Suaeda monoica, Atriplex spongiosa and Spartina x townsendii, are such that the compound could act as a major cytoplasmic osmoticum if it were largely, but not exclusively, concentrated in the cytoplasm and ex-cluded from the vacuole. We have sought to test this assumption by the isolation and purification of vacuoles from red beet storage tissue using a technique

---

Current address [*] School of Biological Sciences, University of Sydney, Sydney, N.S.W., Australia.

[**] Botany School, University of Cambridge, Cambridge, England.

recently developed by Leigh and Branton[3,4]. The results provide strong support for, but not conclusive proof of, the validity of the model in this tissue. Some of the wider physiological implications of the model are also discussed.

MATERIALS AND METHODS

Plant tissues: The plants were grown in a greenhouse with supplemental light in hydroponic culture using Hoagland's medium to which NaCl or polyethylene glycol (PEG) were added incrementally until the desired level was achieved. Details of the individual growth conditions are given by Storey and Wyn Jones[5]. Red beet was obtained from local growers and we are grateful to Mr. Thomas Hughes and Mr. Selwyn Jones for their generosity.

Inorganic analyses: Sodium and potassium were determined by flame photometry. Osmotic pressures were determined on tissue sap extracted using a french pressure cell and hydraulic press in a cryoscopic osmometer (Advanced Instruments Inc., Mass.).

Organic analyses: Betaine was determined by two methods; colorimetrically by a modification of the periodide method at pH 2.0[6] and by semi-quantitative TLC or TL electrophoresis[7]. The former is sensitive but rather non-specific, while the latter is entirely specific but less sensitive and accurate. Precise details are given by Storey and Wyn Jones[5]. Betanin was determined spectrophotometrically at 547 nm in the presence of 1.5% sodium dodecyl sulphate ($\varepsilon = 60,500$)[8]. Proline was measured by the method of Singh et al[9].

Enzyme isolation and assays: Cytosolic malic dehydrogenase was isolated from H.vulgare leaves (cultivar California Mariout) by a minor modification of the method of Greenway and Sims[10] whose assay procedure was also employed. Acid phosphatase was measured using p-nitrophenolphosphate as substrate in 50 mM Na acetate buffer at pH 4.5. Protein was determined on fractions following pre-cipitation in 20% TCA by the Lowry method[11].

Isolation and purification of vacuoles: Beet tissue (450g) was homogenised in a low shear shredder into 2 1. of homogenising medium (Sorbitol 1M; Tris-HCl, 50 mM, pH 7.4; NaEDTA, 1 mM; 2-mercaptoethanol, 2 mM). After filtering the brei and an initial centrifugation, as shown diagrammatically in Fig.1, the pellet was re-suspended in homogenising medium containing 15% Metrizamide (Nyegaard & Co. A/S, Oslo). The sample was overlayed with 10% and 0% metrizamide in homogenising medium and subjected to a density gradient centrifugation. The whole vacuoles were concentrated by floatation as a dense red band at the 0-10% metrizamide interface. Further details of the shredder and the development of the purific-ation method are given by Leigh and Branton[4].

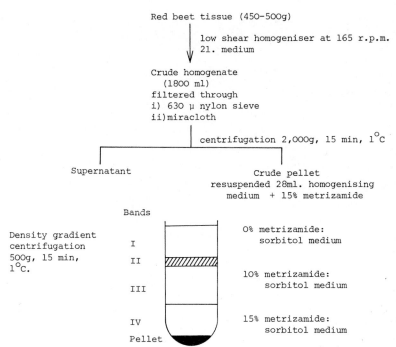

Red beet tissue (450-500g)

low shear homogeniser at 165 r.p.m.
2l. medium

Crude homogenate
(1800 ml)
filtered through
i) 630 µ nylon sieve
ii)miracloth

centrifugation 2,000g, 15 min, 1°C

Supernatant

Crude pellet
resuspended 28ml. homogenising
medium + 15% metrizamide

Bands

Density gradient
centrifugation
500g, 15 min,
1°C.

I

II

III

IV

Pellet

0% metrizamide:
        sorbitol medium

10% metrizamide:
        sorbitol medium

15% metrizamide:
        sorbitol medium

Fig.1. Schematic procedure for isolation and purification of vacuoles

RESULTS AND DISCUSSION

Analytical Data: We have already reported that in many, but not all, halophytes substantial levels of betaine (>10 µmoles g.FW^{-1}) were found even in plants
grown in low salt conditions and that salt and osmotic (PEG) stress caused a
further increase in the level of betaine both in the three halophytes S.monoica,
A.spongiosa and S. x townsendii and in a number of semi-resistant graminaceous
glycophytes e.g. Chloris Gayana, H.vulgare[1, 5,12]. In contrast, little or no betaine was found in the shoot tissue of six salt sensitive species, Phaseolus
vulgaris, Trifolium repens, Raphanus sativus, Pisum sativum, Daucas carota and
Lycopersicum esculentum grown under salt stressed or unstressed conditions.

A close correlation was observed between the osmotic pressure of the shoots
of the plants and their betaine content, both when values are compared in a single
plant grown at various salinities (see Fig.2 - S. x townsendii, p < 0.001, r = 0.98)
and when different plant species are compared under similar growth conditions[5].
In contrast the proline levels only increased significantly in the halophytes at
high, inhibitory NaCl levels and were generally a tenth or less the betaine levels.
Since very high levels of betaine are accumulated in the halophytes, e.g. >150 mM
betaine in extracted sap of S. x townsendii grown at 600 mM NaCl, we have suggested

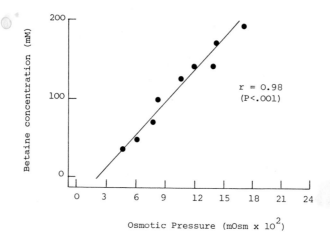

Fig.2. Relation of osmotic pressure to betaine
concentration in extracted sap of S. x
townsendii leaves from plants grown in
Hoagland's medium with incremental increase in NaCl up to 600 mM.

[1,12] that this compound might be involved in osmoregulation. Using the analytical data on osmotic pressure and K^+, Na^+, proline and betaine levels in extracted sap of Spartina grown in Hoagland's medium with incremental increases in NaCl concentration up to 600 mM, a simple model of osmoregulation and ionic compartmentation in this species was constructed (Fig.3)[1]. It was assumed that (i) the cytoplasmic/vacuolar ratio is maintained at 1:4, (ii) the cytoplasmic K^+ level is constant at 100 mM, (iii) betaine and proline are retained in the cytoplasm, (iv) other solutes, e.g. Mg^{2+} salts but not including Na^+ A^-, contribute 75 mOsmol. to the cytoplasmic osmotic potential, (v) the osmotic coefficients of K^+ and Na^+ are 0.9 and those of proline and betaine unity. The rationale for some of these assumptions was discussed previously[1].

Aside from any assumptions about inorganic ion compartmentation, the model requires that betaine be concentrated to very high values in the cytoplasm without any significant deterioration in metabolic functions and largely excluded from the vacuole. We have sought evidence both on the effect of betaine on enzymes and on its compartmentation in cells. This information will be summarised below.

Enzyme Activity: Several authors have noted that the salt sensitivities of cytoplasmic enzymes and functions in glycophytic and halophytic plants are similar and that the cytoplasmic activities of halophytic plant cells appear ill-adapted genetically to high salinity compared with those of some halophytic

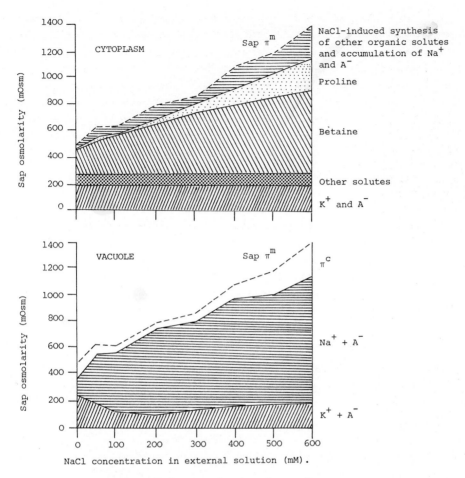

Fig.3. Possible model for cytoplasmic and vacuolar
osmoregulation in shoot tissue of S. x
townsendii.

bacteria[10,13]. We have reinvestigated this problem using, in the main, cytoplasmic malic dehydrogenase from barley leaves as a model system. As reported
by Greenway & Sims[10], this enzyme changes in its salt sensitivity with substrate
concentration. Even at low malate levels (0.5 mM), when the enzyme is readily
inhibited by NaCl and KCl, up to 1M betaine did not affect its activity (Fig.4).
Furthermore, 500 mM betaine partially protected the enzyme against salt inhibition. It should be noted that salt inhibition of the enzyme was not confined to NaCl but occurred with all salts and the order of the effectiveness
followed closely the Hofmeister lyotrophic series for cations; $(CH_3)_4N^+$ <K^+

126

$<NH_4^+$ $<Na^+$ $<Li^+$ $<Mg^{2+}$ $<Ca^{2+}$ and for anions; Cl^- $<NO_3^-$, Br^- $<I^-$ in agreement with previous work on animals[14]. The effects of a number of other organic molecules, some of whom have been reported to act as cytoplasmic osmotica in other organisms, have been tested on malic dehydrogenase. Proline and glycerol did not inhibit activity up to at least 500 mM. They also showed a small protective action in the presence of 300 mM NaCl but were less effective than betaine[1]. Thus it appears that a number of compounds, exemplified by betaine, may act as a nontoxic, possibly protective, cytoplasmic components and further studies are now in progress on their effects on mitochondrial and other activities.

Isolated vacuoles: Two techniques have recently been described for the isolation of whole vacuoles from plant cells[3,15]. One method uses a low shear homogenisation of the tissue and subsequent purification of the released vacuoles by floatation on a metrizamide - sorbitol density gradient (Fig.1). This method has been applied extensively to red beet storage tissue[4]. This tissue is well suited to test the compartmentation of betaine as the cell sap contains betaine (21 mM) and has a fairly high osmolarity (600 mOsm) while the red pigment, betanin, found in the vacuoles, can be used as an internal standard during the density gradient purification of whole, undamaged vacuoles. The whole vacuoles

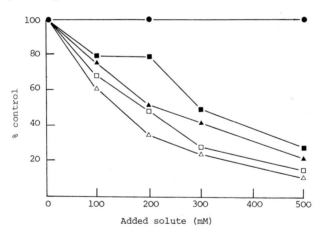

Fig.4 Influence of betaine alone and in presence of NaCl & KCl on the activity of cytosolic malic dehydrogenase from barley leaves.

Enzyme isolated and assayed as described by Greenway & Sims [10].

● betaine;   □ KCl   ■ KCl & betaine (500 mM)   △ NaCl   ▲ NaCl & betaine (500 mM)

are concentrated as a sharp red band at the 0 - 10% metrizamide interface (Band II) on the gradient but, although substantially purified, are apparently contaminated, at a low level, by nuclei and some empty, possibly resealed, vacuoles[4].

TABLE 1.

SPECIFIC ACTIVITIES OF BETANIN, BETAINE, K^+ AND ACID PHOSPHATASE
IN FRACTIONS FROM THE METRIZAMIDE-SORBITOL GRADIENT.

The figures quoted are average values from two separate runs.

Band	Betanin*	p-NPPase**	Betaine*	K^+ *
I	0.74	2.3	31.7	28
II	0.41	1.9	5.4	8.7
III	0.24	1.0	14.0	5.3
IV	0.08	0.5	1.3	2.3
Pellet	0.01	0.1	0.6	0.3

* μmoles mg. protein^{-1} ** ΔOD min^{-1}mg.protein^{-1}

In these experiments vacuoles were prepared and the specific activities of a number of relevant parameters determined in the fractions from the metrizamide gradient (Table 1). A high specific activity of betanin was found in Band II corresponding to the visible concentration of whole vacuoles. A high specific activity was also spasmodically recorded in Band I but this arose from the very low protein content of this fraction and few vacuoles were observed microscopically. The data from this fraction was therefore considered very unreliable. Since betanin is demonstrably in vacuoles, other compounds which are accumulated in vacuoles should be strongly correlated with betanin on the gradient while those not associated with vacuoles should show no such correlation. The correlations between the specific activities of acid phosphatase and betaine and betanin are shown in Figures 5 & 6. Because of the possible errors alluded to above, data on Band I was omitted from the regression analyses although it made little difference to the acid phosphatase: betanin relationship and further diminished the betaine: betanin coefficient.

The excellent correlation of acid phosphatase with betanin (r = 0.98, p <0.001) strongly suggests that this enzyme is located in the vacuole. The absence of a significant betaine: betanin correlation (r = 0.43, p >.2) implies that betaine is not found in the vacuole, in contrast to K^+, which, using the same criterion (r = 0.93; p <0.001; y = 1.83x + 0.07) is found in the vacuole. The extracted sap of these beets contained 75 mM K and 31 mM Na and therefore some K^+ might reasonably be expected to be found in the vacuoles. Since both the water soluble pigment, betanin, and K^+ are found in isolated vacuoles, it appears most unlikely that betaine has leaked out during the isolation procedure and the apparent absence of betaine from isolated vacuoles can be regarded as physiologically significant. Nevertheless, some betaine appears to be carried up in the gradient. The reason for this is not known but could reflect a high concentration of betaine in one of the other organelles contaminating the gradient, e.g. nuclei. Further data from these experiments suggest that some K^+ is lost during vacuole isolation in this medium and this will also require further study. Nevertheless, the evidence strongly favours, but does not completely prove, the hypothesis that betaine is concentrated in the cytoplasm at the expense of the vacuole.   If the betaine were to occupy a cytoplasm itself 5 or 10% of the cell volume, then the betaine sap concentration would mean cytoplasmic concentrations of 420 and 210 mM respectively, which would make significant contributions to the osmotic pressure of 600 mOsm.

These results on enzyme sensitivity and compartmentation provide substantial support for the hypothesis that betaine acts as a cytoplasmic osmotic effector in a number of halophytic plants.

Osmotic and ionic regulation in other organisms: The physiological concept of a fairly constant cytoplasmic ionic strength with K^+ the dominant cation and the accumulation of non-toxic organic molecules to provide osmoregulation is well recognised in animal systems[16,17,18], and is probably best illustrated by

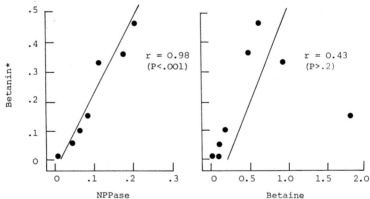

Fig.5. **Relationship** of acid phosphate to betanin specific activities in fractions from gradient and crude pellet.

Fig.6. Relationship of betaine to betanin specific activities in fractions from gradient and crude pellet.

* all units as in TABLE 1.

Steinbach's diagram (Fig.7)[17]. Clearly the model of cytoplasmic regulation for Spartina shown in Fig.3 is essentially identical to the Steinbach model and indeed a knowledge of the comparative situation in animals was a significant factor in the formulation of the Spartina model. The similarity of cytoplasmic ionic relations in eucaryotic cells is clearly revealed in Table 2, which summarises inorganic analyses on a range of plants and animals. In almost all cases intracellular cytoplasmic K^+ is maintained at 100-150 mM regardless of the extracellular environment, while Na^+ is largely excluded except when the extracellular Na^+ is very high, for example, in some marine animals where higher cytoplasmic Na^+ levels are found. The major exceptions to this pattern are some fresh water animals which have adapted to low cytoplasmic K^+ levels e.g. Anodonta cygnea and some of the marine algae e.g. Valonia which are reported to have very high cytoplasmic K^+ contents (400-500 mM) (Table 2). However, since the only analysis using a K^+ - specific microelectrode in such a marine alga revealed a 'normal' K^+ concentration, 153 mM, and conventional analyses are liable to contamination by the large volume of circa 500 mM K^+ in the vacuole, the accuracy of the high cytoplasmic K^+ values may be doubted.

While eucaryotic cells appear alike in maintaining a fairly constant cytoplasmic K^+ concentration and not using massive changes in the total ionic strength of their cytoplasma to achieve osmoregulation, there are also similarities in the compounds employed to bring about intracellular osmoregulation in different

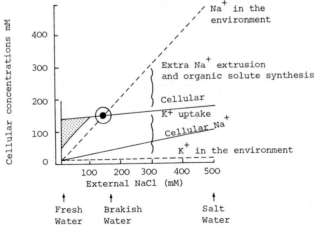

Fig.7. The proposed relationships between cellular K^+ and Na^+ levels and increasing Na^+ in the environment. (Reproduced from Steinbach 1962).

The shaded area represents the extremes of variation of cellular K^+ concentrations in animal cells exposed to dilute environments.

TABLE 2

IONIC COMPOSITION OF BATHING FLUIDS AND TISSUES OF CONTRASTING ORGANISMS

Organism		Fluid/tissue	K^+	Na^+	Cl^-	Comment
Rattus[16] (rat)	Mammal	Plasma	10*	145	116	
		Muscle	185	8	–	
Rana[16] (frog)	Amphibian	Plasma	3	104	74	
		Muscle	126	11	–	
Myxine[16] (Hag fish)	Cyclostome	Serum	10	500	570	
		Muscle	117	122	107	690mM organic solutes
Raja[16] (Skate)	Sea water Elasmobranch	Serum	7	255	241	453mM urea
		Muscle	187	30	–	about 450mM urea
Anodanta[16]	Freshwater Mollusc	Freshwater	0.06	0.5	–	
		Plasma	0.4	15	11	
		Muscle	11	5	11	
Mytilus[16]	Marine Mollusc	Blood	13	490	573	
		Muscle	158	73	56	350-500mM organic solute
Nephrops[16]	Marine Crustacean	Plasma	9	517	527	
		Muscle	167	83	110	520mM organic solute
Avena[19]	Higher Plant (coleoptile)	Ex.soln	10	10	10	
		Cytoplasm	178	15	83	efflux analysis
Hordeum[2]	Higher Plant (Root cortex)	Ex.soln	0.2	1.0	0.2	chemical analysis
		Cytoplasm	90	(10)	–	
		Vacuole	–	70-90	–	
Nitella[20]	Freshwater alga	Ex.soln	0.1	1.0	1.3	
		Cytoplasm	119	14	65	
		Vacuole	75	65	160	
Valonia[20]	Marine alga	'Sea water'	10	470	548	
		Cytoplasm	434	40	138	
		Vacuole	625	44	643	
Griffithsia[20,21]	Marine alga	Cytoplasm	153	30-10	600-600	K^+ electrode
		Vacuole	500-600	30-10	600-600	

* mM

organisms. In most cases either a nitrogen dipole, frequently an amino or imino acid or derivative, such as betaine, or a small polyhydric alcohol or derivative is employed. The major exception to these chemical groupings is urea which is the major organic osmoticum in cartilaginous fish. These animals also differ in that the elevated osmolarity of both the intracellular and extracellular fluids are maintained at that of sea water by urea to which the cell membranes appear freely permeable. Some of the compounds used as osmotic effectors by a range of other organisms are tabulated in Table 3. Clearly it would be of great interest to understand why a particular compound is used in any one organism, as the 'choice', particularly between a polyol or N- dipole, must have profound implications in the nitrogen and carbon economy of the cells and probably the ecology of the organism. As a range of compounds are employed in the various phylla, it might be expected that a number of compounds might be involved in cytoplasmic osmoregulation in higher plants. Undoubtedly the Chenopodiaceae, many Gramineae and probably other families utilize betaine[1]. Data taken from many experiments on the three Chenopods and six Gramineae show a close correlation (r = 0.85) between betaine levels and osmotic pressure. Furthermore, betaine is only accumulated above a minimum osmotic pressure around 250 mOsm which can be compared to the interception point equivalent to 150 mM cell concentration in the Steinbach model (Fig.7).

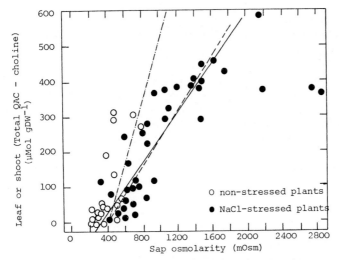

Fig. 8. Relationship of level of total Quaternary ammonium compounds minus choline (effectively betaine see ref.28) to sap osmotic pressure in 9 plants, r = 0.849 p< 0.01; calculated on values up to 2000 mOsm. Values above 2000 mOsm omitted as plants seriously damaged. Concentration of a theoretical cytoplasmic osmoticum calculated assuming,(1) (—·—)F.W: DW ratio of 10:1 and cytoplasm 10% cell volume, (2) (---) FW:DW ratio of 5:1 and cytoplasm occupying 10% cell volume.

There is evidence that other plants, e.g. _Triglochin maritima_ may use proline[25] in place of betaine, although in _A.spongiosa_, _S,monoica_ & _S.x townsendii_ proline is only accumulated when growth is inhibited and is poorly correlated with sap osmotic pressure (see Fig.3). It is arguable whether this represents a difference of degree or kind, as plants must adapt to survive transient stress and proline responses to sudden stress are much quicker than those of betaine[28].

TABLE 3

EXAMPLES OF POSSIBLE OSMOTIC EFFECTORS IN VARIOUS ORGANISMS

Organism	Nitrogen Dipoles	Polyols & Derivatives
Invertebrates[16,22,23]		
Sepia officinalis	α - amino acids, betaine	
Gryphea angulata	α - amino acids, betaine, taurine	
Mytilus edulis	glycine, taurine, betaine	
Aplysia	tauro-choline, betaine, α - amino acids	glycerol
Nephrops	α - amino acids, betaine, TMAO	
various Arthropoda	proline	
various Crustacea, Arachnopmorphs, Mollusca, Annelids, Echinoderms & Sipuncula	amino acids & derivatives	
Higher Plants[5,24,25]		
Chenopodiaceae	betaine	
Spartina townsendii & other Gramineae	betaine (proline)	
Triglochin maritima	proline ?	
Algae[26,27]		
Dunaliella Sp.		glycerol
Ochromonas malhamensis		α - galactosyl glycerol
Monochrysis lutheri		cyclohexane-tetrol
Platymonas suecica		mannitol
Chlorella sp.	proline ?	
Fungi[26,27]		
Dendryphiella salina		mannitol, arabitol
Lichen[26,27]		
Lichinea pygmaea		mannoside-mannitol

The similarities in the osmotic effectors and patterns of cytoplasmic ionic regulation found in higher plants, algae, fungi, invertebrate and vertebrate animals stimulate the speculation that fundamental physical and chemical principles of macromolecular structure underly these observations. As noted previously, the inhibition of cytoplasmic enzymes by NaCl is an example of a general physical chemical phenomenon and not an isolated effect attributable to these specific ionic species. Solutions of monovalent ions up to 100 to 150mM appear to suppress the Donnan diffuse double layer of proteins and other macro-molecules without causing conformational instability[29]. However, higher ionic strength electrolyte solutions decrease protein hydration and promote conformat-ional changes in macromolecules. The efficacy of individual ions in promoting these changes is given by the classical Hofmeister lyotrophic series[14]. This series has been observed in 1. N → D transition of globular proteins, 2. α-helix → coil transitions of homopolypeptides, 3. double helix → coil transitions of polynucleotides and polysaccharides, 4. triple helix → coil transition of collagen to gellatin, 5. monomer → micelle equilibria of surfactants, 6. salting-in and out from aqueous solution, and hypercoil → extended form transitions of synthetic polymers[29] and, of course, enzyme inhibition[14], including barley malic dehydro-genase as described earlier. Thus, the unfolding of proteins and other macro-molecules and the inhibition of enzyme activities by concentrated electrolyte solutions may be regarded as a major constraint in biological systems. It is of great interest to note Eagland's comment[30] 'that it seems certain that an ionic strength of 100 to 150 mM in a solution of a nucleotide or protein has particular significance for both structure and hydration of the macromolecules and this value appears to correspond to the osmotic value of vertebrate species. Just why this is the case remains a matter of speculation, it is, however, not simply a straightforward ionic strength effect although it seems likely to arise from competing effects'.

There is no doubt that the conformational stability of proteins is closely linked to the properties of the solvent system and that alterations in the properties of water, the major biological solvent, by dissolved species will lead to the perturbation of protein conformation[29,30]. In their recent review Franks and Eagland quote approvingly, Hofmeister's original speculation that the lyo-trophic series arose from the power of ions to bind water but a detailed thermo-dynamic analysis of the phenomenon has still not been achieved. Therefore, the interpretation of the biological significance of these phenomena can only be attempted at a superficial level and with great caution.

Since it is observed that the ionic strength of the cytoplasma of eukaryotic cells is remarkably uniform (Table 2) and solutions of Na and K salts above circa 200mM appear to salt-in the quaternary structure of proteins, it may be argued that cytoplasma subjected to high osmotic pressure must accumulate solute molecules which do not have the tendency to salt-in. Further constraints on the molecular species which may be accumulated are indicated by Von Hippel's studies on ribonuclease[14,29] as denaturation is also promoted up the sequence $(CH_3)_4 N^+ <$ $(C_2H_5)_4 N^+ < (C_3H_7)_4 N^+ < (C_4H_9)_4 N^+$ and $CH_2OH.CH_2OH < CH_3OH < CH_3CH_2 OH < CH_3 CH_2 CH_2 OH$. The sequence presumably reflect increasingly hydrophobic interactions with water and imply that molecules interacting strongly with water either due to hydrogen bonding or hydrophobic bonding and capable of salting-in proteins, are undesirable cytoplasmic components. It is reassuring that the molecules most closely related to betaine and glycerol have the least denaturing power. Other physical data imply that betaine is neither a marker water structure-maker or structure-breaker[31] and in similar deduction is contained in Lewin's data on the surface tension of glycerol and related compounds[32]. Thus we tentatively postulate that eukaryotic cells, except in particular cases where specialised proteins have been evolved, accumulate compounds in the cytoplasm to create low osmotic potentials which cause, in conjunction with cytoplasmic inorganic ions, a minimum perturbation of water structure and salting-in of macromolecules.

Two other points should also be noted. Early data suggest that betaine has a high salting-out coefficient[33] and although a non-denaturing precipitation of protein would appear incompatible with an active cytoplasm, it might be conceivable in a dormant cytoplasm. Secondly, although osmotic effectors have been discussed in this paper as 'passive' molecules, there is evidence for taurine and several amino acids changing membrane properties[34] and the potential significance of these effects should not be overlooked.

Some general implications in plant physiology: The application of the general model for cytoplasmic ionic and osmotic regulation to plants has considerable repercussions and is relevant to the elucidation of many of the problems admirably outlined in Cram's recent review[26]. In the space available one can allude to only a few points. The model suggests that there are a restricted number of specifically cytoplasmic components used in the generation of the cytoplasm's contribution to the osmotic and consequently turgor pressure of the cell. As emphasised by Mott and Steward[35] there is great flexibility in the compounds used in vacuoles to generate osmotic and turgor pressure but the cytoplasm may be more fastidious. This could be significant in maintaining growth particularly under stress conditions where the supply of K^+ or the organic cytoplasmic osmoticum, such as betaine, could be limiting. Water stress is common in many environmental stress situations and in many cases an elevated cytoplasmic osmotic pressure would and, probably does, provide some resistance. It would be anticipated both for physical chemical and comparative biochemical reasons that a restricted range of compounds could be employed of which proline, glycerol and betaine are examples. In the specific context of betaine, it is interesting to note that additions of this compound apparently protect plant cells against freezing injury[36] as well as NaCl toxicity[28] and bacterial cells against high molarity sugar media[37] as well as NaCl[38].

A number of tissues during the normal life cycle of a plant undergo dehydration and consequently might be expected to have high sap osmotic pressures or, as in case of stomata are exposed regularly to high vacuolar osmotic pressures. No doubt, specialised proteins stable in high ionic strength media have evolved in some cases, as in halophylic bacteria, but cytoplasmic osmo-compensation may also be essential. It would appear particularly important to consider this possibility in stomata where precisely controlled, rapid fluctuation in osmotic pressure are required. High levels of betaine have been found in _Spartina_ and cotton seeds, in barley anthers[39] and we have recently found significant betaine levels in the aleurone layer and embryo of wheat seed. Proline is also widely distributed in dehydrated tissues[40]. Thus it is not improbable that these observations reflect a general pattern of cytoplasmic behaviour based on a need to minimise the perturbation of protein and other macromolecular structures.

ACKNOWLEDGEMENTS

We are grateful to the Ministry of Agriculture, Fisheries and Food and I.C.I. Plant Protection Ltd.for financial support to R.S.; R.A.L. gratefully acknowledges a Research Fellowship from the Royal Society and N.A. & A.P. similarly acknowledge studentships from the Government of Pakistan and the Science Research Council respectively.

REFERENCES

1. Wyn Jones, R.G., Storey, R. and Pollard, A. (1976) in Transmembrane Ionic Exchanges in Plants. Eds. J. Dainty & M. Thellier.

2. Jeschke, W.D. & Stelter, W. (1976). Planta, _128_, 107-112.

3. Leigh, R.A. & Branton, D. (1975), Plant Physiol. _56_, Supp.p.52.

4. Leigh, R.A. & Branton, D. (1976), Plant Physiol. (submitted).

5. Storey, R. & Wyn Jones, R.G. (1976). Phytochem. (submitted).

6. Speed, D. & Richardson, M. (1968). J.Chromatog. _35_, 497-505.

7. Radecka, C., Genest, K. and Hughes, D.W. (1971). Arzneimittel Forsch. _21_, 548-550.

8. Wilcox, M.E., Wyler, H., Mabry, T.J. & Dreiding, A.S. (1965). Helv. Chim.Acta, _48_, 252.

9. Singh, T.N., Paleg, L.G. & Aspinall, D. (1973). Aust.J.Biol.Sci., _26_, 45-56.

10. Greenway, H. & Sims, A.P. (1974), Aust.J.Plant Physiol. _1_, 15-30.

11. Lowry, D.H., Roseborough, N.J., Farr, A.L. & Randall, R.J. (1951), J.Biol. Chem., _193_, 265-275.

12. Storey, R. & Wyn Jones, R.G. (1975). Plant Sci.Lett. _4_, 161-168.

13. Flowers, T.J. (1975) in Ion Transport in Plant Cells & Tissues, ed. D.A. Baker & J.L. Hall, p. 309-334, North Holland, Amsterdam.

136

14. Von Hippel, P.H. & Schleich, T. (1969) in Biological Macromolecules, ed.
    S.N. Timasheff & G.D. Fasman, p.417-574. Marcel Dekker, New York.

15. Wagner, G.J. & Siegelman, H.W. (1975), Science, 190, 1298-1299.

16. Prosser, C.L. (1973). Comparative Animal Physiology, 3rd ed. W.B. Saunders,
    Philadelphia.

17. Steinbach, H.B. (1962). Perspectives in Biology & Medicine, 5, 338-355.

18. Williams, R.J.P. (1970). Quart.Rev.Biophys. 3, 331-365.

19. Pierce, W.S. & Higinbotham, N. (1970), Plant Physiol. 46, 666-673.

20. Raven, J. (1976) in Ion Transport in Plant Cells and Tissues, ed. D.A. Baker
    & J.L. Hall, p.125-160, North Holland, Amsterdam.

21. Vorobiev, L.N. (1967), Nature, London, 216, 1325-1327.

22. Schoffeniels, E. & Gilles, R. (1970) in Chemical Zoology, Vol. VA, ed.
    M. Florkin & B.T. Scheer, p.255-286. Academic Press, New York.

23. Schoffeniels, E. (1976) in Perspectives in Experimental Biology, Vol.1,
    Zoology, ed. P. Spencer Davies, Pergamon Press, Oxford.

24. Storey, R., Ahmad, N. & Wyn Jones, R.G. (1976). Oecologia (submitted).

25. Stewart, G.R. & Lee, J.A. (1974), Planta, 120, 279-289.

26. Cram, W.J. (1976) in Encyclopaedia of Plant Physiology, Vol.2A, ed. U. Luttge
    & M.G. Pitman. p.284-316, Springer-Verlag, Berlin.

27. Hellebust, J.A. (1976). Ann.Rev.Plant Physiol. 27, 485-505.

28. Storey, R. (1976). Ph.D. Thesis, University of Wales, Cardiff.

29. Franks, F. & Eagland, D. (1975). Crit.Rev.Biochem. 165-219.

30. Eagland, D. (1975) in Water, A Comprehensive Treatise, Vol.5, ed. F.Franks,
    p. 306-516, Plenum Press, New York.

31. Tyrrell, H.J.V. & Kennerley, M. (1968), J.Chem.Soc.A., 2724-2728.

32. Lewin, S. (1974). Displacement of Water and its Control of Biochemical
    Reactions, Academic Press, London.

33. Smith, E.R.B. & Smith, P.K. (1940). J.Biol.Chem. 132, 57-64.

34. Allen, J.A. & Garret, M.R. (1971). Advan.Mar.Biol. 9, 205-253.

35. Mott, R.L. & Steward, F.C. (1972). Ann.Bot. 36, 915-957.

36. Bokarev, K.S. & Ivanova, R.P. (1971). Soviet Plant Physiol. 18, 302-305.

37. Dulaney, E.L., Dulaney, D.D. & Rickes, E.L. (1968). Develop.Ind.Microbial.
    9, 260-269.

38. Rafaeli-Escol, D. & Avi-Dor, Y. (1968). Biochem.J., 109, 687-691.

39. Pearce, R.B., Strange, R.N. & Smith, H. (1976), Phytochem. 15, 953-956.

40. Palfi, G., Bito, M. & Palfi, Z. (1973). Soviet Plant Physiol. 20, 189-193.

Regulation of Cell Membrane Activities in Plants
E. Marrè and O. Ciferri eds.
© 1977, Elsevier/North-Holland Biomedical Press, Amsterdam

PATHOLOGICAL ALTERATIONS IN CELL MEMBRANE
BIOELECTRICAL PROPERTIES*

Anton Novacky and Arthur L. Karr
Department of Plant Pathology, University
of Missouri, Columbia, MO 65201 U.S.A.

SUMMARY

The treatment of mesophyll cells with KCN causes a depolarization of transmembrane potential which is followed by partial, light dependent repolarization. The recovery of transmembrane potential in the presence of KCN is lost as a result of inoculation with leaf spot pathogens or their toxins. The importance of this observation as an early pathological alteration is discussed.

INTRODUCTION

Leaf-spot pathogens cause irreversible damage to critical cellular functions in tissue of the susceptible host plant. This damage leads to cell death and formation of symptoms which are characteristic of the disease. We are interested in identifying these important cellular functions and describing the molecular mechanism by which these functions are altered during successful formation of the host-pathogen complex. Since disease development rapidly leads to general metabolic and structural disorganization in the plant cell, it seemed most profitable to study changes which occur early in the establishment of the complex.

It has been known for some time that alterations in the function of the plasma membrane (PM) occur soon after inoculation[1]. Indeed, the hypothesis-become-dogma of this area of research is that pathogens (or their toxic metabolites) act by directly affecting PM. The membrane is described as becoming "leaky" or "disrupted." This hypothesis is supported by studies showing increased ion leakage from diseased tissue[2,3,4,5] and by ultrastructural observations showing disease caused changes in PM morphology[6,7]. In addition, two groups have shown (in cell free systems) that toxins can cause host-specific alterations in the activity of K^+-stimulated ATPases[8,9].

In our laboratories we are studying disease-caused changes in parameters (transmembrane electropotential (PD), transmembrane electrical resistance (R), coefficient of hydraulic conductivity (Lp) and membrane permeability to non-electrolytes) measured using electrophysiological[10] and plasmometric techniques[11]. Our present results are most simply interpreted by suggesting that the leaf-spot pathogens alter the character of PM by interfering with metabolic systems necessary to the maintenance of a functional membrane and not by direct interaction with the membrane itself. In this paper we will describe the effect of these

* Supported by NSF grant (BMS 74-21347) to A. N.

pathogens (and their toxic metabolites) on the light-dependent ability of the susceptible host cell to recover PD in the presence of KCN (CN).

MATERIALS AND METHODS

Biological materials: Gossypium hirsutum cv. Acala 44 (susceptible) and BV 314-7 (resistant) to Xanthomonas malvacearum provided by L. A. Brinkerhoff were grown in vermiculite irrigated with Hoagland's solution at 24 C - 14h day (9×10^3 μwatts/cm^2) and 21 C - 10h night. Saccharum officinarum clones susceptible and resistant to Helminthosporium sacchari (provided by G. A. Strobel) and Zea mays line B37Tcms, susceptible to Helminthosporium maydis, race T, were grown in a greenhouse. X. malvacearum (Race 10) was grown in nutrient broth for 24 hr at 25 C. Inocula concentration were adjusted with a densitometer after washing in sterile water. Fully developed cotyledons of 14 day old seedlings were inoculated (injection of 10^7 cells/ml of sterile water). Controls were injected with sterile water only. Leaf segments of sugar cane and corn were treated in the flow-through sample cell with the host-specific toxins produced by H. sacchari and H. maydis respectively. The host-specific toxins from H. maydis were purified to homogeneity as described previously[12]. Helminthosporoside (H.S.) was purified from cultures of H. sacchari using a modification (A. L. Karr, unpublished results) of the method of Steiner and Strobel[13].

Electrophysiological measurements: Leaf segments (10x15 mm) were cut with a new razor blade, mounted in plexiglass chambers, and aged overnight (cotton) or for 2 hrs (sugar cane and corn) in aerated bathing solution 1x[14].

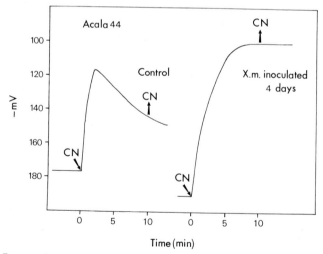

Fig. 1. Transmembrane potential of cotton cotyledon mesophyll cells in healthy tissue (left) and tissue inoculated with X. malvacearum (right). Arrows indicate application and removal of CN. (Same in Fig. 2, 3, 5, 6).

Microcapillaries with glass microfibers were pulled (Tip diameter <0.5 μm; Tip potential < 10 mV; Tip resistance 8-15 MΩ) and filled with 3M KCl. PD and R measurements were made as described previously using a WP Instruments Model 701 Electrometer-Amplifier, a Grass 544 Stimulator, a Keithley 168 Autoranging DMM, and Tektronix Dual Beam Oscilloscope. An illuminator with quartz halogen bulb and goose neck fiber optics was used as a source of light (11.8 x 10^3 μwatts/cm^2 on the surface of experimental tissue segment). Only cells which maintained stable values of PD and R for at least 30 min were used in the experiments. CN was used at a concentration of 1 mM. Ion concentrations were adjusted so $[K^+]$ in the bathing solution and CN solution were identical.

RESULTS AND DISCUSSION

Cells treated with 1mM KCN undergo a rapid depolarization in PD which is

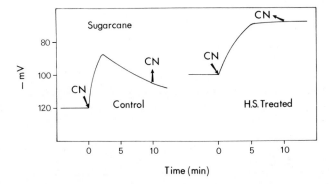

Fig. 2.   Transmembrane potentials in mesophyll cells of healthy sugar cane tissue (left) and tissue treated with helminthosporoside (right).

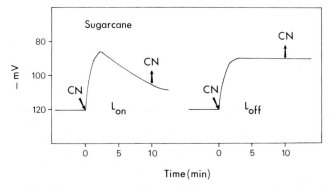

Fig. 3.   Transmembrane potentials in mesophyll cells of sugar cane in light (left) and dark (right) conditions.

complete in 2-4 min. No change in R accompanies this first depolarization of PD
to CN. Cells undergo a second, slower, depolarization of PD when treated with CN
for periods longer than 10 min. This second depolarization is accompanied by a
progressive increase in R and a large decrease in Lp. The change in Lp has been
interpreted to result from a closing of "apparent polar routes" for transmembrane
water movement[11]. Treatment of sugar cane with H.S. or inoculation of cotton with
X. malvacearum causes neither a marked decrease in the magnitude of the first
depolarization of PD to CN nor the increase in R associated with the second
depolarization of PD to CN. These results suggest that interruption of respiratory energy supplies (or their utilization) is not severe in the early stages of
disease development. The data in the remainder of this paper will deal with
disease caused changes which do occur in the first depolarization of PD to CN.
Inherent in this discussion is the assumption that the observed CN effects
result from the blockage of respiratory energy[15].

Cotton (Fig. 1, left) or sugar cane (Fig. 2, left), treated with CN, undergo
the rapid depolarization in PD described above. The initial depolarization in PD
is followed by a partial recovery in PD which is complete about 10 min after the
addition of CN. This recovery in the presence of CN (CNR) is completely absent
in cotton cotyledons inoculated with X. malvacearum (Fig. 1, right) and sugar
cane leaf pieces treated with H.S. (Fig. 2, right). CNR is a light dependent

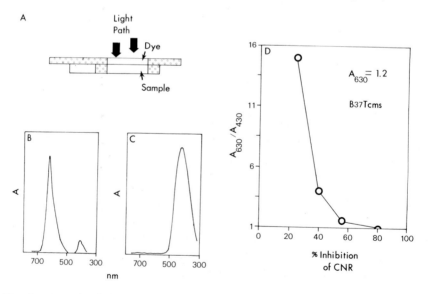

Fig. 4. A: Diagram of the sample chamber with the dye compartment (3 mm)
B, C: Absorbance spectra of blue (630 nm) and yellow (430 nm) dyes in a 1 mm
cell. A_{630} was arbitrarily set at 1.2. D: % inhibition of CN recovery at
different ratios of blue and yellow dyes.

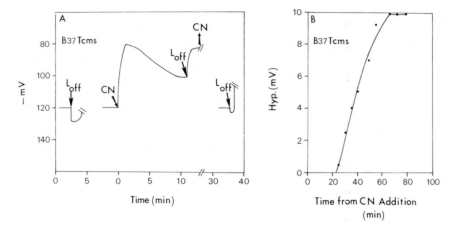

Fig. 5. A. Transient hyperpolarization of transmembrane potential (corn mesophyll cells) caused by a light-dark transition. The response to "light off" is reversed after CN treatment. B. Recovery of the amplitude of hyperpolarization as a function of time after initial CN depolarization.

process and is absent in unilluminated tissue.(Fig. 3, right). The magnitude of CNR is dependent on light intensity. CNR is absent in darkgrown seedlings (corn) and appears after the plants are transferred to light and become green. Present data are not sufficient to either define the efficiency of CNR or to relate its appearance to the formation of functional chloroplasts.

We constructed the sample chamber shown in Fig. 4A to test the wavelength dependency of CNR. This chamber permits introduction of dyes into the light path of the illuminator. In the experiments summarized in Fig. 4., two dyes, one blue (Fig. 4B) and one yellow (Fig. 4C) were introduced separately or in combination into the dye compartment of the sample cell. Neither the blue or the yellow dye alone inhibited CNR. When yellow dye was added to blue dye inhibition of CNR increased and reached a maximum at a dye ratio of 1:1. As would be expected with a chlorophyll mediated process, green light was inefficient in driving CNR.

The suggestion that chloroplasts can alter PD by pumping ions into or out of the cytoplasm or by providing energy to PM-bound ion pumps is not unique (e.g.16). What is unique is the suggestion that leaf spot pathogens interfere with this light dependent component of PD and do so early in the infection process (CNR in sugar cane mesophyll cells is lost within 10 min after treatment with H.S. Cotton cotyledon cells exhibit this loss 3 h and 12 h after inoculation with avirulent (hypersensitivity) or virulent (disease) races of X. malvacearum respectively). It seems likely that loss of such processes would eventually be lethal in a green cell.

We have also investigated disease caused changes in the chloroplast-dependent transients of PD observed after light-dark and dark-light transitions[16] and the

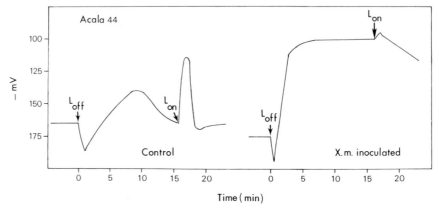

Fig. 6. Change of transmembrane potential induced by light-dark, dark-light transitions in mesophyll cells of cotton cotyledons: healthy (left), and inoculated with X. malvacearum (right).

relationship of these transients to CNR. The first part, the initial hyperpolarization (1st H) of PD, of the light-dark transient is shown in Fig. 5A. When the same cell is treated with CN the typical pattern of CNR is observed. Following CNR the 1st H is lost. When CN is removed from the system the amplitude of the hyperpolarization signal increases and becomes normal 60-70 min after the initial depolarization to CN (Fig. 5B). The hyperpolarization is first detectable in corn 20-25 min and in sugar cane 35-40 min after the first depolarization to CN. Because of this observation we have considered the possibility that CNR and the first hyperpolarization of PD are different manifestations of the same physical process. With this hypothesis we would consider that CNR represents the maximum possible chloroplast caused change in cytoplasmic ion concentration under conditions where PM-pumps are inhibited. The curve in Fig. 5B would result because the progressive activation of PM-pumps begins changing cytoplasmic ion concentrations back to normal and permits renewed chloroplast effects on PM-ion gradients. Curve (Fig. 5B) would represent a titration of the reactivation of PM-ion pumps. This explanation appears likely since curve 5B is identical to the PD tracing obtained when cells are treated with CN in the dark.

One would expect to see a loss or modification in the chloroplast-dependent transients of PD when CNR is lost in diseased or toxin treated tissue if CNR and these transients are closely related. The normal light-dark and dark-light transients are shown in Fig. 6 (left). These transients are grossly modified in cotton cotyledon tissue inoculated with X. malvacearum (Fig. 6, right). While the amplitude of 1st H is not changed, the general shape of this response is changed. (The amplitude of 1st H is reduced in corn treated with the toxins produced by H. maydis, race T). The depolarization which follows 1st H is increased in magnitude and the cell has lost the ability to recover back to normal PD in the dark. The normal dark-light transients have also been modified. The depolarization and recovery of PD observed in healthy tissue (Fig. 6, left) has nearly disappeared in diseased tissue (Fig. 6, right). If CNR and the chloroplast-dependent transients are due to the transport of ions across the chloroplast envelope, the results presented in this paper suggest that an early effect of leaf spot pathogens on host tissue is to disrupt these transport phenomena. This disruption could result from modification of the energy supplies available to the chloroplast ion pumps or from modification of the structure of the chloroplast membranes.

Finally we must consider the possibility that CNR and transients result from different processes under common cellular control. The transients could result from the pumping of some ion, perhaps H^+, across the chloroplast envelope[17] while CNR might result from the pumping of another ion into or out of the chloroplast (The magnitude of CNR is uneffected by H^+ concentration in the bathing solution in the range from pH 5 to 7). Alternatively, it is possible that CNR results from the activation of other cellular ion pumps which are capable of using chloroplast derived energy supplies. In this case one could postulate that it is the common control of ion transport processes and not the processes themselves which is damaged during the early stages of disease development. Such a control apparatus would include a thermodynamically-determined sensor(s)[18,19] of total PM ion gradients and would be of interest since workers have suggested that PD is under strict cellular control.

REFERENCES

1. Wheeler, H. and Hanchey, P. (1968) Annu. Rev. Phytopathol. 6, 331-350.
2. Wheeler, H. and Black, H. S. (1963) Am. J. Bot. 50, 686-693.
3. Gardner, J. M., Mangour, I. S. and Scheffer, R. P. (1972) Physiol. Plant Path. 2, 197-206.
4. Cook, A. A. and Stall, R. E. (1968) Phytopathology 58, 617-619.
5. Burkowicz, A., and Goodman, R. N. (1969) Phytopathology 59, 314-318.
6. Luke, H. H., Warmke, H. E. and Hanchey, P. (1966) Phytopathology 56, 1178-1183.
7. Goodman, R. N. (1972) in Phytotoxins in Plant Diseases (R. K. S. Wood, A. Balio and A. Graniti eds), pp. 311-329. Academic Press.
8. Tipton, C. L. Mondal, M. H. and Benson, M. J. (1975) Physiol. Plant Path. 7, 277-286.
9. Strobel, G. A. (1975) Scientific American 232, 80-88.
10. Novacky, A. Karr, A. L. and Van Sambeek, J. W. (1976) Bio Science 26, 499-504.
11. Turner, J. G. (1976) The nonelectrolyte permeability of tobacco cell membranes in tissue inoculated with Pseudomonas pisi and Pseudomonas

144

tabaci. University of Missouri Ph. D. Thesis.

12. Karr, A. L., Karr, D. B. and Strobel, G. A. (1974) Plant Physiol. 53, 250-257.

13. Steiner, G. W., and Strobel, G. A. (1971) J. Biol. Chem. 256, 4350-4357.

14. Etherton, B. (1963). Plant Physiol. 38, 581-585.

15. Higinbotham, N. and Anderson, W. P. (1974) Can. J. Bot. 52, 1011-1021.

16. Pallaghy, C. K. and Lüttge, U. (1969) Z. Pflanzenphysiol. 61, 58-67.

17. Bentrup, F. W., Gratz, H. J. and Umbehauen, H. (1973) in Ion Transport in Plants (W. P. Anderson ed.), pp. 171-182, Academic Press, London and New York.

18. Melchior, D. L. and Steim, J. M. (1976) Annu. Rev. Biophys. Bioeng. 5, 205-238.

19. Zimmermann, U. and Steudle, E. (1974) J. Membr. Biol. 16, 331-352.

EFFECTS OF AUXIN AND FUSICOCCIN
ON ION TRANSPORT

Regulation of Cell Membrane Activities in Plants
E. Marrè and O. Ciferri eds.
© *1977, Elsevier/North-Holland Biomedical Press, Amsterdam*

H^+ ION TRANSPORT IN PLANT ROOTS

M. G. Pitman, W. P. Anderson* and N. Schaefer
School of Biological Sciences, University of Sydney
N.S.W., Australia 2006

*Research School of Biological Sciences
Australian National University
Canberra, A.C.T., Australia 2601

INTRODUCTION

Plant roots can be major sites for H^+ transport. Release of H^+
ions from the roots has been known for many years to be important in
plant nutrition and in the process of ion accumulation. For example,
uptake of Al^{3+}, and hence the sensitivity of roots to Al^{3+}, has been
shown to be related to the acidification produced in the soil by the
roots, accounting for varietal differences in Al tolerance (1).
Incipient Fe-deficiency in sunflower plants has been shown to acidify
the solution around the roots and addition of Fe then reduces the
activity (2). Hydrogen transport also may be involved in reduction
of Fe^{3+} to Fe^{2+} in maize roots (3). On a more general scale,
reduction of NO_3^- to amino acids can lead to net fluxes of H^+ (or OH^-)
across the root surface, depending on the overall cation/anion
balance in the plant (4).

Early studies with low-salt roots (grown on $CaSO_4$ solution only)
showed that H^+ release could accompany net ion uptake (5). This
process was shown to be related to synthesis of organic acids such
that there was an almost 1:1 stoichiometry between organic acid
production and H^+ release, small differences arising from H^+
associated with the buffering capacity of the cell (6, 7). The
release of H^+ ions from the cell was shown to be against the electrochemical potential gradient for the ion, and to require metabolic
energy (8). Inhibitors of energy metabolism such as carbonyl cyanide
m-chlorophenylhydrazone (CCCP) and arsenite reduced H^+ transport and
ion uptake. The suggestion that H^+ and OH^- ions are involved in salt
uptake to higher plant cells has had a long history (9, 10) which
pre-dates the recent fashion for involvement of H^+ in almost all

processes.

Cells of roots grown in nutrient solutions or soils containing adequate K^+ can be in flux equilibrium for H^+/OH^-. Barley roots, for example, show little net uptake of H^+ at the natural pH of dilute salt solutions (about 5.3 - 6.0). It was exciting to find that the phytotoxin fusicoccin (FC) stimulated H^+ efflux in these roots. The stimulation of H^+ efflux was accompanied by increased K^+ uptake and hyperpolarisation of cell PD (11) as previously reported in other tissues (12, 13, 14, 15) but there appeared to be no cell enlargement in roots. The stimulation of ion transport was interpreted as electrogenic coupling between H^+ efflux and cation influx (16). The H^+ release stimulated by FC is nearly equivalent to organic acid synthesis, with a small component derived from the buffering capacity of the cell (N. Schaefer, unpublished results).

Hydrogen ion release in stem tissues (oat coleoptiles) responds spectacularly to auxin (IAA) concentrations (17) and can be inhibited by abscisic acid (ABA) (18). Effects of hormones on H^+ transport in roots is less impressive. We have found no effect of IAA or ABA on H^+ transport in barley roots, nor interaction of ABA with the FC-stimulated efflux (16). Shaner et al. (19), using maize roots, have reported that ABA leads to a small inhibition of H^+ release and K^+ uptake and depolarises cell PD by about 10 mV.

Ion uptake to low-salt barley roots can be stimulated by the addition of Ca^{2+} (Viets effect (20, 21)). The accompanying H^+ release is also sensitive to Ca^{2+} and changes rapidly if the level of Ca^{2+} in solution is changed (22). From the speed of the response it appears that Ca^{2+} is acting externally, possibly by its effect on the surface potential of the membrane or on the H^+ concentration of the Donnan free space.

Hodges and others have suggested that the ion-stimulated ATP-ases (24) associated with membrane fractions of root cells may be H^+ transporting systems such that H^+ export can be coupled with cation movement and OH^- remaining in the cytoplasm can either be coupled with Cl^- or neutralised by HCl diffusing inwards. It is natural to think that the ATPase may therefore be stimulated by FC, if it is involved in transport at the plasmalemma, but this does not appear to be the case (R. A. Wildes, unpublished results).

The model for action of the membrane ATPase emphasises that H^+ transport cannot be thought of in isolation from OH^-, which in turn can be related to organic acid synthesis and possibly to other transport processes. Measurements of cytoplasmic pH (25, 26) in tissues other than roots show that it is maintained at a relatively stable level despite large changes in external pH. Regulation of the cytoplasmic pH is thus another process that has to be considered in interpreting the overall H^+/OH^- traffic across the cell membranes.

An unresolved problem for root cells is the extent to which diffusion of H^+ along its electrochemical gradient might contribute to the H^+ traffic. Large diffusive fluxes of H^+ would require corresponding active secretion of H^+, if the regulatory system were to maintain constant cytoplasmic pH. The results described in this paper allow estimation of a passive permeability to H^+ ions for root cells and show that diffusive H^+ fluxes can be appreciable at low pH (about 4).

MATERIALS AND METHODS

Material: Roots of barley seedlings (Hordeum vulgare cv Abyssinian) were used in these experiments. The seedlings were germinated on 0.5 mM $CaSO_4$ and transferred to aerated solution of 5 mM KCl + 0.5 mM $CaSO_4$ 24 h before use. When used the seedlings were about 5 days old and the K^+ content of the roots was about 65 μmol g_{FW}^{-1}.

H^+ transport: Rates of H^+ transport were determined from the amount of HCl or KOH (NaOH) that had to be added to maintain the pH at the selected level. It should be noted that net H^+ loss from solution could be due to H^+ uptake or OH^- efflux, and this proviso applies to the results given below. For convenience H^+/OH^- transport is usually referred to as "H^+ transport".

Ion fluxes: Rates of ^{86}Rb, ^{36}Cl and ^{22}Na uptake were estimated from the tracer content of the tissue at the end of a period of about 1 h, during which time pH was maintained by titration. The tissue was rinsed in unlabelled solution for 10 minutes and hence the tracer uptake estimates the net flux ϕ_{ov}, which can be described in terms of the component unidirectional fluxes as $\phi_{ov} = \phi_{cv}.S_c = \phi_{cv}.\phi_{oc}$ ($\phi_{cv} + \phi_{co} + \phi_{cx}$) where subscripts show direction relative to solution (o), cytoplasm (c) and vacuole (v) (27).

Potential measurements: The potential difference of the vacuole relative to the solution (PD) was estimated from microelectrodes filled with 3 M KCl inserted perpendicular to the axis of the root into the 2nd or 3rd cortical cell from the root surface. The PD was often found to be stable over periods of about 1 h, allowing several changes in external solution to be made.

RESULTS

Effect of external pH on H^+ efflux and ion transport: Measurements were made of net H^+ uptake by barley roots in solutions of 5 mM KCl + 0.5 mM $CaSO_4$ when pH was varied. The pH range was from about 4 to 5.5. At the low end of this range (pH 4.1) the tissue was more or less at H^+ equilibrium when FC was present and at the upper end of the range (pH 5.3) there was zero net H^+ flux when FC was absent.

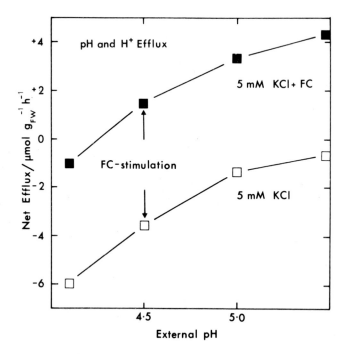

Fig. 1. Uptake (-) or efflux (+) of H^+ by barley roots in 5 mM KCl (■); (□) are results with the addition of 10 μM FC.

The effect of varying pH is shown in Figure 1. In the absence of FC, as pH was decreased there was increased H^+ uptake to 6.4 µmol g_{FW}^{-1} h^{-1} at pH 4.1. In 10 µmol FC the H^+ efflux was 4.4 µmol g_{FW}^{-1} h^{-1} at pH 5.3, falling with reduced pH to an uptake of 1.7 µmol g_{FW}^{-1} h^{-1} at 4.1. Though net H^+ transport responds to external pH in both sets of data, the stimulation of efflux due to FC was nearly independent of external pH.

Measurements were also made of ^{86}Rb and ^{36}Cl uptakes from 5 mM KCl at pH 4.1 and 5.3 (Table 1). As shown earlier (16), FC stimulated both ^{86}Rb and ^{36}Cl uptakes, which in these experiments were net movement of tracer into the vacuoles of the cells and hence were complex estimates of fluxes (27). In general there was no effect of pH on ^{36}Cl uptake, but ^{86}Rb uptake was reduced at lower pH. This pattern was found in several repeats of the measurements though absolute values varied by about 20% between experiments.

TABLE 1

pH AND ION FLUXES

Measurements of net H^+ flux, ^{86}Rb and ^{36}Cl uptakes from 5 mM KCl

	pH	H^+ efflux*	^{86}Rb uptake*	^{36}Cl uptake*
Control	4.1	−3.8	3.7	3.5
+ µM FC	4.1	0	6.8	4.9
FC-stimulation		3.8	3.1	1.4
Control	5.3	0.8	6.3	3.5
+ µM FC	5.3	4.8	12.2	5.0
FC-stimulation		4.0	5.9	1.5

* µmol g_{FW}^{-1} h^{-1}

TABLE 2

pH AND CELL PD

Solution	pH	PD (mV)	ΔPD (mV) *
5 mM KCl	4.1	-84	8.25 $\pm$ 1.7 (4)
5 mM KCl	5.3	-92	
5 mM NaCl	4.1	-182	24.75 $\pm$ 4.6 (4)
5 mM NaCl	5.3	-207	

* Difference between PD in pH 4.1 less PD in pH 5.3; mean $\pm$ SEM.

Measurements of cell PD: Cells were set up in 5 mM KCl + 0.5 mM CaSO$_4$ at pH 5.3 and a microelectrode was inserted into a cell in the 2nd or 3rd layer of the cortex. When the PD was steady the solution was changed to one of pH 4.1. There was a small depolarisation (Table 2) which was reversible when pH was changed back to 5.3. Similar measurements (Figure 2) were made using 5 mM NaCl + 0.5 mM CaSO$_4$ but now the PD's were more electronegative than in 5 mM KCl. The change in PD between pH 4.1 and 5.3 also was larger in 5 mM NaCl than in 5 mM KCl. In both KCl and NaCl, the effect was highly significant on a paired comparison t-test (P<0.001).

If the difference in PD between pH 4.1 and 5.3 is due to a diffusion potential then an estimate of the ratio of permeability of the membrane to K^+ (L_K) and to Na^+ (L_{Na}) can be calculated. Using the Goldman equation (28) the difference in PD between two situations where external H^+ concentrations are H_{01} and H_{02}, external K^+ is held constant at K_0 and the ratio L_H/L_K is ε, is given by

$$\Delta PD \quad = \quad RT\ln \frac{K_0 + \varepsilon H_{01}}{K_0 + \varepsilon H_{02}}$$

Substituting the data from Table 2 for ΔPD gives:

$$\varepsilon \quad = \quad 20.6 \pm 5.3$$

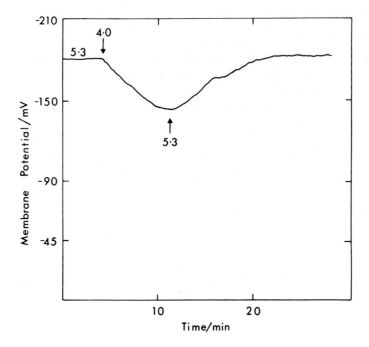

Fig. 2.   Time course of PD change in 5 mM NaCl on transfer from pH 5.3 to 4.0 and back.

A similar calculation can be used to estimate the ratio L_H/L_{Na} as

$$L_H/L_{Na} \quad = \quad 96 \pm 8$$

and hence

$$L_{Na}/L_K \quad = \quad 0.21 \pm 0.08$$

This value compares favourably with the estimates of L_{Na}/L_K determined from flux measurements ($L_{Na}/L_K = 0.3$ (29)) and from electrical measurements ($L_{Na}/L_K = 0.15$ to 0.3 (16)). In other sets of data (e.g. Table 3) the difference in PD between pH 5.3 and 4.1 may be larger by about 10 mV. The effect is to increase the estimate of ε by a factor of 2, but the difference between KCl and NaCl remains.

The PD in 5 mM KCl at pH 5.3 was more negative than the equilibrium PD for K^+ (-67 mV based on the measured K^+ content of the tissue). The PD in 5 mM NaCl at pH 5.3 also was much more negative than the PD estimated (-105 mV) by substituting concentrations of K^+, Na^+ and H^+ in the Goldman equation, taking $L_{Na}/L_K = 0.2$, and ε as given above. The hyperpolarisation of cell PD relative to these Goldman PD's (25 mV in KCl; 102 mV in Na Cl) may be due to some other factor such as electrogenic potentials or accumulation of K^+ in the cytoplasm to a higher concentration than the measured tissue average. Measurement of the effect of CO on PD in light and dark showed that a rapid change in PD of 60 mV could be produced. This effect has been interpreted as due to a change in electrogenic as opposed to diffusion PD (30).

<u>Calculated values for H^+ fluxes</u>: Estimates have been made of the K^+ permeability of barley root cell membranes as 3×10^{-9} m s^{-1} (31), based on K^+ efflux from the cells and assuming the Goldman flux equation (28). Hence the permeability to H^+ can be inferred as 62×10^{-9} m s^{-1}.

On this basis H^+ fluxes into the cells in 5 mM KCl can be estimated as follows (assuming 0.1 m^2 cell surface/g_{FW} (32).

pH 4.1:   17 nmol m^{-2} s^{-1} = 6 µmol g_{FW}^{-1} h^{-1}

pH 5.3:   1.1 nmol m^{-2} s^{-1} = 0.4 µmol g_{FW}^{-1} h^{-1}

A similar calculation assuming the cytoplasm is at pH 7, estimates the diffusive H^+ efflux at 0.2 nmol g_{FW}^{-1} h^{-1}. The estimates of net H^+ transport therefore agree extremely well with the data in Figure 1, and Table 1.

The more negative potential in 5 mM NaCl than in 5 mM KCl predicts that the H^+ influx should be larger in NaCl than in KCl. For example in 5 mM NaCl, H^+ influx is:

pH 4.1:   36 nmol m^{-2} s^{-1} = 12.8 µmol g_{FW}^{-1} h^{-1}

pH 5.3:   2.4 nmol m^{-2} s^{-1} = 0.9 µmol g_{FW}^{-1} h^{-1}

TABLE 3

H^+ FLUXES AND ION UPTAKE IN KCl AND NaCl*

Solution	pH	H^+ efflux	^{22}Na uptake	^{86}Rb uptake	^{36}Cl uptake
5 mM NaCl	5.3	-0.6	0.6		5.5
5 mM KCl	5.3	+0.8		2.9	2.2

* all fluxes are µmol g_{FW}^{-1} h^{-1}

Some data comparing H^+ and ion fluxes in 5 mM KCl and NaCl are given in Table 3.

The predicted difference in H^+ flux was found but the agreement between observed and calculated values is not as good as with the data for Figure 1. The range of values for ε make this comparison only qualitatively satisfying. The high values of H^+ efflux predicted by large ε may not be realised because the internal hydrogen ion concentration changes.

The H^+ efflux from low-salt roots in 5 mM KCl at pH 5.3 was the same as in 5 mM NaCl at pH 5.3. The lack of any difference agrees with the observed ratio of L_K/L_{Na} in these roots being close to 1 (16).

Interactions of FC and external pH on PD: Measurements of cell PD were made at pH 4.1 and 5.3 with the addition of 10 µM FC. The PD was first measured in 5 mM KCl (or NaCl) at pH 4.1 and 5.3 to establish the pH-induced change. Using the same root (or in some cases the same cell) solution containing FC was then introduced and after the initial hyperpolarisation when the PD had stabilised, the pH response was determined again (Table 4). Paired comparison t-test of the pH-induced changes with and without FC showed no significant difference due to FC. This implies that the passive permeability to H^+ was unaffected by FC and supports the view that FC acts by increasing active H^+ efflux.

TABLE 4

PD CHANGE WITH FC BETWEEN pH 4.1 AND 5.3

Solution	ΔPD (-FC)	ΔPD (+FC)
5 mM KCl	30 (5)	25 (5)
5 mM NaCl	38 (4)	32 (4)

CONCLUSIONS

At the normal pH of the solutions (about 5.3) the net flux of H^+ across the membranes is near zero in 5 mM KCl. The diffusive flux of H^+ into the cells is small (about 0.5 μmol g_{FW}^{-1} h^{-1}) and this sets a lower limit on the active H^+ efflux under these conditions. However, there could be a diffusive efflux of HCO_3^- of about 1 μmol g_{FW}^{-1} h^{-1} (8) based on estimated HCO_3^- concentration in the cytoplasm (the CO_2 efflux from respiration is about 15-20 μmol g_{FW}^{-1} h^{-1}). This efflux of HCO_3^- would appear as an OH^- efflux and lead to a revised estimate for active H^+ efflux of 1.5 μmol g_{FW}^{-1} h^{-1} at flux equilibrium at pH 5.3. The point we wish to establish is that there could be appreciable active H^+ efflux even though the diffusive influx at pH 5.3 appears to be small. Such an active efflux is required for the electrogenic hyperpolarisation of the membrane PD (see above).

The stimulation of H^+ efflux by FC at pH 5.3 is large compared with H^+ traffic at that pH and so requires internal mechanisms for dissipation of OH^- if the cytoplasmic pH is to be maintained near 7. At pH 4.0 in FC the diffusive influx is adequate to balance the active efflux.

The relatively large permeability of the membrane to H^+ (L_H/L_K = 21) raises the question of whether H^+ can leak back into the cell following active export. Such a diffusional flux would reduce the efficiency of the coupling between H^+ efflux and cation uptake. Ions transported across the membrane and appearing in the solution outside the membrane (at concentration H_m) will have the possibility of migrating to the solution or back into the cell. Treating these as unidirectional fluxes we have

To cell: $\phi_{in} = L_H \cdot \dfrac{FE_m/RT}{1 - \exp(FE_m/RT)} \cdot H_m$

To solution: $\phi_{out} = \dfrac{D}{\ell} \cdot \dfrac{FE_o/RT}{1 - \exp(FE_o/RT)} \cdot H_m$

where D is the diffusion coefficient of H^+ in solution (10^{-9} m^2 s^{-1}), ℓ is the effective path length to the solution (about 150 μm), E_m is the membrane potential and E_o is the potential difference from outer surface to the solution (possibly near zero). Substituting values it can be calculated that

$$\phi_{in}/\phi_{out} \simeq 1/30$$

Thus only a small proportion of H^+ export (about 3%) should re-enter the cell over and above the diffusional flux of H^+ across the membrane.

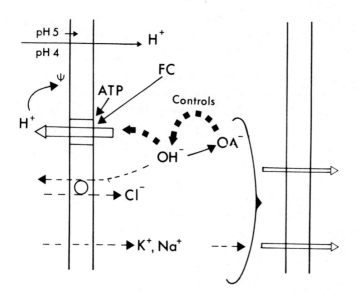

Fig. 3.   Interactions of fluxes of H^+ in the cell with other processes.   Other active K^+ and Na^+ fluxes are not shown.

It was pointed out above that opposing H^+ and OH^- movements are often indistinguishable. One effect of H^+ export from the cell is assumed to be an increase in cytoplasmic OH^-. This can account for greater Cl^- uptake in FC if active Cl^- uptake is due to an OH^--Cl^- antiport system making use of the electrochemical potential of OH^- inside the cell (Figure 3). The increased Cl^- uptake in NaCl compared with KCl solutions (Table 3) is difficult to interpret in relation to this scheme unless it is related to the decreased Cl^- efflux in NaCl due to the hyperpolarisation of cell PD.

The mechanism by which fusicoccin stimulates H^+ efflux at present is obscure. There seems to be no effect on energy metabolism since there is no change in oxygen consumption (16) and we have found no change in ATP level in barley roots. The content of FC treated roots was 76.6 ± 1.9 nmol g_{FW}^{-1} compared with 82.8 ± 1.6 nmol g_{FW}^{-1} in controls.

The results shown here differ from those reported by Cocucci et al. (33) for the effect of pH on PD in maize roots. They found reduction in pH hyperpolarised the PD instead of depolarising it (as shown in Figure 2). We are unable to explain this behaviour of the maize roots in relation to pH, unless it relates to the presence of Mg^{2+} and phosphate in the solution used, or was due to a change in active transport.

According to the data of Figure 1 and Table 4, FC does not alter the passive permeability of the membrane to H^+. Similar results were reported for maize (33), where the pH induced changes in PD were the same with and without FC. We consider these results rule out the possibility that FC acts as an H^+-ionophore.

Acknowledgements: This work was supported by grants from the Australian Research Grants Committee and University of Sydney (M.G.P. & N.S.). We are grateful to Dale Wellfare and Barbara J. Wright for technical assistance, and Mrs S. Titt for setting out the manuscript.

REFERENCES

1. Foy, C. D., Gerloff, G. C. and Gabelman, W. H. (1973) J. Amer. Soc. Hort. Sci. 98, 427-432
2. Marschner, H. (1974) in Mechanisms of Regulation of Plant Growth (Bieleski, R. L., Ferguson, A. R. and Cresswell, M. M. eds), pp. 99-109, Bulletin 12, The Royal Society of New Zealand, Wellington
3. Chaney, R. L., Brown, J. C. and Tiffin, L. O. (1972) Plant Physiol. 50, 208-213
4. Smith, F. A. and Raven, J. A. (1976) in Encyclopedia of Plant Physiology, New Series, Vol. IIA, (Lüttge, U. and Pitman, M. G. eds), pp. 317-346, Springer, Berlin-Heidelberg-New York
5. Hoagland, D. R. and Broyer, T. C. (1940) Am. J. Bot. 27, 173-185
6. Hiatt, A. J. (1967) Plant Physiol. 42, 294-298
7. Hiatt, A. J. and Hendricks, S. B. (1967) Z. Pflanzenphysiol. 56, 220-232
8. Pitman, M. G. (1970) Plant Physiol. 45, 787-790
9. Briggs, G. E. (1930) Proc. Roy. Soc. B. 107, 248-269
10. Jackson, P. C. and Adams, H. R. (1963) J. Gen. Physiol. 46, 369-386
11. Pitman, M. G., Schaefer, N. and Wildes, R. A. (1975) Plant Sci. Lett. 4, 323-329
12. Lado, P., Rasi-Caldogno, F., Pennachioni, A. and Marrè, E. (1972) Planta 110, 311-320
13. Marrè, E., Colombo, R., Lado, P. and Rasi-Caldogno, F. (1974) Plant Sci. Lett. 2, 139-150
14. Marrè, E., Lado, P., Ferroni, A. and Ballarin Denti, A. (1974) Plant Sci. Lett. 2, 257-265
15. Marrè, E., Lado, P., Raso-Caldogno, F., Colombo, R. and De Michelis, M. I. (1974) Plant Sci. Lett. 3, 365-379
16. Pitman, M. G., Schaefer, N. and Wildes, R. A. (1975) Planta 126, 61-73
17. Cleland, R. (1973) Plant Physiol. Suppl. 51, 2
18. Rayle, D. L. and Johnson, K. D. (1973) Plant Physiol. Suppl. 51, 2
19. Shaner, D. L., Mertz, S. M. and Amtzen, C. J. (1975) Planta 122, 79-90
20. Viets, F. G. (1944) Plant Physiol. 19, 466-480
21. Epstein, E. (1961) Plant Physiol. 36, 437-444

22. Neirinckx, L. and Stassart, J. M. (1975) Bull. Soc. Roy. Bot. Belg. 108, 65-77

23. Marschner, H. and Mengel, K. (1966) Z. Pflanzenernähr. Düng. Bodenk. 112, 39-49

24. Hodges, T. K. (1976) in Encyclopedia of Plant Physiology, New Series, Vol. IIA (Lüttge, U. and Pitman, M. G. eds) pp. 260-283 Springer, Berlin-Heidelberg-New York

25. Davis, R. F. (1974) in Membrane Transport in Plants (Zimmermann, U. and Dainty, J. eds), pp. 197-201, Springer, Berlin-Heidelberg-New York

26. Walker, N. A. and Smith, F. A. (1975) Plant Sci. Lett. 4, 125-132

27. Walker, N. A. and Pitman, M. G. (1976) in Encyclopedia of Plant Physiology, New Series, Vol. IIA (Lüttge, U. and Pitman, M. G. eds), pp. 93-127, Springer, Berlin-Heidelberg-New York

28. Briggs, G. E., Hope, A. B. and Robertson, R. N. (1961) Electrolytes and Plant Cells, Blackwell, Oxford

29. Pitman, M. G. and Saddler, H. D. W. (1967) Proc. Nat. Acad. Sci. 57, 44-49

30. Anderson, W. P., Hendrix, D. L. and Higinbotham, N. (1974) Plant Physiol. 54, 712-716

31. Pitman, M. G. (1976) in Encyclopedia of Plant Physiology, New Series, Vol. IIB, (Lüttge, U. and Pitman, M. G. eds), pp. 95-128, Springer, Berlin-Heidelberg-New York

32. Pitman, M. G., Anderson, W. P. and Lüttge, U. (1976) in Encyclopedia of Plant Physiology, New Series, Vol. IIB, (Lüttge, U. and Pitman, M. G. eds), pp. 57-69, Springer, Berlin-Heidelberg-New York

33. Cocucci, M., Marrè, E., Ballarin Denti, A. and Scacchi, A. (1976) Plant Sci. Lett. 6, 143-156

Regulation of Cell Membrane Activities in Plants
E. Marrè and O. Ciferri eds.
© 1977, Elsevier/North-Holland Biomedical Press, Amsterdam

HORMONAL CONTROL OF H[+]-EXCRETION FROM OAT CELLS

Robert E. Cleland and Terri Lomax
Botany Department, University of Washington
Seattle, WA 98195, USA

SUMMARY

There are three ways in which hormone-induced H[+]-excretion might occur: by
an electrogenic H[+]-pump coupled with passive K[+]-uptake (I), by an electroneutral
K[+]/H[+] exchange (II), or by an electrogenic K[+]/H[+] exchange (III). In order to
discriminate between these three we have measured the effect of IAA and FC on
the membrane potential and K[+]-uptake of oat coleoptiles. Both compounds cause
a hyperpolarization of the membrane potential starting at about the same time as
the H[+]-excretion. FC causes a large, rapid stimulation of K[+]-uptake, while auxin
increases K[+]-uptake only slightly after a long lag. We conclude that auxin acts
via mechanism I while FC acts by III. Attempts to demonstrate FC-activation of
any of the coleoptile ATPases have been uniformly negative.

FC also induces H[+]-excretion and K[+]-uptake by oat root sections after a lag
of about 2 minutes. Again, FC may be acting via mechanism III. Oat leaf protoplasts excrete protons in response to FC even without the addition of any ions
to the external medium, suggesting that cation uptake is not an absolute requirement for FC-induced H[+]-excretion.

INTRODUCTION

According to the acid-growth theory[1,2,3] the growth-promoting compounds
indoleacetic acid (IAA) and fusicoccin (FC) initiate cell enlargement in stem and
coleoptile cells by causing the cells to excrete hydrogen ions. The lowered wall
pH activates some as yet unidentified enzyems which cleave load-bearing bonds in
the wall and thus permit cell enlargement to occur. The evidence for auxin and
FC-induced H[+]-excretion is now extensive[4,5] and attention has now shifted to the
question as to how these compounds induce the H[+]-excretion.

There are a number of ways in which hormone-induced proton release might
take place[6] but three mechanisms deserve special consideration. First, the
H[+]-excretion might occur via some sort of electrogenic H[+]-pump (I) such as a
plasma membrane-associated ATPase[2]. The proton excretion would be accompanied
by a hyperpolarization of the membrane potential (PD) which would, in turn,
increase the uptake by passive diffusion of cations such as K[+]. The increase in
K[+]-uptake would be proportional to the exponential of the hyperpolarization. A
second possibility is an electroneutral K[+]/H[+]exchange[7] (II). This would cause no
change in the PD and would result in a ratio of K[+]-uptake to H[+]-excretion
(K/H ratio) of 1 under all conditions. Finally, the H[+]-excretion might occur via
an electrogenic K[+]/H[+] exchange[8] (III); i.e., it might be mediated by a carrier

162

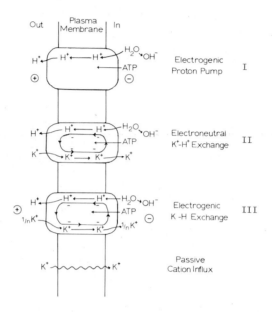

Fig. 1. Three possible mechanisms for H⁺-excretion. Details are described in
the text.

that exports protons as it moves from inside to the outside of the plasma membrane,
and then recycles either as an anion, resulting in a hyperpolarization of the PD,
or carrying a cation such as K^+, in which case there would be no change in the PD.
In this case the K/H ratio would depend on the external K^+ concentration, but
would usually be less than 1, and the rate of K^+-uptake would be inversely proportional to the amount of hyperpolarization of the PD.

   To distinguish between these three possibilities we need to know the effects
of auxin and FC on the PD and the uptake of K^+, and we must know the K/H ratio.
Both the speed with which the hormone effects are initiated and the magnitude of
the effects need to be determined. Some of this information has been available
[9-13], but for no tissue was the information sufficient to permit a decision to
be made between these three mechanisms (although after this study was initiated
Marrè & coworkers[12,14] presented evidence that the FC-induced H⁺-excretion from
corn roots is via III). This study was conducted in order to obtain the
necessary information about auxin and FC-induced H⁺-excretion from oat
coleoptiles, and some information about FC-induced H⁺-excretion from oat root
and leaf protoplast cells.

MATERIALS AND METHODS

   Avena sativa, cv. Victory sections were obtained as described by Cleland[15];

all sections were deleafed and had their epidermis removed by peeling. Oat roots were grown as per Hodges & Leonard[16], and the roots were cut into either 0.5 or 1-cm sections starting from the tip. Avena leaf protoplasts were prepared from 7-day old light-grown leaves by the methods of Brenneman & Galston.[17]

Membrane potentials were measured with pairs of 3 M KCl-filled glass micro-electrodes and the electronic equipment described in Pitman et al.[18] Segments were mounted horizontally in 2.5-3.5 ml of 1 mM MES-tris buffer, pH 6.0 + 0.1 mM $CaCl_2$ and either 0.1 or 1 mM KCl, and the measuring electrode was inserted through the external longitudinal wall of a cell in the outer mesophyll layer. When a stable PD had been achieved sufficient IAA or FC was added to make 10 μM and the recording of the PD was continued until a stable PD was again reached.

K^+-uptake was determined either by measurement of the loss of K^+ from the medium by atomic absorption spectometry or by uptake of $^{86}Rb^+$ from solutions containing 0.3-3 μc/ml $^{86}RbCl$ (1 μM) plus carrier KCl[8]. The two techniques were not expected to give identical results since the former technique measures net movement of K^+ into and out of the tissue while the latter measures only the Rb^+ influx rate. Furthermore oat coleoptiles appear to take up Rb^+ preferentially over K^+ with the result that use of $^{86}Rb^+$ overestimates the K^+-uptake rate by at least 25%.[19]

H^+-excretion was determined by measurement of the amount of NaOH needed to maintain the pH at the initial value of 6.0, with titrations every 20 min.[20]

Membrane fractions were prepared and ATPase was assayed using the techniques of Hodges & Leonard.[16]

RESULTS AND DISCUSSION

Effect of Auxin and Fusicoccin on the Membrane Potential of Oat Coleoptiles.

A study of the effect of IAA and FC on the membrane potential of oat coleoptile cells was carried out at Washington State University in cooperation with H.B.A. Prins, J.R. Harper & N. Higinbotham.[21] In the absence of hormones a PD that averaged -109 mV was found in the presence of both 0.1 and 1 mM KCl (the lack of response of the PD to the external K^+ concentration has already been noted for Avena coleoptiles when Ca^{++} is present[22]). IAA caused a hyperpolarization of the membrane potential starting after a lag of 7-8 minutes (Fig. 2) and reaching an average of -26 mV. The lag is shorter than that reported[23] for H^+-excretion or growth of peeled coleoptiles (ca. 12 min), but this is probably due to the fact that the PD determinations were run at 27-29C rather than 25C. FC induced a hyperpolarization of the PD that started after a lag of less than 30 seconds and reached -49 mV in the presence of 1 mM KCl (Fig. 2). When the KCl level was reduced to 0.1 mM the amount of hyperpolarization increased to -75 mV (Table 1). Addition of FC to auxin-pretreated tissus caused a hyperpolarization of -22 mV in addition to the already existing auxin-induced hyperpolarization. Clearly both

164

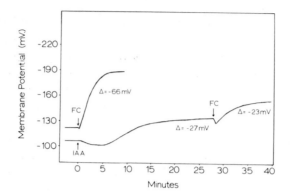

Fig. 2. Time-course of effect of IAA and FC on the membrane potential of oat
coleoptile sections. Peeled coleoptiles incubated 1-3 hrs in 1 mM MES-tris,
pH 6.0 + 1 mM KCl and 0.1 mM $CaCl_2$, then PD measurement started. At first arrow
10 μM IAA or FC added. In lower curve FC was added in addition to IAA after 28
min. Amounts of hyperpolarization of the PD are given.

TABLE I

EFFECT OF K^+ LEVEL ON FC-INDUCED MEMBRANE HYPERPOLARIZATION

K^+-Concentration	FC-induced Hyperpolarization
1 mM	-49 mV
0.1 mM	-74 mV

Conditions same as in Fig. 2 except KCl present at 1 or 0.1 mM.

IAA and FC activate some electrogenic process with lags which agree well with those
for H^+-excretion, suggesting that it is the H^+-excretion which is electrogenic and
responsible for the hyperpolarization of the PD.

Effect of Auxin and FC on K^+-uptake into Oat Coleoptiles.

Under the conditions employed in these experiments (1 mM MES-tris buffer, pH
6.0 + 0.1 mM $CaCl_2$ + 1 mM KCl) K^+-uptake, in the absence of hormones, occurred at
a low, constant rate which was about one-fourth the apparent rate of H^+-excretion
(Table 2). Auxin had no effect on this uptake until after a lag that usually
exceeded 30 minutes, then it enhanced the uptake a maximum of 30% after 90-120
minutes[8] (Fig. 3). As H^+-excretion was increased over 2-fold by auxin there was
a reduction in the K/H ratio.

FC, on the other hand, stimulated K^+-uptake by 5-10 fold after a lag of only

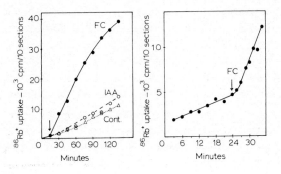

Fig. 3. Time-course of action of FC and IAA on K^+-uptake into oat coleoptiles. Peeled sections incubated in MES-tris, pH 6.0 + 1 mM $CaSO_4$, 0.5 mM K_2SO_4 and 2.1 μC $^{86}RbCl$. At arrow 10 μM FC or IAA added. From Cleland[8].

TABLE 2

K/H RATIO FOR AVENA COLEOPTILES IN THE PRESENCE OF FC AND IAA

Treatment	K^+ uptake	H^+ excretion	K/H Ratio
	μMoles/hr/10 sections		
Control	0.04	0.16	0.25
IAA, 10 μM	0.04	0.37	0.11
FC, 10 μM	0.35	0.58	0.60

Groups of 10 sections incubated in buffer + 1 mM KCl. H^+-excretion determined by titration to pH 6 every 20 min. K^+-uptake is loss of K^+ from solution as determined by atomic absorption spectometry.

90 seconds (Fig. 3) and within 2 hours over 35% of the K^+ had been taken up from the solution. This uptake is not specific for K^+-Rb^+ in oat coleoptiles.[8] For example Na^+ competed nearly as efficiently as K^+ for $^{86}Rb^+$ uptake, and Mg^{++} was only slightly less effective. Ca^{++} and Li^{++} were without effect. This contrasts with the situation in maize roots where Na^+ if ineffective in competing for K^+ uptake[14].

The uptake of K^+ as measured by $^{86}Rb^+$ uptake was proportional to the KCl concentration so that a plot of log uptake vs log KCl gave a nearly straight line (Fig. 4). H^+-excretion, on the other hand, was less sensitive to the external K^+-concentration with the result that the K/H ratio varied greatly with the KCl concentration. At 10 μM KCl the ratio appeared to be 1, but this is an over-

estimate due to the preference of the tissue for $^{86}Rb^+$ over K^+; the actual K/H
ratio was closer to 0.8. At 1 mM KCl the ratio was 0.60 (Table 2) and the 0.01
mM KCl the ratio had fallen to 0.02.

A comparison of FC-induced K^+-uptake with FC-induced H^+-excretion showed that
the two processes are remarkably similar in several ways. Both responses were
induced by FC within 90 seconds.[8,23] Both show a similar sensitivity to the
external pH, with the FC effect being at a minimum a low pHs and increasing as the
pH rose until a maximum was reached at pH 6.5-7.5.[8,24] Both responses were
inhibited by metabolic inhibitors such as CCCP and KCN, but were unaffected by
either protein synthesis inhibitors such as cyloheximide or osmotic inhibitors
such as mannitol.[8,24] It is tempting to conclude, then, that the FC-induced
K^+-uptake and H^+-excretion are linked processes.

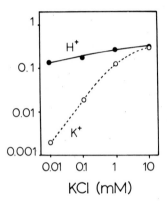

Fig. 4. Effect of external KCl on FC-induced H^+-excretion and K^+-uptake.
Solutions contained 0.1 mM $CaCl_2$, 1 µM $^{86}RbCl$ and 0 to 10 mM KCl. H^+-excretion
determined by titration and K^+-uptake by uptake of $^{86}Rb^+$ into the tussue.

The Mechanism of IAA and FC action in Oat Coleoptiles.

Some tentative conclusions concerning the mechanism of IAA and FC action in oat
coleoptiles can be reached in light of the information just presented. Neither IAA
or FC can be acting via electroneutral K^+/H^+ exchange (Mechanism II, Fig. 1) as
both hormones cause a hyperpolarization of the membrane potential and the K/H
ratio is less than 1 for at least the first 2 hours following addition of either
compound.

Auxin may activate an electrogenic H^+-pump (I) as suggested by Hager et al.[2]
In that case the stimulation of K^+-uptake observed in auxin-treated tissues would
actually be in response to the increased negativity of the PD rather than to any
direct action of auxin on K^+-uptake. The kinetics of the apparent auxin-stimula-
tion of K^+-uptake are compatible with such a mechanism. But FC cannot also act by
the same mechanism. It must be remembered that with this mechanism the increase in

K^+ uptake will be proportional to the exponential of the hyperpolarization of the PD. Since FC causes a two-fold greater hyperpolarization than auxin it should cause at most a three-fold greater increase in K^+-uptake. But in fact the stimulation by FC is at least 10-fold greater than that induced by auxin, far too great to be explained by mechanism I. Furthermore the close correlation between the kinetics of FC-induced K^+-uptake and H^+-excretion would not be expected for mechanism I.

The data for FC fits well with an electrogenic K^+/H^+ exchange mechanism (III). As predicted by this mechanism the hyperpolarization decreases as the K^+ in the medium increases. As predicted, K^+-uptake is proportional to the external K^+ level but is inversely proportional to the amount of hyperpolarization. As expected the kinetics of the FC-induced K^+-uptake and H^+-excretion are similar, and the sensitivity of the two processes to inhibitors and external pH are identical. Auxin might also act via mechanism III, although this seems unlikely because of the difference in the timing of the auxin-induced H^+-excretion and K^+-uptake. We conclude, then, that the evidence is best explained by assuming that auxin acts via an electrogenic H^+-pump while FC activates an electrogenic K^+/H^+ exchange process.

Since the FC-induced K^+/H^+ exchange is inhibited by all inhibitors of ATP synthesis[8,19] it is tempting to believe that the exchange is mediated by a plasma membrane-associated ATPase. In that case we might expect to be able to demonstrate FC activation of isolated coleoptile ATPases. In cooperation with Dr. T. K. Hodges we have fractionated coleoptile membranes on a discontinuous sucrose gradient, and have tested each fraction, with and without FC, for its ATPase activity. Each assay was run both in the presence and absence of KCl, and the ATPases of each fraction were also tested after solubilization with Triton X-100. A typical experiment is illustrated in Table 3. We failed to find any stimulation of ATPase activity by FC in any fraction under any set of conditions.

TABLE 3

LACK OF EFFECT OF FC ON AVENA COLEOPTILE ATPases.

Membrane Fraction	ATPase Activity	
	− FC	+ FC
	ug Pi/mg protein/hr	
20-30% sucrose	9.8	11.0
30-34% sucrose	18.0	16.8
34-45% sucrose	19.8	17.6

Membrane fractions were prepared from oat coleoptiles and were tested for ATPase activity by the procedures of Hodges & Leonard.[16] FC (10 μM) was present only during the ATPase assay (25 min).

Using a different isolation and fractionation procedure Tepfer & Cleland[25] obtained similar negative results, and Hodges[26] has failed to find any stimulation of oat root ATPases with FC. While it is premature to conclude that FC does not activate ATPase activity _in vivo_, we must consider the possibility that FC does not enhance ATPase activity, but rather couples the ATPase to the actual carrier mechanism. If so, K^+-export from ATP-treated membrane vesicles should be enhanced by FC. Alternatively, the action of FC may be indirect; FC may cause a change in the cytoplasmic pH to occur through enhanced uptake of CO_2 or HCO_3^- or by formation of malic acid, and the change in pH may, in turn, enhance the ATPase activity. We have found that some of the coleoptile ATPase fractions are quite sensitive to the external pH with an optimal activity at ca. pH 7. Unfortunately we have no information as yet as to the cytoplasmic pH in coleoptiles or the way in which it might change in response to FC.

FC Action in Oat Root and Leaf Cells

FC is known to cause H^+-excretion from a wide range of tissues including roots, stems, leaves and coleoptiles.[3,14,27] Is its mode of action the same in each of these tissues; i.e., does FC always act via an electrogenic K^+/H^+ exchange? Pitman et al.,[13,28] who demonstrated FC-induced H^+-excretion and K^+-uptake into barley root tissues, proposed that FC acts via mechanism I instead, although the data are compatable with mechanism III as well. We have examined the response of oat root sections to FC with the intention of seeing if we could obtain an indication as to the mechanism by which FC act in this tissue.

The rates of both H^+-excretion and K^+-uptake vary with the position along the oat root (Fig. 5). The rates of both are low in the apical 5 mm, where all of the

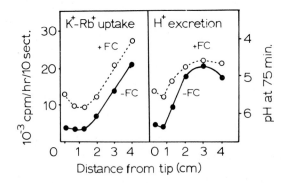

Fig. 5. H^+-excretion and K^+-uptake in oat root sections as a function of FC and of position along the root. Solutions contained 1 mM MES-tris, pH 6.0, 1 mM KCl, 1 μM [86]RbCl, ± 10 μM FC. Twenty 0.5 or ten 1-cm sections were used.

growth occurs. H^+-excretion begins to increase about 1 cm from the tip and reaches
a maximum at 3 cm. K^+-uptake begins to increase at 1.5 cm and the rate steadily
increases with distance at least to 4 cm. FC stimulates H^+-excretion and K^+-uptake
in all parts of the root. The absolute increase in K^+-uptake is nearly the same in
all parts of the root, with the result that the percent increase is greatest at the
tip where the control rate is low. The maximum increase in H^+-excretion, both in
percent of control and on an absolute basis is greater at the tip. We have therefore concentrated our attention on the apical 1 cm of the oat root.

FC stimulates both H^+-excretion and K^+-uptake into these apical root sections
after a lag of about 2 minutes (Fig. 6). Both processes are enhanced by over 3-
fold, and the K/H ratio is 0.5-0.6 both in the presence and absence of FC. The
similarity in the kinetics of the two processes suggests that FC is acting via
mechanism III here, too, but in the absence of the necessary electrophysiological
data a more definite conclusion cannot be reached.

Finally, a preliminary study has been initiated on the ability of FC to induce
H^+-excretion from oat leaf protoplasts. This system has the advantage that the
ionic concentration of the outside medium can be more accurately known and
controlled because the absence of the cell wall removes a major reservoir for
cations. Oat leaf protoplasts excrete protons rapidly in response to FC, with a

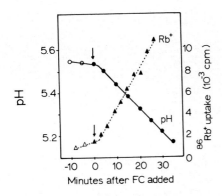 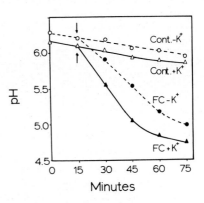

Fig. 6 (left). Time-course of FC-induced H^+-excretion and K^+-uptake by apical 1-
cm oat root segments. Conditions same as in Fig. 5. FC added at arrows.

Fig. 7 (right). H^+-excretion from oat leaf protoplasts as a function of FC and
external KCl. Protoplasts were in unbuffered 0.6 M mannitol ± 1 mM KCl and the
pH was measured every 15 min. FC (10 μM) was added at the arrows.

delay of less than 2 minutes before the rate of excretion is accelerated (Fig. 7).

The presence of 1 mM KCl in the medium increases the rate of FC-induced H^+-excretion, but it will be noted that considerable proton excretion occurs in the absence of any ions added to the external medium. Of course the medium is not completely free of ions, as some must be released from both intact protoplasts and by bursting of other protoplasts, but these results provide additional evidence that FC-induced H^+-excretion does not have an obligitory counter-movement of K^+, although the rate of H^+-excretion is enhanced when the uptake of K^+ can occur. This system appears to be ideal for further studies on the effect of external cations on hormone-induced H^+-excretion.

This research was supported by Contract E(45-1)-2225-T19 from the U.S. Energy Research & Development Administration. The assistance of T. K. Hodges, N. Higinbotham, H.B.A. Prins & J. R. Harper in portions of this research is gratefully acknowledged.

REFERENCES

1. Cleland, R.E. (1971). Ann. Rev. Plant Physiol., 22, 197.
2. Hager, A., Menzel, H. & Krauss, A. (1971). Planta, 100, 47.
3. Marrè, E., Lado, P., Rasi Caldogno, F. & Colombo, R. (1973). Plant Sci. Letters, 1, 179.
4. Davies, P.J. (1973). Bot. Rev., 39, 139.
5. Rayle, D. L., & Cleland, R. E. (1976). Current Topics Devel. Biol., in press.
6. Thomas, R. C. (1976). Nature, 262, 54.
7. Haschke, H. P. & Luttge, U. (1973). Zeit. Naturforsch. 28c, 555.
8. Cleland, R. E. (1976). Biochem. Biophys. Res. Com., 69, 333.
9. Etherton, B. (1970). Plant Physiol., 45, 527.
10. Nelles, A. (1975). Biochem. Physiol. Pflanzen, 157, 182.
11. Marrè, E., Lado, P., Ferroni, A. & Ballarin Denti, A. (1974). Plant Sci. Letters, 2, 257.
12. Cocucci, M., Marrè, E., Ballarin Denti, A. & Scacchi, A. (1976). Plant Sci. Letters, 6, 143.
13. Pitman, M.G., Schaefer, N. & Wildes, R.A. (1975). Plant Sci. Letters, 4, 323.
14. Lado, P., De Michelis, M. I., Cerana, R. & Marrè, E. (1976). Plant Sci. Letters, 6, 5.
15. Cleland, R. E. (1972). Planta, 104, 1.
16. Hodges, T. K. & Leonard, R. T. (1974). Methods Enzymology, 32B, 392.
17. Brenneman, F. N. & Galston, A. W. (1975). Biochem. Physiol. Pflanzen, 168, 453.
18. Pitman, M.G., Mertz Jr., S. M., Graves, J.S., Pierce, W.S. & Higinbotham, N. (1972). Plant Physiol., 47, 70.

19. Cleland, R. E. Unpublished data.

20. Cleland, R.E. (1973). Proc. Nat. Acad. Sci. U.S., _70_, 3092.

21. Prins, H.B.A., Harper, J.R., Higinbotham, N. & Cleland, R.E. (1976).
    Plant Physiol., _57_, S3.

22. Higinbotham, N., Etherton, B. & Foster, R.J. (1964). Plant Physiol.,
    _39_, 196.

23. Cleland, R.E. (1976). Plant Physiol., _58_, in press.

24. Cleland, R. E. (1975). Planta, _127_, 233.

25. Tepfer, M. & Cleland, R.E. Unpublished data.

26. Hodges, T. K. Personal communication.

27. Marrè, E., Lado, P., Rasi Caldogno, & F., Colombo, R. (1974). Plant
    Sci. Letters, _2_, 139.

28. Pitman, M.G., Schaefer, N. & Wildes, R.A. (1975). Planta, _126_, 61.

Regulation of Cell Membrane Activities in Plants
E. Marrè and O. Ciferri eds.
© 1977, Elsevier/North-Holland Biomedical Press, Amsterdam

THE STOMATAL TURGOR MECHANISM AND ITS RESPONSES

TO CO_2 AND ABSCISIC ACID:

OBSERVATIONS AND A HYPOTHESIS

Klaus Raschke

MSU/ERDA Plant Research Laboratory

Michigan State University

East Lansing, Michigan   48824   U.S.A.

INTRODUCTION

Stomata admit CO_2 into the intercellular spaces of plants when conditions favorable for the assimilation of CO_2 prevail. Stomata also moderate water loss if water stress develops in the plant. They accomplish this dual task by a variation in the turgor of the guard cells which in turn leads to opening and closing of the stomatal pore. Pressure and volume changes of guard cells are the osmotic effects of import and export of inorganic ions and the metabolism of organic acids. These processes of ion transfer and metabolism are controlled in direction and magnitude by the $[CO_2]$ in the intercellular spaces and by the level in the plant of (+)-abscisic acid*, the plant hormone formed under stress. CO_2 and ABA are the most important messengers we know in the stomatal feedback system controlling gas exchange.

I shall first describe how stomata respond to each one of the two feedback messengers, separately and in combination. Then I shall summarize the present knowledge of the stomatal turgor mechanism. Both sets of information will then be used to develop a hypothesis explaining how CO_2 and ABA possibly affect processes occurring in the cytoplasm and in the membranes of guard cells.

More information on the stomatal mechanism than can be supplied in this paper is available from recent reviews[1,2]. These reviews are, however, obsolete in their account of the metabolism of organic acids and carbohydrates in guard cells. I have attempted to incorporate into this paper recent advances in the elucidation of the carbon metabolism of epidermal tissue.

In general, stomata open when the level of the intercellular $[CO_2]$ drops and close when it increases. The stomatal response to CO_2 follows saturation kinetics; half saturation occurs near 200 µl 1^{-1} in Zea mays. There are, however, two groups of observations which make a modification of these statements necessary.

---

*Abbreviations: ABA, (+)-abscisic acid; FC, fusicoccin; PEP, phosphoenol-
pyruvate.

(i) In a number of plant species, stomata do not open to their widest apertures if they are completely closed and then are brought into an atmosphere
completely free of CO_2. An increase of the $[CO_2]$ from 0 to about 100 µl 1^{-1} is
necessary to obtain maximal opening. If the $[CO_2]$ is increased further, stomata
become narrower again, now obeying the general rule. When the change in $[CO_2]$
is reversed, an effect of hysteresis becomes evident. Stomata continue to open
even when the $[CO_2]$ decreases below 100 µl 1^{-1}. Maximal opening is now observed
in air completely devoid of CO_2[3].

(ii) Stomata of Xanthium strumarium, Gossypium hirsutum, and possibly other
species, lose their sensitivity to CO_2 if grown in an environment which prevents
the development of water stress[4,5]. Once open, these stomata cannot be closed
again by an increase in $[CO_2]$. An increase of the ABA level in the leaf, however,
returns to the stomata their ability to respond to CO_2; ABA sensitizes stomata
to CO_2[4]. The presumed messenger of water stress in leaves, ABA, appears to act
in stages, depending on the degree of water stress; low levels of ABA induce
stomata to adjust their apertures in synchrony with variations in CO_2 uptake;
high levels produce stomatal closure, even if the requirement for CO_2 of the
photosynthetic apparatus is not met. Then water relations take priority over
photosynthesis.

Similar to the sensitization of stomata to CO_2 by ABA, a requirement of CO_2
can be observed for the action of ABA on stomata[5]. Stomata of Xanthium strumarium
may be in a state in which they respond to ABA only if CO_2 is present in the air.
This enhancement of the response to ABA by CO_2 follows saturation kinetics; half
saturation occurs near 200 µl 1^{-1}. The enhancement of stomatal responses to ABA
by CO_2 has been observed in several species. The degree of this interaction
varies widely between species.

The relationship between the amount A of ABA needed to elicit a stomatal response and the [ABA] in the transpiration stream can be described by

$$A = \text{coefficient} \times [ABA]^n$$

In this expression, A stands for the amount of ABA supplied to the leaf until the
stomatal conductance is reduced by 5%. The exponent n was found to be $0.6 < n < 0.9$.
This relationship resembles a Freundlich adsorption isotherm, with repulsion
occurring between the adsorbed molecules. These results are based on experiments
conducted with leaves of Xanthium strumarium[5], Gossypium hirsutum and with epidermal strips of Commelina communis (unpublished).

The inhibition of stomatal opening by ABA can be overcome by a simultaneous
presentation of fusicoccin[6]. This toxin can also inhibit stomatal closing in
response to CO_2[7].

Since stomatal regulation of gas exchange occurs through a variation of guard cell turgor, we have to expect that CO_2 and ABA control (i) the transfer of inorganic osmotica within the epidermis and (ii) the production and catabolism of organic osmotica within the guard cells. We turn now to the mechanism of this controlled system.

THE STOMATAL TURGOR MECHANISM[2]

Solute requirement for opening and the elastic properties of the guard cell walls: The volume of the guard cells (magnitude approximately 10 pl per pair) roughly doubles when stomata open from the closed to the fully open state. The relationship between stomatal aperture and guard cell volume has been found to be linear for stomata of Vicia faba, and this was true also for the relationship between aperture and solute content. This unexpected relationship turned out to result from the unusual elastic properties of the guard cell walls. The relation between hydrostatic pressure, p, of a cell, coefficient of elasticity, α, and the relative volume deviation from the resting state, $v = (V-V_0)/V_0$, can be described by $p = \alpha v^n$. For most plant cells n>1, for guard cells of Vicia faba, n was found to be <1, viz. n = 2/3. An n of 2/3 happens to result in a nearly constant solute requirement for a unit change in stomatal opening over a wide range of stomatal apertures ($0 \leq v \leq 1$). With ψ standing for the water potential, and s for the solute content of the guard cells, one can write

$$\psi = p-\pi = \alpha v^n - sRT/\{V_0(1 + v)\}$$

Rearrangement and differentiation with respect to v yields

$$\frac{ds}{dv} = \frac{V_0\alpha}{RT} \{nv^{n-1} + (n + 1)v^n - \frac{\psi}{\alpha}\}$$

Substitution of 2/3 (or a slightly larger value) for n results in fairly constant sums of the two first terms in the brackets. Data obtained by other authors on Vicia faba[8], Nicotiana tabacum[9] and Phaseolus vulgaris[10] confirm a linear relationship between stomatal aperture and solute content of guard cells. This linearization can be looked upon as an evolution facilitating the regulation of stomatal apertures independent of the degree of stomatal opening.

The elastic properties of the guard-cell walls (and the pressure exerted on the guard cells by the other epidermal cells) determine the solute requirement for stomatal movement.

The osmotica in guard cells[1,2]: It is now well established that in all plant species salts of K^+ are the osmotica which are used by the plant to increase the pressure in guard cells above the pressure prevailing in guard cells of closed stomata. Some halophytes are able to use Na^+ in addition to K^+. Electron-probe microanalysis conducted simultaneously with determinations of the volumes of guard cells of Vicia faba and their total solute contents demonstrated (i) that the amounts of K^+ imported into guard cells (4 peq $stoma^{-1}$) are sufficient to account for the observed changes in guard-cell volume and pressure[11]. As a matter of fact, the amounts were too high if K^+ was associated with a monovalent anion, but about right when divalent anions provided the counter charges for K^+. The microprobe work further showed that (ii) neither the contents of guard cells of N, S, or P changed during stomatal movement, and that of Cl only little. Inorganic anions can therefore be unimportant as counter ions for K^+ in guard cells of Vicia faba; organic anions must perform this function. Malate was thought the best candidate[11] and indeed was shown by Allaway[12] to accumulate in the epidermis of Vicia faba when stomata opened and to disappear when they closed. About one-half of the K^+ was balanced by malate in his material. At the same time, experiments conducted with K^+ salts of non-absorbable acids led to the conclusion that malate was most probably formed within the guard cells during K^+ import and that electroneutrality was maintained by release of H^+ from guard cells[13]. This exchange became manifest through an acidification of the medium on which epidermal strips were exposed. By automatic titration it could be shown that the amounts of H^+ released by guard cells were of the same order of magnitude as the amounts of K^+ taken up.

The lack of participation of Cl^- in providing countercharges for K^+ is, however, not general. Depending upon availability of Cl^- and on some unknown factors, Cl^- can contribute considerably to the anion content of guard cells, even in Vicia faba. Under experimental conditions providing Cl^-, 75% of the negative charges can be represented by this anion (unpublished).

The participation of Cl^- in stomatal turgor variation is very striking in grasses. This has been shown in Zea mays[14]. In this species, subsidiary cells of the stomatal complex store large amounts of K^+ and Cl^-. During stomatal opening, both ions migrate rapidly into the guard cells, during closure they rapidly return to the subsidiary cells. On the average, 40% of the K^+ were paired by Cl^-, the remainder by organic anions. In some individual guard cells, K^+ was completely balanced by Cl^-.

Pallaghy[15] measured negative transmembrane potentials in guard cells. It seems therefore probable that during stomatal opening, K^+ is taken up passively into the vacuole, possibly as a consequence of an anion uptake into the vacuole and a simultaneous expulsion of H^+ from the guard cells into their environment. Independent of what the initiating transport process may be, pH regulation would lead to an import of Cl^- into the guard cells and a production of organic acids. The fact that Cl^- seldom balances all the K^+ in guard cells, and often does not participate as an important osmoticum in stomatal movement, makes it unlikely that Cl^- uptake is the primary mechanism driving stomatal opening.

The important role of a K^+/H^+ exchange mechanism or an active expulsion of H^+ from guard cells during their inflation is indicated by the dramatic increases in stomatal opening observed after application of fusicoccin. Graniti[16] reported that plants treated with this toxin wilt easily, Turner and Graniti[17] showed that this effect was due to stomatal opening and that opening was correlated with a proportional increase in the K^+ content of the guard cells[10]. It was suspected that (i) FC would enhance proton expulsion in guard cells and (ii) that systems exhibiting a proton release in response to FC would show a K^+ dependence. The K^+ dependence of FC stimulated H^+ excretion was meanwhile confirmed by Marrè et al.[18] on pea internode segments. The enhancement of H^+ excretion from guard cells was shown in a series of experiments conducted at East Lansing. Since the results are not published yet, I give a brief summary here. (i) An addition of 10^{-4}M FC to 10 mM KCl doubled stomatal opening (3.5 versus 7.5 µm) within 4 h in epidermal strips of Tulipa gesneriana. The hydrogen ion excretion also doubled, from 2.7 to about 5×10^{-11} eq stoma^{-1} (tulip has very large guard cells). (ii) Proton excretion into a 10 mM solution of KCl from epidermal strips of Vicia faba, as recorded with a flat-surface combination pH electrode, was accelerated by the addition of FC (10^{-5}M). The response began in a time <1 min. (iii) Automatic titration of the release of H^+ from strips of Commelina communis yielded a similar result: H^+ excretion accelerated within 50 sec after the addition of FC. (iv) The enhancement of the excretion of H^+ by FC depended on the presence of K^+ (Table 1).

In conclusion, the increase in osmotic pressure of guard cells involves a H^+/K^+ exchange mechanism, which in turn, sets in motion an uptake of Cl^- (in exchange for OH^-?) and a production of organic acids whose H^+ are used to obtain K^+ and whose anions are used to balance the positive charges of K^+. In the case of a complete neutralization of K^+ by Cl^- there will be, of course, no net exchange of H^+. At this point it remains open whether H^+/K^+ exchange occurs stoichiometrically at a ratio 1:1 or not, and whether this exchange is self-triggered, or

TABLE 1

HYDROGEN ION EXCRETION BY EPIDERMIS

Titration of H^+ excreted within 4 hours by epidermal strips of Tulipa gesneriana floating in CO_2-free air on various solutions.

Solution		n	Excretion of H^+
KCl	Fusicoccin		neq mm^{-2}
0	0	1	1.0
0	0.1 mM	2	1.1
10 mM	0	2	0.8
10 mM	0.1 mM	4	1.8
100 mM	0	3	1.6

initiated by a signal from a turgor sensor in the guard cells, or from a system
regulating the pH of the cytoplasm.

Clearly, the change in osmotic pressure in guard cells is linked not only to
the transfer of inorganic ions between cells but also to the metabolism of
organic acids. One can suspect that CO_2 participates through the carboxylation
of PEP to oxaloacetate which subsequently is reduced to malate, or transaminated
to aspartate.

Acid metabolism of guard cells: The early autoradiograms of Shaw and Maclachlan[19] demonstrated that the power of the epidermis to assimilate [14]CO_2 resides in
the guard cells. This was taken as evidence for the guard cells' ability to reduce CO_2 photosynthetically. Recent determinations of the assimilation pattern
of epidermal samples of Tulipa gesneriana and Commelina communis however showed
that this is not correct[20]. The main fixation products of CO_2 are malic and
aspartic acids, independent of whether stomata are in the light or in darkness,
and, this is important, whether they are open or closed. Earlier findings of
phosphoglyceric acid among the fixation products in epidermis[21] must be ascribed
to contamination of the epidermal samples by adhering mesophyll chloroplasts.

Incubation of epidermal strips with ^{14}C-labeled sugars and glucose-1-phosphate
yielded radioactive malate and aspartate[20]. Epidermal samples possess PEP carboxylase activity proportional to their stomatal density[22]. Guard cells are,
therefore, able to produce malate by the carboxylation of PEP. The latter can be
obtained by glycolysis (explaining the frequently observed disappearance of starch
during stomatal opening). Production of organic acids in guard cells should be
self-limiting through feedback on PEP carboxylase[23].

During stomatal closing, malate is disposed of by three processes occurring
simultaneously[20]: (i) A specific leakage, together with K^+, from the guard cells,
(ii) oxidation in the tricarboxylic acid cycle, and (iii) gluconeogenesis within
guard cells after decarboxylation. K^+ and Cl^- leave the guard cells during
stomatal closing. If subsidiary cells are present, they receive and store the
K^+ and Cl^- released from the guard cells[14].

Stomatal closure initiated by malic acid: The velocity of stomatal closing in
response to CO_2 followed saturation kinetics. In Zea mays, half saturation
occurred at a $[CO_2]$ near 200 µl l^{-1}, which is equivalent to the K_M (CO_2) of PEP
carboxylase[2]. The possibility exists that closing is mediated by the formation
of malic acid. This possibility was tested by the exposure of epidermal samples
with open stomata to solutions of malic acid. In most experiments, the pH of
the solution was adjusted to the first pK of malic acid in order to have 50% of
the acid present in the protonated and therefore absorbable form. There were no
metal cations in the solution (adjustments in pH were made with small amounts of
bis tris propane) and a solution of the non-absorbable acid MES served as control.

TABLE 2

STOMATAL CLOSING IN RESPONSE TO MALIC ACID

Epidermal strips with open stomata were placed on 0.1 M solutions at time zero.

Species	Acid	pH	Stomatal aperture in μm after exposure for t min				
			t = 0	15	30	50	125
C. comm.[*]	MES	3.4	16.7	21.8	28.3		
	1-Malic	3.4	16.3	4.6	4.0		
V. faba[**]	MES	3.4	open			28.5	
	1-Malic	3.4	open			4.2	
V. faba	H_2O	4.0	11.9	12.2	12.1	12.0	12.0
	MES	4.0	10.7	12.2	11.8	11.3	12.0
	1-Malic	4.0	8.9	9.3	8.6	6.8	3.5

[*]Commelina communis L.

[**]Vicia faba L.

Stomata closed in epidermal strips floating on the malic acid solution at low pH, they did not close on the solution of MES (Table 2) nor did they close on malic acid solutions of high pH. It can be assumed that the protonated malic acid entered (the cytoplasm of) the guard cells and caused stomatal closure. Experiments are under way to find out whether this closing effect is merely due to an increase in the acidity of the guard cells or due to a specific response to malic acid. The results of the experiments done so far indicate that both kinds of reactions contribute to closing.

THE HYPOTHESIS

Involvement of turgor regulation: I consider guard cells not to differ from other plant cells in their ability to establish and regulate turgor. In non-stressed plants, stomata may stay open all the time. CO_2 and ABA, each alone or in combination, could modulate the functioning of this regulator, for instance by affecting, directly or indirectly, the mechanical and electrical properties of the cell membranes that are involved in turgor regulation. (Zimmermann[24] has indicated how cell membranes could be involved in this control process.) Uptake or loss of ions and metabolism of organic acids would follow such membrane changes. Alternatively, one can envisage that the primary responses to CO_2 and ABA occur in the cytoplasm, for instance, changes in the chemical potential of H^+ and perhaps other ions, which would then lead to reactions overriding the commands given by the turgor regulator. It is very likely that responses occur at both locations, at the membranes and in the cytoplasm, and that they affect each other.

The dual role of malic acid and the sensing of CO_2: CO_2 is required to make malate for stomatal opening (if Cl^- is not available as counter ion for K^+).

Yet CO_2 is known to cause stomatal closure. This contradiction can be resolved if malate content and pH of the cytoplasm determine direction and magnitude of the ion fluxes. I propose that at low $[CO_2]$ the processes of expulsion of H^+ from the cytoplasm of the guard cells and removal of the malic anion into the vacuoles will keep pace with acid formation; stomata will open. At high $[CO_2]$, the level of H^+ and malate in the cytoplasm will increase, cause a slow-down of the production of acids and ultimately lead to a leakage of K^+ and Cl^-, and to the dissipation of malate by the three mechanisms listed before. The level of malate will, of course, also be determined by the rate at which deacidification takes place. PEP carboxylase and the malic dehydrogenases appear to form by themselves already a push-pull system which adjusts the intracellular malate level (and pH) in some relationship to the intercellular $[CO_2]$[2]. This combination of mechanisms may serve as the CO_2 sensor of the stomata. If this is true, guard cells should be assimilating CO_2 into malic acid in the light and in darkness, whether stomata are closed or open. This was indeed found[20]. If a low pH in the cells leads to closure, then exposing open stomata to protonated acids should lead to closure; this was found, too. Effects of organic acids on roots provide an analogy[25]. If FC leads to enhanced expulsion of H^+ from guard cells, then application of this substance should prevent CO_2 from causing stomatal closing; also this was observed[7]. The hysteresis in the relationship between $[CO_2]$ and stomatal opening (as mentioned earlier in this paper) can be interpreted: When stomata open in CO_2-free air they cannot obtain CO_2 from respiration fast enough to produce malate for maximal stomatal opening; an increase in exogenous CO_2 to an optimal level will allow malate formation to keep pace with malate removed into the vacuoles: maximal stomatal opening occurs between 100 and 200 μl CO_2 l^{-1}. If, however, stomata are brought from a high into a low $[CO_2]$ they start with an ample supply of CO_2. As the exogenous $[CO_2]$ declines, endogenous CO_2 seems to be sufficient to produce the increment in malate necessary to obtain maximal stomatal opening.

Stomata which have lost their sensitivity to CO_2 may either possess strong "pumps" removing malate and H^+ from the cytoplasm of the guard cells, or possess membranes whose properties have changed when strained by high turgor[24].

ABA enhances acidification? Stomata insensitive to CO_2 can be sensitized by endogenous or exogenous ABA[4], and vice versa, stomatal responses to ABA are enhanced by the presence of CO_2. In some species, like Xanthium strumarium, stomata may be in a state in which they are able to respond to ABA only if CO_2 is also present[5]. Stomatal responses to ABA could be overcome by an application of fusicoccin[6]. These observations indicate that ABA prevents stomatal opening and causes closure by much the same mechanism by which CO_2 acts. Apparently, ABA accelerates the build-up of high levels of H^+ and malate in the cytoplasm, either by enhancing malate formation or by inhibiting acid removal, including expulsion of H^+, from the cytoplasm of the guard cells. Observation on stomata favor the latter explanation. It is likely that ABA attaches for action to very specific sites of a guard-cell structure from where it detaches with relative ease when the medium bathing the cells is free of ABA.

FACTS AND CONJECTURE SUMMARIZED

Facts: The turgor of guard cells increases as a result of an import of K^+ and Cl^- into the vacuoles, with Cl^- in some cases completely balancing the positive charges of K^+; in others only to a small extent. The K^+ that are not paired by Cl^- are neutralized by organic anions, mainly malate ions. These organic acids are produced within the guard cells after the carboxylation of PEP. The anions of these acids are transferred into the vacuole, the H^+ are used to obtain K^+ from outside. Fusicoccin enhances expulsion of H^+ from guard cells, uptake of K^+ and stomatal opening.

The solute content of guard cells is linearly related to cell volume and stomatal aperture. This unexpected relationship is due to the unusual elastic properties of the guard-cell walls. In Vicia faba, turgor is proportional to the 2/3 power of the relative increase in cell volume.

During stomatal closure, K^+ and Cl^- are released from the guard cells. Malate is partially also released; another part is used for gluconeogenesis, a third finds its way into the tricarboxylic acid cycle.

CO_2 and ABA cause stomatal closure. In some species, CO_2 and ABA are required simultaneously to initiate stomatal closure; a high level of CO_2 alone will not narrow stomata.

Conjecture: Guard cells are able to function as regulators of gas exchange through the evolution of sensitivities to CO_2 and ABA. These sensitivities were acquired through modification of two control mechanisms found in most plant cells: turgor regulation and regulation of pH. Apparently, these mechanisms are still serving their original purpose in guard cells of non-stressed plants of many species when stomata are insensitive to CO_2; slight stress suffices to induce a sensitivity to CO_2 through the formation and action of ABA.

CO_2 and ABA appear to enhance the accumulation of organic acids in the cytoplasm; CO_2 serves as substrate, ABA inhibits the expulsion of H^+ (or enhances acid synthesis). The acid level in the cytoplasm determines whether the production of acids will be accelerated or decelerated, whether H^+ will be exchanged for K^+, and Cl^- for OH^-. Acid metabolism is controlled by the dependence of enzyme activity on the concentrations of H^+, metal ions and organic anions in the cytoplasm, while the exchange of H^+ for K^+ and Cl^- for OH^- may be affected by alterations of membrane properties through pressure, distribution of charges, and the absence or presence of malate anions. High malate content and low pH lead to the leakage of K^+, Cl^-, and malate from guard cells. Changes in the ion

182

content of the cytoplasm are transmitted to the vacuole through transport mechanisms situated at the tonoplast.

In the presence of CO_2 and ABA, these modifications of the two control systems no longer serve homeostasis with respect to pH and turgor but rather help to moderate gas exchange, perhaps even to optimize the relationship between water loss and CO_2 uptake through the leaf's epidermis.

ACKNOWLEDGMENT

My research was supported by the United States Energy Research and Development Administration under Contract E(11-1)-1338 as well as by the John Simon Guggenheim Memorial Foundation and the Deutsche Forschungsgemeinschaft.
I thank my collaborators Carol Van Kirk and Robert Saftner for providing unpublished data.

REFERENCES

1. Hsiao, T. C. (1976) Transport in plants. Vol. 2, pt. B (U. Lüttge and M. G. Pitman, eds.), Springer-Verlag, Berlin-Heidelberg.
2. Raschke, K. (1975) Ann. Rev. Plant Physiol. 26:309.
3. Dubbe, D., Farquhar, G. D., and Raschke, K. (unpublished).
4. Raschke, K. (1974) Plant Growth Substances, 1973, Proc. VII. Internat. Conf. on Plant Growth Substances, Hirokawa Publ. Co., Tokyo.
5. Raschke, K. (1975) Planta 125:243.
6. Squire, G. R. and Mansfield, T. A. (1972) Planta 105:71.
7. Squire, G. R. and Mansfield, T. A. (1974) New Phytol. 73:433.
8. Fischer, R. A. (1972) Aust. J. Biol. Sci. 25:1107.
9. Sawhney, B. L. and Zelitch, I. (1969) Plant Physiol. 48:1350.
10. Turner, N. C. (1973) Amer. J. Bot. 60:717.
11. Humble, G. D. and Raschke, K. (1971) Plant Physiol. 48:442.
12. Allaway, W. G. (1973) Planta 110:63.
13. Raschke, K. and Humble, G. D. (1973) Planta 115:47.
14. Raschke, K. and Fellows, M. P. (1971) Planta 101:296.
15. Pallaghy, C. K. (1968) Planta 80:147.
16. Graniti, A. (1964) Host-parasite relations in plant pathology (Z. Király and G. Ubrizsy, eds.) Research Institute for Plant Protection, Budapest.
17. Turner, N. C. and Graniti, A. (1969) Nature 223:1070.
18. Marrè, E., Lado, P., Rasi-Caldogno, F., Colombo, R. and DeMichelis, M. I. (1974) Plant Sci. Lett. 3:365.
19. Shaw, M. and Maclachlan, G. A. (1954) Can. J. Bot. 32:784.
20. Dittrich, P. and Raschke, K.; Raschke, K. and Dittrich, P. Planta (in press).
21. Willmer, C. M. and Dittrich, P. (1974) Planta 117:123.
22. Willmer, C. M., Pallas, J. E. Jr. and Black, C. C. Jr. (1973) Plant Physiol. 52:448.

23. Osmond, C. C. (1976) Transport in plants. Vol. 2, pt. A (U. Lüttge and M. G. Pitman, eds.), Springer-Verlag, Berlin-Heidelberg.

24. Zimmermann, U. (1974) Membrane transport in plants (U. Zimmermann and J. Dainty, eds.) Springer-Verlag, Berlin-Heidelberg-New York.

25. Jackson, P. C. and Taylor, J. M. (1970) Plant Physiol. 46:538.

Regulation of Cell Membrane Activities in Plants
E. Marrè and O. Ciferri eds.
© *1977, Elsevier/North-Holland Biomedical Press, Amsterdam*

EFFECTS OF FUSICOCCIN AND HORMONES ON PLANT CELL
MEMBRANE ACTIVITIES : OBSERVATIONS AND HYPOTHESES

Erasmo Marrè

Centro di Studio del C.N.R. per la Biologia Cellulare e
Molecolare delle Piante, Istituto di Scienze Botaniche
Università di Milano, 20133 Milano (Italy)

SUMMARY

The available evidence concerning the effects of FC and natural hormones on ion transport is interpreted as indicating the presence in plant tissues of a system catalysing the electrogenic proton/monovalent cation exchange. This hormone-regulated system seems to play an important rôle in the control of electric potential difference and of proton and ion intracellular and extracellular concentration. The analogies and differences between the effects of FC and those of auxin, and of other hormones, on this system are interpreted as depending on the different modes of action: relatively direct in the case of FC, complex and comprehending various intermediated steps in that of the natural hormones. The relationships are discussed between the effects of FC and auxin on proton extrusion, cation uptake, potential difference and other apparently secundary responses.

INTRODUCTION

In recent years considerable attention has centered on the possibility that several important aspects of the action of plant hormones and also of various plant activity regulating agents may depend on their capacity to influence ion transport. Thus, large effects on the electrogenic transport of H^+, K^+ and monovalent cations have been reported by several laboratories to be induced by auxins, by the growth-promoting toxin fusicoccin (FC)[1-9], by cytokinin[10] and by ABA[11, 12, 13]. The regulation of stomata opening by light seems mediated by the activation or inhibition of a H^+/K^+ exchange mechanism[11]. Recent evidence suggests that also the phytochrome-mediated effects of red and far red light on cell elongation and seed germination are accompanied by, and possibly at least in part depend on, definite changes in the activity of ion transport systems[14-16].

The understanding of the mechanism of hormone action on ion transport is thus becoming a very important field of plant hormonology. Conversely, the possibility of influencing ion transport by hormonal treatment provides a new, powerful tool for the study of the ion transport processes.

The aim of the present paper is to discuss some central problems arising from the evidence available up to now in this field, concerning : a) the nature of the transport system activated by fusicoccin; b) its relationship with the system(s) influenced by natural hormones; c) the mechanism of action of fusicoccin and of natural regulators on the ion transport system(s).

The physiological implications of the effects of hormones and FC on ion transport have been discussed in previous articles[3].

I) NATURE OF THE FUSICOCCIN-ACTIVATED SYSTEM

a) Peculiarities of the effects of FC

The effect of FC on ion transport shows some unique features, that emphasize the interest of its study. Firstly, FC induces large and fundamentally homogeneous changes of ion transport not only in all the higher plants species so far tested, from Characeae and mosses to mono- and dicotyledons, but, also, within each species, in a very large variety of tissues (stems, coleoptiles, roots, leaf mesophyll, stomata guard cells, seed embryos and cotyledons, tuber storage tissues). Secondly, the activating effect of FC on some important parameters of ion transport such as H^+ extrusion, K^+ uptake and hyperpolarization of the transmembrane electric potential (PD), are much larger than the maximal effects of auxin or any other known natural or exogenous phytoregulator. Thirdly, FC seems to influence an area of cell physiology in general, and of ion transport in particular, much better defined than that influenced by the natural hormones and regulating factors. For example, FC does not significantly influence (or influences only as a remote consequence of its primary action on ion transport) a number of processes known to be characteristically controlled by natural plant hormones, such as senescence, cell division, differentiation, and, in general, biosynthetic processes.

b) Hierarchy of FC effects on ion transport and related processes

The effects of FC on ion transport may be subdivided into major and minor. Major effects are those quantitatively larger and reproducible in all tissues so far investigated; among these are the stimulation of H^+ extrusion, the increase of uptake of K^+ and of other monovalent cations, the increase of selectivity for K^+ and the hyperpolarization of PD. Minor effects are the increases of anion, glucose and amino acid uptake. A third category of effects includes those on respiratory metabolism (increase of respiration, of dark CO_2 fixation into malate, of pyruvate and glucose-6-P levels, decrease of the C_1/C_6 ratio) and on cell wall extensibility (and thus on cell enlargement). The major effects are interdependent and seem to represent the primary response of the transport system to FC, while the effects on anion, amino acid and glucose uptake, on metabolic parameters and on cell wall characteristics may be explained as secondary consequences of the activation of the electrogenic H^+/K^+ exchange.

c) Energetics of FC-stimulated electrogenic H^+/K^+ exchange

FC-induced increases of H^+ extrusion, K^+ uptake and PD are markedly depressed by inhibitors of the respiratory metabolism such as CO, cyanide, DCCD, by the phosphorylation uncouplers DNP, CCCP and by low temperature [1, 2, 7, 17]. These results, while showing that activation of H^+/K^+ exchange by FC depends on the integrity of the metabolic system, provide only circumstancial evidence for its dependence on metabolic energy. In fact, any severe injury of oxidative processes obviously leads to large changes of the metabolic pattern, which might affect the mechanism of FC action on ion transport at the level of various, undefined steps. This appears also true for phosphorylation uncouplers, which are known to aspecifically affect the lipid component of the cell membrane, making it permeable to protons [18], and might thus counteract FC-stimulated H^+ release by letting protons reenter the cell, rather than by decreasing the ~P level in the cell.

More convincing evidence for the energy-dependence of the FC-stimulated process comes from simple thermodynamic considerations. In various tissue, e.g. in maize root segments, FC-induced H^+ accumulation in the medium is still detectable at an external pH of 4, while the pH of the cell sap is of ca.6, and that of cytoplasm presumably higher by about 1 unit. Under the same conditions, PD is of ca. -120 mV, or even more negative. Thus, FC-stimulated H^+ extrusion occurs against a very steep electrochemical gradient, the neutralization of which requires that it be coupled to some exoergonic process providing some 5000 cal per mole of H^+ extruded. The finding is important, in this connection, that the activation of H^+/K^+ exchange by FC is constantly associated with an increase of the inside negative PD, and thus with an increase of the electrochemical gradient against which the protons are secreted [2, 8, 9, 17].

By extrapolation from what known for other biochemical systems, the energy for active H^+ extrusion across a membrane may be provided by: a) a redox reaction, such as that postulated in chloroplasts and in mitochondria by the Mitchell theory; b) an ATPase-mediated reaction coupled to proton transport, such as that catalysed by chloroplast and mitochondria Mg^{++}-dependent ATPase; c) an antiport or a symport of the protons against, or together with, some other ion or solute whose transport across the membrane would be enough exoergonic to overcompensate the endergonicity of H^+ extrusion.

The latter hypothesis seems ruled out by the finding that under appropriate conditions (for example in root segments and at low K_2SO_4 concentrations) FC-induced H^+ extrusion is not accompanied by a significant accumulation of anions in the medium, or by important changes in the net transport of ions other than K^+, whose translocation for external K^+ concentrations close to Nernst equilibrium (namely, for most tissues, between 10^{-3} and 10^{-2} M), does not involve large free energy changes. In conclusion, theoretical considerations as well as the available evidence strongly suggest that the energy for FC-induced proton extrusion must be supplied either by an oxidation-reduction process or by the hydrolysis of ~P bonds .

The latter possibility is somewhat supported by recent results obtained in our laboratory, showing that FC can stimulate "in vitro" by ca. 20-30% the K^+, Mg^{++}-dependent ATPase activity of plasmalemma-enriched membrane preparations from maize coleoptiles and from spinach leaf tissues [19]. Further characterization of this "in vitro" effect is needed, however, before drawing definite conclusions about its significance in explaining the mechanism of action of FC on H^+ extrusion.

d) Relationship between H^+ extrusion, and cation and anion uptake

In all cases investigated FC-induced H^+ extrusion appears strictly coupled to an increase of the uptake of K^+ or other monovalent cations [3, 8, 9, 20]. In fact both responses are evident almost immediately (there is practically no lag period, and the maximal effect is observed within 5-10 minutes), and both are inhibited by the same metabolic poisons. Moreover, H^+ extrusion is markedly stimulated by increasing the concentration of K^+, Rb^+ or Na^+ in the medium, and the extent of this stimulation is roughly proportional to the rate of uptake of the different cations in the 1-100 mM range [3, 21, 22]. On the other hand, the data reported under various experimental conditions for the stoichiometry of the processes (namely the titratable ratio H^+ extruded/K^+ uptaken, $(-\Delta H^+/\Delta K^+)$. are far from homogeneous, ranging from values close to 1 to values lower than 0.2 [22]. As previously suggested, these wide fluctuations may be explained by assuming that part of the H^+ extruded by the FC-activated system might disappear from the medium (and thus escape titration) by either reassociating with ^-OH (secreted as such or as HCO_3^-) or by reentering the cell. Electrophysiological considerations make likely that both the postulated ^-OH secretion and/or the H^+ reentry into the cell depend upon electroneutral antiport or symport with anions rather than on passive diffusion. In fact, large differences in Cl^- and, in general, in anion uptake, such as induced by changing the anion species and their concentrations in the medium, have no significant effects of PD, either in the presence or in the absence of FC [17]. On the other hand, the finding that FC-induced proton extrusion in root segments is quantitatively almost unaffected by changes in the pH of the medium in the pH 6 – pH 4 range [23, 24] suggests that passive permeability of the plasmalemma to protons must be relatively low. On the basis of these considerations, if the ratio of the amount of protons really extruded by the FC-activated mechanism to the amount of K^+ taken up were close to unity, one would expect a value close to 1 for the ratio (titratable H^+ extruded + Cl^- (or A^-) uptaken)/ K^+ uptaken, rather than for the $-\Delta H^+/\Delta K^+$ ratio, as the amount of extruded protons disappearing because either reassociation or symport would be equal to the amount of anion taken up. This prediction is confirmed by a long series of results recently obtained in this laboratory under very different experimental conditions, showing that in FC-stimulated plant tissues the $(-\Delta H^+ + \Delta anion)/\Delta K^+$ ratio is within the 0.8 – 1.1 range, while the $-\Delta H^+/\Delta K^+$ ratio varies from 0.1 to 0.9 (Table I).

It is interesting to observe that according to this hypothesis the value of the measured $-\Delta H^+/\Delta K^+$ ratio would be largely influenced by any change in the rate of anion uptake induced by treatment with FC or other factors. In fact, low rates of anion uptake would correspond to a relatively high $-\Delta H^+/\Delta K^+$ ratio and this ratio would progressively decrease with the increase of anion uptake.

e) Nature of the coupling of H^+ extrusion with K^+ uptake, and of the electrogenic effect of FC

An important problem concerning FC-stimulated H^+/K^+ exchange is that concerning the nature of the coupling between the two processes. Two main mechanisms have been proposed : a) chemical coupling, in which a H^+/K^+ antiport would be mediated by a single carrier (as in the case of electrogenic K^+/Na^+ antiport proposed for animal systems [25]); b) electrogenic coupling, in which K^+ uptake would be driven by the increase of PD consequent to the activation of a " pure", and thus strongly electrogenic, proton pump [3, 8]. In the absence of comparatively relevant flux changes of ions other than H^+ and K^+, the ratio of proton extrusion to K^+ uptake would always be close to 1, independently of which of these two coupling mechanisms would be operating. Also the constant finding that FC-induced stimulation of H^+/K^+ exchange is associated with a simultaneous hyperpolarization of the PD [3, 8, 17] may be satisfactorily interpreted both in the case of the electrogenic coupling and in that of the chemical coupling, by assuming that the stoichiometry of the H^+/K^+ antiport tends to favour H^+ extrusion (see Poole[26]).

TABLE I - Values of the ratios $- \Delta H^+/\Delta K^+$ and $(- \Delta H^+ + \Delta anion)/\Delta K^+$ in pea internode and in maize root segments treated $\pm$ FC or IAA under different conditions

		Conditions							$\dfrac{-\Delta H^+}{\Delta K^+}$	$\dfrac{(-\Delta H^+ +\Delta A^-)}{\Delta K^+}$
		Buffer	pH		ions in the medium					
			initial	final	anion	m M	cation	mM		
PEA INTERNODES	Control	Succinate 0.5 mM	5.70	5.54	Cl⁻	10	K⁺	10	0.37	0.85
	IAA	" 0.75 "	"	5.45	"	"	"	"	0.41	0.89
	FC	" 1.0 "	"	5.46	"	"	"	"	0.36	0.82
	Control	Succinate 0.5 mM	5.70	5.56	SO₄²⁻	5	K⁺	10	0.73	1.00
	IAA	" 0.75 "	"	"	"	"	"	"	0.84	1.08
	FC	" 1.0 "	"	5.43	"	"	"	"	0.85	1.12
	Control		5.90	5.29	Cl⁻	5	K⁺	5	0.13	0.97
	FC		"	4.34	"	"	"	"	0.27	0.94
	Control		5.90 a)		Cl⁻	5	K⁺	5	0.55	1.26
	FC		"		"	"	"	"	0.51	1.09
	Control		5.90 a)		Benz.S.	10	K⁺	10	0.50	0.79
	FC		"		"	"	"	"	0.73	0.96
MAIZE ROOTS	Control		5.70	6.05	Cl⁻	1	K⁺	1	(b)	1.13
	FC		"	5.20	"	"	"	"	0.11	0.97
	Control		5.70	5.85	Cl⁻	60	K⁺	60	(b)	1.08
	FC		"	4.53	"	"	"	"	0.13	0.89
	Control		5.65	5.50	SO₄²⁻	30	K⁺	60	0.04	0.66
	FC		"	4.60	"	"	"	"	0.42	0.87
	Control	Succinate 0.5 mM	5.82	5.97	Cl⁻	10	K⁺	10	(b)	0.87
	FC	" 1.0 "	5.80	5.60	"	"	"	"	0.30	1.06

a) pH was maintained constant in the $\pm$ 0.02 units range by continoous titration
b) Negative ratio value, due to apparent H^+ influx.

2 mm long segments from the growing region of the distal internode of 7 days old etiolated pea seedlings were prepared as described by Lado et al.[21]. Subapical 8 mm long root segments were prepared after removal of the apical part (2 mm) from maize seedlings germinated for 3 days [23]. General condition of treatment of the materials and determinations of ΔH^+ by titration and of K^+(over a period of 1 h) were as previously described [21,23].

In conclusion, at the present moment, no clear cut evidence seems to allow the discrimination between these two mechanisms of coupling.

f) Other effects of FC on transport

Recent results show that FC stimulates in pea stem and in maize root segments an energy-dependent uptake of anions (chloride, sulfate and succinate), of amino acids (leucine, α -aminoisobutyric acid) and of sugars (glucose, sucrose [27]). These effects, although much smaller than the major ones observed for H^+ extrusion and K^+ uptake, are still quite significant, ranging from +30 to +80%, and are not influenced by increasing the osmolarity of the medium to values close to the plasmolysis point. Thus these effects are independent of those of FC on cell enlargement and water uptake. The data however are still incomplete and more work is required to approach the problem of whether they represent secondary consequences of the activation of the H^+/K^+ - PD system. Tentatively, the stimulation of Cl^- uptake might be interpreted as depending on the creation of a H^+ gradient across the membrane, as both the hypothesized H^+-anion symport or the ^-OH/anion antiport (see above) would be favoured by low external and high internal pH values. As a matter of fact, some stimulation of anion uptake by low external pH has been reported in various materials [27, 28, 29].

The relatively large effect of FC on glucose uptake (as 3-0- methylglucose) might be suggestively interpreted as depending on a glucose $-H^+$ symport utilizing the FC-induced increase of the electrochemical proton gradient, according the mechanism beautifully worked out in Neurospora [30] and in Chlorella [31]. Unfortunately, the results till now available in this regard are not very encouraging, since we have not been able to show well defined effects of changing pH and salt concentration in the medium on both normal and FC-induced 3-0- methylglucose uptake in our materials.

However, the problem remains open, and intensive work is being carried out on this interesting point, as well as on that of the mechanism of the FC-induced increase of amino acid uptake.

g) Relationship between the effects of FC on ion transport and on metabolism

FC has been shown to markedly enhance dark CO_2 fixation in oats coleoptile [32] and in pea internode segments [33]. In both cases, most of the fixed radioactivity is localized into malate. The simplest interpretation of this result is that alkalinization of the cytoplasm due to the stimulation of H^+/K^+ exchange increases the rate of PEP carboxylation, as postulated by the pH-stat theory of Raven and Smith [34] and of Davies [35]. On the other hand, the observed changes of other metabolic parameters suggest that FC induces a complex pattern of possibly far-reaching effects on respiratory metabolism (Table II). The strong increase of pyruvate level in pea internode segments might be interpreted as depending on the increased rate of NADP-dependent decarboxylation of malate to pyru- vate; the consequent increase of the NADPH/NADP ratio might explain the inhibition of Glc-6-P oxidation to Ru-5-P, and thus the increase of Glc-6-P and the decrease of the C_1/C_6 ratio. A difficulty arising from this interpretation is that, in order to explain the simultaneous FC-induced activation of PEP carboxylation and of NADP-dependent malate decarboxylation, one has to postulate a network of allosteric effects about which almost no data are available.

h) Model for FC action

The model of Fig. 1 represents an attempt to coordinate the main known responses of higher plant tissues to FC, according to the considerations and hypotheses outlined in this section. In this model, all FC-induced effects are interpreted as consequences of a single primary activation of an energy-dependent system catalizing electrogenic H^+ extrusion or H^+/K^+ exchange.

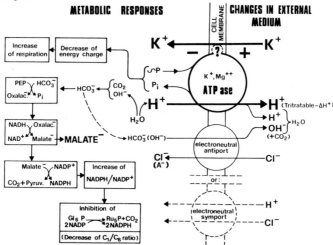

Figure 1 - Model for the FC-activated electrogenic H^+/K^+ exchange system.Activation of the proton pumping ATPase by FC would determine the hyperpolarization of PD,only partially neu- tralized by simultaneous K^+uptake. The coupling between H^+extrusion and K^+uptake might be either electrogenic or due to electrogenic H^+/K^+ countertransport. Part of the extruded pro- tons would disappear from the medium because of either reentry with Cl^-(electroneutral sym- port),or association with secreted OH^- or HCO_3^- (electroneutral countertransport).

TABLE II - Similarities between the effects of IAA and FC in pea internode segments.

	Controls	IAA	FC
Cell enlargement	5	15	28
Q_{O_2}	180	225	260
C_1/C_6 ratio	0.99	0.93	0.73
Glucose-6-P	0.53	0.59	0.68
Pyruvate	0.060	0.088	0.350
Dark CO_2 fixation	27.5	33.0	43.6
K^+ uptake (1 mM)	0.48	0.96	2.20
Na^+ uptake (1 mM)	0.13	0.16	0.26
K^+/Na^+ uptake ratio	3.7	6.0	8.5
Cl^- uptake (10 mM)	1.7	2.6	3.9
SO_4^{2-} uptake (5 mM)	0.59	0.67	1.31
Titratable H^+ extrusion (1 mM KCl)	0.16	0.40	0.89
H^+ extrusion activation by 10 mM KCl	+187%	+273%	+317%
H^+ extrusion activation by 10 mM NaCl	+37%	+67%	+54%
3-0-methylglucose uptake (5 mM)	0.37	0.44	0.51
α-aminoisobutyric acid uptake (5 mM)	0.16	0.19	0.24
Transmembrane electric potential	-60	-77	-83

Cell enlargement as percent increase of fresh weight in 2 hours. Q_{O_2} as μl CO_2 taken up x g fr.w.$^{-1}$ x h^{-1}. Glucose-6-P and pyruvate as μmoles x g fr. w.$^{-1}$ at 2 h of treatment. CO_2 fixation as n moles CO_2 incorporated in the acid soluble fraction x g fr. w.$^{-1}$ in 30 min. H^+ extrusion and uptake of ions, 3-0-methylglucose and α-aminoisobutyric acid as μeq x g^{-1} x h^{-1}. Transmembrane electric potential as mV. From Marrè [37].

II) DO AUXIN AND OTHER HORMONES INFLUENCE ION TRANSPORT BY REGULATING THE SAME H^+/K^+ TRANSPORT SYSTEM ACTIVATED BY FC ?

This question is important both from a theoretical and from an operational point of view. If a same system mediates the effects of FC, of IAA and of other hormones and regulating factors on ion transport, then the available evidence would lead to the con-clusion that practically all higher plant tissues contain a main H^+/K^+ exchange system, whose rate of operation is regulated by hormonal or, in general, by internal physiological stimuli. On the other hand, if the H^+/K^+ exchange system activated by FC is different from that activated by IAA (and/ or by other hormones) then FC would represent a powerful tool to understand various aspects of ion transport, electrogenesis, relationship between pH in the wall space and cell enlargement, etcetera, but the analysis of its mode of action would only marginally improve our knowledge about the mode of action of hormones on ion transport.

In my opinion, the available data favour the view that the FC-activated system is the same influenced by auxins, and probably also by other hormones. In fact, the data of Table II concerning pea internode segments (the material most thoroughly investigated in this respect) show that the process activated by IAA shares with the one sensitive to FC the following characteristic : a) simultaneous activation of H^+ extrusion, K^+ uptake and PD hyperpolarization ; b) dependence of H^+ extrusion on the uptake of monovalent cations; c) higher selectivity for K^+ as compared to Na^+ and other cations as far as ion uptake and the effect on H^+ extrusion are concerned; d) minor activating effects on the uptake of other solutes such as Cl^-, SO_4^{2-}, amino acids and glucose; e) stimulation of carboxylative malate synthesis accompanied by an increase of pyruvate; f) apparent inhibition of the oxidative part of the hexosephosphate shunt, indicated by the decrease of the C_1/C_6 ratio and, perhaps, also by the modest but significant increase of Gl_6P. It may be observed that quantitatively the various responses to each of the two phyto-regulators are significantly different, but this may be easily explained as a consequence of the much lower activity of IAA as compared to FC. Other ion transport systems not directly influenced by either IAA or FC are presumably operating in the untreated tissues. Therefore one would expect that this " background " activity is quantitatively affected in a somewhat different way by the different degrees of activation of the ion transport mechanism by IAA or, respectively, by FC.

In conclusion, FC and IAA actions on ion transport and related responses appear to be mediated either by a single system, or by two different systems extremely similar and so far almost indistinguishable. The hypothesis of a single mechanism is simplest, and thus seems preferable, were it only for methodological reasons.

Substantially similar considerations may be drawn concerning the system(s) mediating the effects of other natural hormones and physical regulating factors on ion transport, even if in these cases the evidence is still by far too incomplete to allow a detailed analysis. The positive effects of cytokinin on H^+ extrusion and PD in isolated cotyledons[10], the inhibitory action of ABA on H^+/K^+ exchange in stomata guard cells (easily reversed by FC)[13], the inhibition by red light of auxin-stimulated H^+ extrusion in coleoptiles[14] and possibly, some important effects of GA_3 and light on ion transport development in germinating seeds[16] may be conveniently explained - at least in the absence of new data - as depending on the capacity of these factors to influence in some way the same system activated by FC.

On the other hand, there are good reasons to believe that the FC-sensitive H^+/ K^+ exchange system normally operates in all physiologically conditions, even in the absence of treatments with hexogenous regulating factors. Mechanisms controlling and coordinating electrogenesis, proton extrusion and selective cation uptake must operate in any cell, if cell omeostasis has to be maintained. The identification of one of these basal mechanisms with the FC-sensitive system is suggested by the finding that H^+ extrusion, selective cation uptake and the PD value usually respond quantitatively to changes in the experimental conditions in the same way independently of the presence or the absence of FC or of IAA (see, for example, in Fig. 2, the similar effects of cations on H^+ extrusion with and without FC or IAA).

It is true that plant materials, when incubated at pH close to neutrality in the absence of stimulating factors usually show little extrusion, or even some uptake of protons, while they actively absorb K^+ and Cl^-. This finding, however, is not in contrast with the operation of a H^+/ K^+ exchange mechanism, if one accepts that a consistent fraction of H^+ secreted by the proton extruding system desappears from the medium because of either Cl^-/ OH^- exchange or $H^+ - Cl^-$ symport, as postulated in the model of Fig. 1.

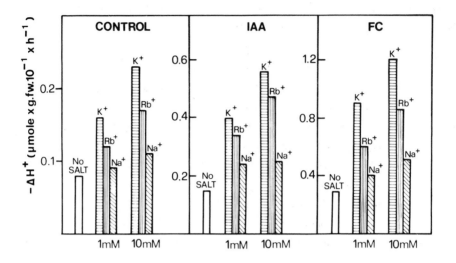

Figure 2 - Activating effect of K$^+$, Rb$^+$ and Na$^+$ (as chlorides) on basal and on IAA- or FC-stimulated proton extrusion ($-\Delta H^+$) in pea internode segments. 2 mm long segments from the growing part of the distal internode of etiolated pea seedlings were prepared, and ΔH^+ was measured as described by Lado et al.[21].

III) MODE OF ACTION OF FC, AUXIN AND OTHER HORMONES ON THE H$^+$/ K$^+$ EXCHANGE SYSTEM

As discussed in the above section, the analogies between the effects of FC and those of IAA and other natural regulatory factors on ion transport suggest that the same transport system is involved in all cases. This conclusion, however, leaves completely open the possibility of different mechanisms of action of FC, IAA and other factors on this system. The available evidence indeed strongly favours the hypothesis of different primary receptors and different modes of action for the different classes of factors regulating ion transport.

For FC and for IAA - on the action of which more data are available -- this conclusion is clearly supported by the findings summarized in Table III, showing that:

a) FC is active on almost all plant tissues and organs, while IAA is much more tissue-specific in its action on ion transport; b) the effects of IAA on ion transport occur after a measurable lag, while those of FC are practically immediate; c) IAA induces "in vivo" a series of morphogenetic and biochemical responses (cell division, differentiation, root development, ethylene synthesis) which are not induced by FC under the same experimen-tal conditions; d) the effects of IAA on ion transport (as well as on growth) are practi-cally suppressed by protein synthesis inhibitors, while a consistent part of FC action is still detected even under conditions of 95% inhibition of protein synthesis.

TABLE III - Differences between the effects of IAA and FC

	IAA	FC
Tissue specificity (a)	Stems and coleoptiles	Stems and coleoptiles roots, leaves, cotyledons, seed embryos, stomata guard cells
Kinetics (b)	Lag period of minutes	No lag period
Ethylene synthesis (c)	Stimulation	No effect
Cell multiplication in tissue cultures (d)	Stimulation	No effect
Rooting (e)	Stimulation	No effect
Sensitivity to protein synthesis inhibitors (f)	High	Low
Site of binding "in vitro" (g)	Endoplasmic reticulum-enriched fraction	Plasmalemma-enriched fraction

(a) For the effects on H^+ extrusion and K^+ uptake

(b) For the effects on electric potential difference and H^+ extrusion [3, 4, 8, 9, 17]

(c) For concentrations of IAA and FC above 10^{-4}M (pea internode segments [36, 37])

(d) Sycamore cell culture [38]

(e) Vine stems and bean hypocotyls [39]

(f) Effects of actinomycin, cordycepin, cycloheximide, puromycin and amino acid analo-logues on cell enlargement, H^+ extrusion and K^+ uptake (pea internode and maize coleoptiles [40, 41])

(g) Membrane preparations from etiolated maize coleoptiles [42].

Another less direct indication suggesting different primary receptors comes from recent work on the capacity of auxin and of FC to bind to cell fractions obtained by differential centrifugation and density gradient fractionation : maximum IAA binding occurs in the ER-enriched fraction, while FC binds to the plasmalemma-enriched fraction, and no cross competition is observed between the two compounds .

The simplest interpretation of these findings is that FC and IAA interact with different primary receptors and that they influence the system responsible of H^+/K^+ exchange through different pathways. For FC, a relatively direct, and specific action on ion transport and PD is suggested by the lack of a lag period, the very wide range of tissue-specificity, the apparent independence on protein synthesis and the preliminary evidence of the capacity to activate " in vitro " the K^+-dependent ATPase from plasmalemma - enriched preparations.

In contrast, the effect of IAA on ion transport seems mediated by a relatively long and complex chain of steps, possibly involving protein synthesis, and including branching points leading to sequences of reactions able to regulate physiological functions other than H^+/K^+ exchange. This interpretation would satisfactorily account for : a) the lack of action of IAA on ion transport in a number of higher plant tissues (due to the absence in these cases of some terminal link in the chain); c) the capacity of IAA (and not of FC) to induce effects different from those on ion transport (due to branching points in the chain); d) the much greater sensitivity of IAA action to protein synthesis inhibitors (due either to the involvement of some short lived protein(s) in the chain, or, alternatively, to the induction of the synthesis of specific protein(s)).

A similar interpretation may by given also to the effects of other hormones, such as GA_3 ABA and cytokinin, and of light. Here again, the indirect, complex nature of the mechanism of the action on ion transport is indicated by the finding that all these natural regulating factors show some tissue-specificity in their action on H^+/K^+ exchange and influence other phenomena than ion transport (see, for a more detailed discussion [43]).

On the basis of the above-reported considerations, a working hypothesis on the mode of action of FC and of natural regulators on ion transport is schematically presented in Fig. 3.

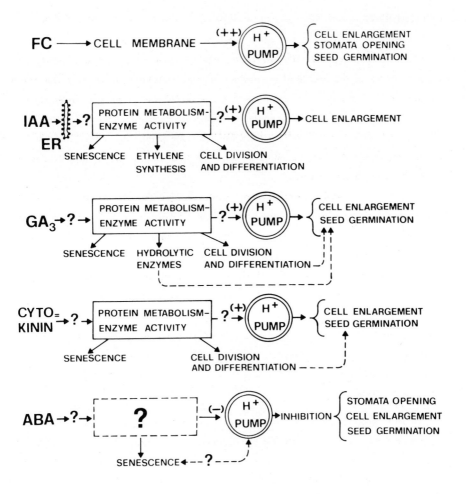

Figure 3 - Model for the different modes of action of FC, IAA and other hormones on the H^+/K^+ exchange system and associated responses. Among the hormonal effects those only are considered in the scheme which : a) may be interpreted as direct consequences of H^+/K^+ exchange activation or inhibition; or, b) may be induced in at least some mate-rials without a simultaneous detectable effect on H^+/K^+ exchange.

CONCLUSIONS

The evidence discussed in this paper suggests that in all higher plant tissues there exists a system that catalizes the metabolic, energy-dependent electrogenic extrusion of protons coupled to the uptake of monovalent cations, and preferentially K^+. The rate of operation of such system is an important factor in the control of transmembrane electric potential and of intra- and extracellular pH and ionic composition, and, consequently, of all other cell activities in any way dependent on these parameters. Among these are cell turgor, wall plasticity, organic acid metabolism and, probably, various aspects of solute (included glucose) transport.

The system is normally operating at a suboptimal rate, and can be either activated or inhibited by internal stimuli, natural hormones and physical factors. The action of hormones on the H^+/K^+ exchange system appears rather indirect, and may explain only a limited part of their specific physiological activities.

The further understanding of the physiological implications of this mechanism depends primarily upon the recognition of its biochemical characteristics, including its structural localization, the source of energy, the nature of the coupling between H^+ extrusion, cation uptake and electrogenesis and the rôle of substrates and allosteric factors in its regulation.

V) AKNOWLEDGEMENTS

The data reported here represent the combined efforts of a group of individuals including A. Ballarin-Denti, N. Beffagna, R. Cerana, M. Cocucci, M.C. Cocucci, S. Cocucci, R. Colombo, M.I. De Michelis, P. Lado, G. Lucchini, P. Pesci, M.C. Pugliarello, F.Rasi-Caldogno and A. Scacchi.

Thanks are due to Prof. A. Ballio of the Department of Biochemistry, University of Rome, for the generous gift of fusicoccin.

REFERENCES

1. Marrè,E., Lado,P., Rasi-Caldogno,F. and Colombo, R.(1973) Plant Sci. Lett. 1, 185-192.

2. Marrè,E., Lado,P., Ferroni, A. and Ballarin-Denti,A. (1974) Plant Sci.Lett. 2, 257-265.

3. Marrè, E., Lado,P., Rasi-Caldogno,F, Colombo,R., Cocucci,M.and De Michelis M.I. (1975) Physiol. Vég. 13, 797-811.

4. Cleland,R. and Lomax,T. (1977) in Regulation of Cell Membrane Activities in Plants (Marrè,E. and Ciferri,O. Eds.) Elsevier North-Holland, Amsterdam.

5. Yamagata,Y. and Masuda, Y. (1975) Plant Cell Physiol. 16, 41-52.

6. Rayle,D.L. and Cleland,R. (1976) in Current Topics of Developmental Biology (Moscona,A.A. and Monroy,A. Eds.) in press, Academic Press, New York.

7. Cleland, R. (1976) Biochem.Biophys. Res. Commun. 69, 333-338.

8. Pitman, M.G., Schaefer,N. and Wildes, R.A.(1975) Planta 126, 61-73.

9. Pitman, M.G., Schaefer,N. and Wildes, R.A.(1975) Plant Sci.Lett. 4, 323-329.

10. Marrè,E., Colombo,R., Lado,P. and Rasi-Caldogno,F.(1974) Plant Sci. Lett,2,139-150.

11. Squire,G.R. and Mansfield, T.A. (1972) Planta 105, 71-78.

12. Raschke, K. (1975) Ann. Rev. Plant Physiol. 26, 309-340.

13. Raschke,K. (1977) in Regulation of Cell Membrane Activities in Plants (Marrè, E. and Ciferri,O. Eds.) Elsevier North-Holland, Amsterdam.

14. Lürssen, K. (1976) Plant Sci. Lett. 389-399.

15. Lado,P., Rasi-Caldogno,F.,and Colombo, R.(1975) Physiol.Plant. 34, 359-364.

16. Cocucci,S. and Cocucci,M. (1976) Paper-Demonstration at "The 9[th]Inter.Conference on Plant Growth Substances",Lausanne 1976.

17. Cocucci,M., Marrè,E., Ballarin-Denti,A. and Scacchi, A.(1976) Plant Sci. Lett. 6, 143-156.

18. Hoffer,V.,Lehninger,A.L. and Leunarz,W.G.(1970) J.Membrane Biol. 3, 142-155.

19. Beffagna,N.,Cocucci,S.and Marrè,E.(1977) Plant Sci. Lett. 8, in press.

20. Marrè,E., Lado,P., Rasi-Caldogno,F.,Colombo,R. and De Michelis, M.I. (1974) Plant Sci. Lett. 3, 365-379.

21. Lado,P., Rasi-Caldogno,F., Colombo,R., De Michelis, M.I. and Marrè, E. (1976) Plant Sci. Lett. 7, 199-209.

22. Marrè,E. (1977) in Proceedings of C.N.R.S. International Workshop on "Transmembrane Ionic Exchanges in Plants"(Ducet,R., Heller,R. and Thellier,M.Eds)in press.Paris

23. Lado,P., De Michelis,M.I., Cerana,R.and Marrè, E. (1976) Plant Sci.Lett. 6, 5-20.

24. Pitman,M.G. (1977) in Regulation of Cell Membrane Activities in Plants (Marrè, E. and Ciferri, O. Eds.) Elsevier North-Holland, Amsterdam.

25. Ritchie,M.G. (1971) Curr.Topics in Bioenergetics, Ed. by D.R. Sanardi,4,327-356.

26. Poole, R.J. (1974) Can. J. Bot. 52, 1023-1025.

27. Lado, P., Rasi-Caldogno, F., Colombo, R. and De Michelis, M.I. (1976) Paper-Demon-stration at "The 9th Intern.Conference on Plant Growth Substances, Lausanne 1976.

28. Rubinstein, B. (1974) Plant Physiol. 54, 835-839.

29. Pitman, M.G.(1970) 45, 787-790.

30. Slayman, C. and Slayman, C.W. (1974) Proc. Nat. Acad. Sci. U.S.A. 71, 1935-1939.

31. Komor, E. and Tanner, W. (1974) J. Gen. Physiol. 64, 568-581.

32. Johnson, K.D. and Rayle, D.L. (1976) Plant Physiol. 57, 806-811.

33. Lucchini, G. (1976) Giorn. Bot. It., 110, in press.

34. Raven, G.A. and Smith, F.A. (1974) Can. J. Bot. 52, 1035-1048.

35. Davies, D.D. (1973) Symp. Soc. Exp. Biol. 27, 513-529.

36. Lado, P., Rasi-Caldogno, F., Pennacchioni, A. and Marrè, E. (1973) Planta 110, 311-320.

37. De Michelis, M.I. and Lado, P. (1974) Rend. Accad. Naz. Lincei, 56, 808-813.

38. Rollo, F., Nielsen, E. and Cella, R. (1977) in Regulation of Cell Membrane Activi-ties in Plants (Marrè, E. and Ciferri, O. Eds.) Elsevier North-Holland, Amsterdam.

39. Lado, P. et al., unpublished data.

40. Lado, P., Rasi-Caldogno, F. and Colombo, R. Plant. Sci. Lett., in press.

41. Cocucci, M.C. et al., unpublished data.

42. Dohrman, U. et al., unpublished data.

43. Marrè, E. (1976) Proceedings of "The 9th Intern. Conference on Plant Growth Sub-stances, (Pilet, P.E. Ed.) in press, Lausanne.

Regulation of Cell Membrane Activities in Plants
E. Marrè and O. Ciferri eds.
© *1977, Elsevier/North-Holland Biomedical Press, Amsterdam*

AUXIN EFFECTS ON THE IONIC RELATIONS

OF PETROSELINUM CELL CULTURES

Friedrich W. Bentrup, Wolfgang Gutknecht and Helmut Pfrüner

Abteilung Biophysik der Pflanzen
Institut für Biologie I der Universität
D-7400 Tübingen

INTRODUCTION

Control of cell extension in a higher plant tissue by auxins, like indole-3-acetic acid(IAA) or 2,4-dichlorophenoxyacetic acid (2,4-D), presumably involves ion transport regulation across the plasma membranes[1]. An analysis of this mode of hormone control requires that transmembrane fluxes of individual ion species, as well as electrical potential and conductance can be studied at the cellular level. For this purpose we have adopted heterotrophically growing suspension culture cells derived from a root callus of the Parsley, Petroselinum sativum. Plate 1A shows a sample of these cells which, as a rule, grow in clusters of 4 to 8 cells.

Previously we have reported that IAA affects the influx of Cl^-, but not of K^+, into Petroselinum cells[2]. This paper communicates an effect of 2,4-D upon the influx of Na^+, and of IAA upon the membrane potential(E_{vo}).

MATERIALS AND METHODS

Axenic suspension cultures of Petroselinum sativum were grown in an aerated nutritional medium of 135 mM total osmolarity including 20 mM K^+, 0.22 mM Na^+, 3 mM Ca^{2+}, 5.9 mM Cl^-, 10^{-5}M IAA, and $5 \cdot 10^{-7}$ M 2,4-D; pH was 4.9, temperature 20°C. For experiments either this medium (Table 1, Fig.2) or the tenfold diluted one (Fig.1) was used. The cells were transferred to the hormone-free medium 24 hours before the experiments. Details of culture and of the employed tracer ion flux technique are given in Ref.3, of the electrophysiological technique in Ref.4. Plate 1B shows a cell with a microelectrode inserted apparently into the vacuole which is partly concealed by cytoplasmic particles.

RESULTS AND DISCUSSION

1. Effects of 2,4-D upon Na$^+$ Uptake. Continuous growth of Petroselinum cell cultures depends upon the presence of auxin in the
medium. Analysis of the intracellular, i.e. by volume essentially
vacuolar, ion content of cells equilibrated in the tenfold diluted
culture medium yields 362 mM Na$^+$, 115 mM K$^+$, and 13.7 mM Cl$^-$. Gran-
ted a reasonable increase of the Na$^+$ concentration in the Donnan
Free Space of the cell wall, i.e. from 0.022 to 0.1 mM Na$^+$, a value
of E_{Na} of -206 mM is obtained which by far exceeds the average value
of E_{vo} of -50 mV. Thus Na$^+$ is accumulated against a large electrochemical gradient by these cells. This could be accomplished by a
H$^+$/Na$^+$ exchange through the putative ubiquitous ATP-fueled proton
export pump in the plasmalemma. If this pump exist in Petroselinum
and is chemically controlled by auxin[5], a stimulation of the Na$^+$
influx by auxin should occur. Fig.1 shows that 10^{-5}M of the synthe-

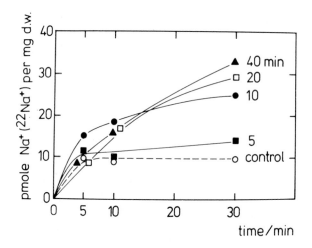

Fig.1. Time-course of the uptake of ^{22}Na$^+$ by Petroselinum cells
preincubated with 10^{-5}M 2,4-D. The nuclid was added at t = 0, the
hormone at times before t = 0, given at the curves.

Plate 1. (A) Light micrograph of Petroselinum sativum cells which
have been grown heterotrophically in a suspension culture.
(B) Single cell with a glass microelectrode inserted from the right
hand side. The dark vertical stripe indicates the glass barrier
holding the cell during impalement.

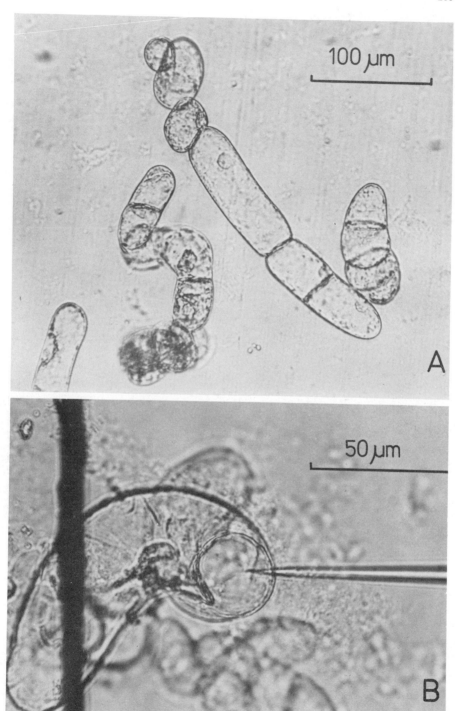

tic auxin 2,4-D increases the net uptake of $^{22}Na^+$ by <u>Petroselinum</u>. This stimulation appears as early as 20 min after addition of the hormone: if 2,4-D is added 10 min before $^{22}Na^+$, the increased uptake of the latter can be measured already after another 10 min. We have also found a stimulation of Na^+ uptake by IAA. Other experiments indicate, however, that IAA and 2,4-D might not always act identically upon Na^+ fluxes, whereas they so far do upon Cl^- fluxes.

Although presently no reliable conclusion is possible, a chemically coupled H^+/Na^+ exchange might be invoked similar to the H^+/K^+ exchange proposed for <u>Helianthus</u> hypocotyl cells[6], and <u>Avena</u> coleoptile cells[7]. (A cotransport of Na^+ with H^+ seems unlikely: A rough calculation yields that the inward proton motive force across the plasmalemma, $E_H + E_{co}$, might reach but not notably exceed the force driving Na^+ outward, $E_{Na} - E_{co}$.)

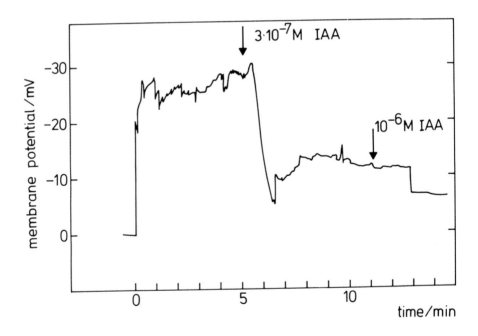

Fig.2. Time course of the membrane potential, E_{vo}, of a <u>Petroselinum</u> cell upon addition of IAA as indicated. Impalement was commenced at t = 0 and finished at t = 13 min.

2. IAA-dependent Changes in Membrane Potential. An electrophysiological study of the Petroselinum cell reveals that the membrane
potential, E_{vo}, is governed by diffusion of K^+ and Cl^-, but only
insignificantly of Na^+ and H^+; the putative electrogenic H^+ export
pump appears to be shunted by the diffusional regime[8]. Explicitely,
a permeability ratio, P_{Cl}/P_K, of about one can be derived from the
Goldman equation for E_{vo}. Similarly low values of E_{vo} have been recorded from suspension culture cells of Acer[9].

Addition of IAA depolarizes the membrane as shown by the trace
from a representative cell in Fig.2. The magnitude, rate and latency
time of depolarization depends upon the concentration of IAA. Contrary to the magnitude, the rate of depolarization and the latency
time turned out to be clearly related to the concentration of IAA
added (Table 1). Obviously, at 10^{-4}M IAA a response of E_{vo} is already recorded after 7 sec. But there is no evidence from these ex-

TABLE 1

Depolarization by Indole-3-acetic Acid of the Plasmalemma
of Petroselinum sativum Cells

IAA concentration (M)	Latency period[§] (sec)	Rate of depolarization ($\mu V \cdot sec^{-1}$)
10^{-11}	98 ± 8	6 ± 2
10^{-9}	71 ± 5	5 ± 6
10^{-7}	46 ± 5	12 ± 5
10^{-5}	16 ± 2	26 ± 9
10^{-4}	7 ± 1	31 ± 5

[§]after addition of IAA; figures include the S.E.M.

periments that IAA triggers this response by the same molecular
mechanism as it does in cell wall extension. Since E_{vo} does not
significantly respond to the pH of the test medium, IAA unlikely
acts through the pH. In fact, the IAA-dependent depolarization
slightly decreases, if the pH is by one unit either lowered or increased. A similar response to the pH has been noted for the IAA-
dependent decrease in Cl^- uptake[2].

It must be stressed that the interpretation of IAA-induced changes in membrane potential require appropriate ion flux and electrical conductance measurements- particularly in <u>Petroselinum</u>, because IAA may also hyperpolarize this membrane in media of high ionic strength and auxin concentrations below 10^{-7}M. Special attention is to be focussed on Ca^{2+}: In <u>Zea</u> coleoptile cells IAA interferes with the action of Ca^{2+} upon E_{vo} and may also cause either hyper- or depolarization[10]. A key role of Ca^{2+} in the primary action of auxins is suggested by the notion that IAA-induced acidification depends upon the presence of Ca^{2+} in the medium of suspension cultures of <u>Acer</u> cells[11] and <u>Avena</u> coleoptiles[12].

This work was supported by the Deutsche Forschungsgemeinschaft.

REFERENCES

1. Van Steveninck, R.F.M. (1976) Encyclopedia of Plant Physiology New Series (A. Pirson and M.H. Zimmermann, editors) Vol. 2B, pp. 307-342. Springer Berlin-Heidelberg-New York.
2. Bentrup, F.W., Pfrüner, H., Wagner, G. (1973) Planta <u>110</u>, 369-372.
3. Pfrüner, H. (1972) Diplomarbeit Universität Tübingen.
4. Bentrup, F.W. (1970) Planta <u>94</u>, 319-332.
5. Hager, A., Menzel, H., Krauss, A. (1971) Planta <u>100</u>, 47-75.
6. Marrè, E., Lado, P., Rasi-Caldogno, F., Colombo, R., De Michelis, M.I. (1974) Plant Science Letters <u>3</u>, 365-379.
7. Haschke, H.-P., Lüttge, U. (1975) Plant Physiol. <u>56</u>, 696-698.
8. Gutknecht, W., Bentrup, F.W., in preparation.
9. Heller, R., Grignon, C., Rona, J.-P. (1974) Membrane Transport in Plants (U. Zimmermann and J. Dainty, editors) pp. 239-243. Springer Berlin-Heidelberg-New York.
10. Nelles, A., Müller, E. (1975) Biochem. Physiol. Pflanzen <u>167</u>, 253-260.
11. Fisher, M.L., Albersheim, P. (1973) Plant Physiol. Suppl. <u>51</u>, 2.
12. Cohen, J.D., Nadler, K.D. (1976) Plant Physiol. <u>57</u>, 347-350.

Regulation of Cell Membrane Activities in Plants
E. Marrè and O. Ciferri eds.
© *1977, Elsevier/North-Holland Biomedical Press, Amsterdam*

AUXIN, PROTON PUMP AND CELL TROPHICS

V.V. Polevoy and T.S. Salamatova
Department of Plant Physiology and Biochemistry
Leningrad State University
Leningrad I99I64, USSR

SUMMARY

It has been shown experimentally, that auxin prevents acidosis in tissues of maize coleoptile and mesocotyle sections, which is accompanied by a delay of RNA, DNA and protein destruction, delay of the respiration rate decrease, delay of acid-soluble SH-group content decrease and by increased inclusion of ^{14}C-2-uracil into RNA.

Suggestion has been made of a scheme of auxin action mechanism on the active H^+ ion transport. According to this scheme, auxin increases throughput of the reduced redox chain in the plasmalemma, which interacting with the redox chain of mitochondria, is functioning as a proton pump, directed outward. This process is accompanied by the reversion of electron transport in the mitochondria. This results in a decreased tissue acidity, increased reducing power, and entering of cations into the cytoplasm, in exchange for H^+ ions. All these factors stabilize intracellular structures, activate metabolism, supporting a favourable balance between anabolism and catabolism. We believe that the proton pump appeared when life came into being, eliminating H^+ ions produced in the primary cell in the process of anaerobic fermentation of organic substances.

INTRODUCTION

In this paper we should like to draw attention to some processes which develop not outside but inside the plant cells with activated auxin-dependent proton pump.

It is known that ion balance in the cells is a determining factor in many aspects of metabolism (I-3). Of great importance in this balance are H^+ ions, which are present in the cytoplasm in concentrations comparable to those of microelements (about 10^{-7} M). H^+ ions exercise a strong influence on the biocolloid state and enzyme activity. Doubtless, functioning in the cells are mechanisms controlling pH level within some definite allowable limits. We believe, that among those mechanisms, of special importance is the active transport of H^+ ions through biological membranes. This process, according to Mitchell's chemiosmotic theory, is also the basis of membrane transformation of energy, ATP formation, and transport of substances (4,5). The possibility of a proton pump functioning in the plasmalemma of the plant cells has been indicated by Mitchell (4). At first, this idea was put forward for algae cells (6). At present, it becomes to be commonly accepted for higher plants as well (7-I2).

In our laboratory we put forward a hypothesis that the activation

210

of membrane transport of H^+ and other ions is the basis of both the
trigger mechanism starting up the functioning activity of the plant
cell (I3-I8) and its trophics (9,I5,I6,I8-20). The latter manifests
itself in the slower ageing of auxin-dependent isolated tissues.
Trophics, as used in this paper, is a complex of processes responsible for the balance between anabolism and catabolism and thus ensu-
ring structural integrity and functional readiness of the cell.

MATERIALS AND METHODS

Experiments have been carried out with I0 mm coleoptile and mesocotyle sections of 4-day old etiolated maize seedlings (Bucovinsky 3).
As auxin we used methyl esters of indoleacetic acid (mIAA) or 2,4-
dichlorophenoxyacetic acid (m2,4-D). The incubation medium was distilled water. In experiments with inhibitors, 2,6-dichlorophenol in-
dophenolate (2,6-DCPIP) was introduced into the medium simultaneously
with auxin; 2,4-dinitrophenol (2,4-DNP) and $K_3Fe(CN)_6$ 2-3 hours prior
to IAA (2I). For tissue pH determination, use has been made of
homogenate obtained by grinding 20 sections in I5 ml of distilled
water (9,I9). Nucleic acid was made by Schmidt and Tannhauser technique, proteins - by biuret reaction, RNAse activity - according to
Shannon et al. (I9,20). Standard deviation are given in the figures.

RESULTS AND DISCUSSION

Study of the mechanism of IAA action on the proton pump. Study
of the nature of slow biopotential oscillations, induced by auxin in
maize coleoptile sections (I3,I4,I7) has shown that this auxin action depends on the presence of calcium, but not magnesium, and is eliminated by H^+, but not K^+ ions. Taking into account structural similarity between auxin and serotonin, the neuromediator, and the fact
that IAA, like the neuromediator, influences electrophysiological
processes and the functional activity of cells (cytoplasm streaming,
growth by elongation), a conclusion has been made that IAA is able
to induce ion shifts in cells. This general concept has been presented in a scheme (Fig. I) according to which IAA influences metabolism
and growth due to its specific action on transmembrane ion fluxes,
Ca^{++} and H^+ transport first of all (I3,I4,I7).

Then (9,I5-I7,I9-2I), in experiments with maize coleoptile sections, it has been shown that mIAA and m2,4-D does induce H^+ ion excretion into the distilled water (incubation medium), showing specific
dependence between auxin concentration, acidification of incubation
medium and elongation of sections (Fig. 2). Beginning with I973,

Fig. I. Scheme of IAA action on metabolism and cell growth through activation of ion fluxes (I3, I4).

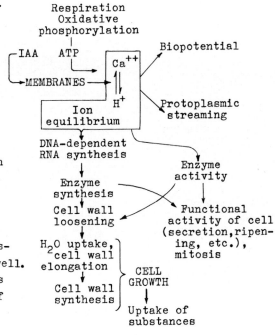

this auxin action was demonstrated with different objects, and it was proved that auxin induces active transport of H^+ ions (IO-I2). But, as seen from Fig. 2, auxin action results not only in the acidification of the external medium, but in the decrease of acidity in tissue sections (9,I5-20),as well. It were these observations that led us to the idea of the role of the proton pump in the regulation of the intracellular pH.

For further studies of the mechanism of auxin action on H^+ ion transport we used different inhibitors (2I). Table I shows, that 2,4-DNP, in concentration O.OI mM, gives 34 per cent increase of O_2 uptake by maize coleoptile sections. This is indicative of the fact that the uncoupler penetrates into the cells and acts on the mitochondria. But 2,4-DNP,when used in the above concentration, does not yet show any inhibitory action on growth and respiration, induced by auxin (2I). Thus, probably, there is no close correlation between ATP generation in the cells and auxin action on the growth (on the proton pump?). On the other hand, 2,6-DCPIP (I.7 mM) and ferricyanide (50 mM) result in a very strong inhibition of mIAA action on the coleoptile section growth, without influencing respiration in the control sections. This allows a consideration that neither electron acceptor penetrates into the cells and both are acting on the outer surface of the plasmalemma. It is known, that 2,6-DCPIP and ferricyanide can take electrons on the flavoproteid level of the reduced redox chain (22). Comparison of our results with Rehm's scheme explaining the mechanism of active H^+ ion transport in the stomach (23), led us to conclusion (2I) that IAA is, probably, increasing the throughput of the reduced redox chain in plasmalemma, which

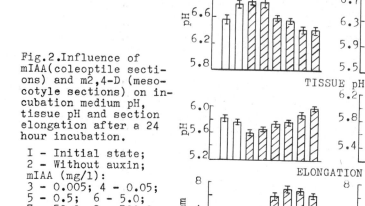

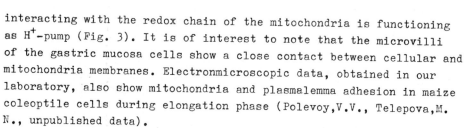

Fig.2.Influence of
mIAA(coleoptile secti-
ons) and m2,4-D (meso-
cotyle sections) on in-
cubation medium pH,
tissue pH and section
elongation after a 24
hour incubation.

I - Initial state;
2 - Without auxin;
mIAA (mg/l):
3 - 0.005; 4 - 0.05;
5 - 0.5; 6 - 5.0;
7 - 50.0; 8 - I00.0;

m2,4-D (mg/l):
3 - 0.0I5; 4 - 0.I5;
5 - I.50; 6 - I5.0;
7 - 50.0

interacting with the redox chain of the mitochondria is functioning
as H^+-pump (Fig. 3). It is of interest to note that the microvilli
of the gastric mucosa cells show a close contact between cellular and
mitochondria membranes. Electronmicroscopic data, obtained in our
laboratory, also show mitochondria and plasmalemma adhesion in maize
coleoptile cells during elongation phase (Polevoy,V.V., Telepova,M.
N., unpublished data).

   Inhibitory action of auxin on cell acidosis and ageing processes.
Fig. 2 shows, that aged sections of maize coleoptiles and mesocoty-
les develop acidosis. This process is delayed or fully prevented by
auxin (mIAA and m2,4-D), in concentrations increasing up to a definite
level. Simultaneously, in the external medium there occurs accumulation of H^+ ions. These ions are unlikely to flow into the external medium from the vacuolar sap. Probably, the proton pump, localized in
the plasmalemma, pumps H^+ from the cytoplasm.

   We observed maximum activity of RNAse in the homogenate from mesocotyle and coleoptile sections at pH 5.3 (9,20). Consequently, it
is possible to think that keeping the intracellular medium from acidification will decrease the activity of acid RNAse and other acid
hydrolases and in this way delay destruction processes.

   And, in fact, according to our data, mIAA (30-I00 mg/l) and m2,4-D
(I-I5 mg/l) inhibit destruction of RNA and DNA in coleoptile and

TABLE I

Effect of IAA (I0 mg/l), 2,4-DNP (0.0I mM), 2,6-DCPIP (I.7 mM),
and $K_3Fe(CN)_6$ (50 mM) on O_2 uptake (µl/I5 sections/4 hours)
and elongation of coleoptile maize sections (mm/I section/5 hrs)

Inhibitors	IAA	O_2	Elongation
2,4-DNP	− −	3I3 ± 30	I.0 ± 0.02
	− +	425 ± I3	2.6 ± 0.I0
	+ −	4I8 ± 33	0.9 ± 0.09
	+ +	572 ± I5	2.5 ± 0.I4
2,6-DCPIP	− −	309 ± 23	0.6 ± 0.08
	− +	383 ± I9	I.9 ± 0.34
	+ −	336 ± I8	0.3 ± 0.I3
	+ +	379 ± I8	0.7 ± 0.29
$K_3Fe(CN)_6$	− −	450 ± 2I	I.2 ± 0.09
	− +	570 ± 28	2.9 ± 0.22
	+ −	475 ± 35	0.2 ± 0.I3
	+ +	440 ± 49	0.5 ± 0.I5

mesocotyle sections (9,I5-20). In mesocotyle sections, m2,4-D (I5 mg/l) delay also the decrease of acid-soluble SH-group content and the respiration rate. Auxin supported a higher level of label incorporation in RNA. But, as seen from Fig. 4, the supporting influence of m2,4-D on the RNA and protein content is connected not so much with the increase of their synthesis, but with the retardation of their decay (I9,20).

It is important to note that in aged sections, which are characterised by increased total (potential) RNAse activity, m2,4-D does not decrease this activity;on the contrary,it increases it (I9,20). It seems that, together with the increased acid RNAse content, auxin creates such physico-chemical conditions in the cells which inhibit its activation. We believe, that such conditions are created by functioning of the proton pump, excreting H^+ ions from the cytoplasm.

Activation of cation uptake. The effect of retarding action of auxin on RNA destruction in the mesocotyle sections may be increased by introducing Mg^{++} or Mn^{++} into the incubation medium. Together with the latter, the effect of m2,4-D is especially strong, and RNA destruction in the sections is almost completely prevented during 24 hours (9,20). Mn^{++} and Mg^{++} show a strong inhibitory effect on RNA destruction in section homogenate as well. K^+ and Ca^{++} used in similar concentrations, do not show such an effect. These data allow a conclusion that divalent cations, entering into the cells in exchange for H^+ ions (antiport, according to Mitchell), favour the stabilization of intracellular structures, ribosome in particular, and activate metabolism.

Maintenance of cell reduction level. As has been already pointed

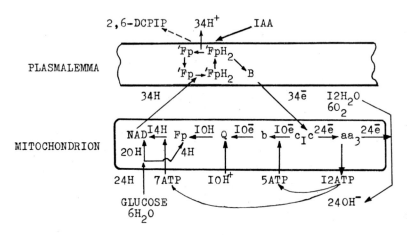

Fig. 3. Scheme of auxin action on the proton pump.

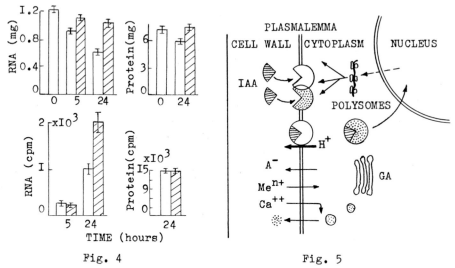

Fig. 4   Fig. 5

Fig. 4. Action of m2,4-D (I5 mg/l) on the RNA content (20 sections) and protein content(40 sections) and on ^{14}C-2-uracil inclusion
into RNA and ^{35}S-methionin inclusion into proteins of mesocotyle
sections. Dark columns - with auxin, light columns - without auxin.

Fig. 5. General scheme of auxin action on the cell growth (28).
A$^-$ - Anions, Me^{n+}- Cations, GA -Golgi apparatus.

out, in the presence of m2,4-D, the decrease of acid-soluble SH-group content in mesocotyle sections has been delayed. This phenomenon
can be understood from the scheme of IAA action on the proton pump
presented in Fig. 3. According to this scheme, the activation of the

proton pump is connected with the reversion of electron transport in the redox chain of mitochondria. This must result in reduction of NAD, NADP, and other compounds.

RNA and protein synthesis. Activation of RNA and protein synthesis due to auxin is a long established fact (24,25). The decreased acidity of the cytoplasm, uptake of cations, some of which (Mn^{++}, Mg^{++}) are activators of RNA polymerase and stabilizers of ribosomes, increased reduction of the intracellular medium must favour synthetic processes.

It has been shown recently that, due to IAA action, cell membranes release some factor ("transcription factor"), which, together with IAA, can activate RNA synthesis in chromatin (26,27). It is not excluded that the "transcription factor" is connected with the receptor starting H^{+}-pump and released from the membrane only by the degree the proton pump is activated (Fig. 5).

Proton pump and life. Data are available suggesting that different adverse factors and dying tissues cause cell acidosis both in animals (29) and plants (Fig. 2). We think, that the proton pump of the cellular membranes, eliminating H^{+} ions from the cytoplasm, is supporting life in the cell.

The proton pump is functioning in bacteria, mitochondria and chloroplasts, in plasmalemma of the plant cells, in the gaster etc. It seems to be a very old physiological mechanism, appearing, probably, in protobionts for the elimination, from the primary cell, of H^{+} ions produced due to anaerobic fermentation of organic substances. We believe the proton pump to be the very mechanism due to which primary protobionts developed non-equilibrium between internal and external medium; it gave the beginning of life.

REFERENCES

I. Dixon, M. and Webb, E. (1964) Enzymes. 2nd Ed., Longmans.

2. Davies, D.D. (1973) Symp.Soc.Exp.Biol., 27, 513-529.

3. Polevoy, V.V. (1975) Vestnik Leningr.Univ.,Ser.Biol.,3, 105-108.

4. Mitchell, P. (1970) In: Organization and Control in Prokaryotic and Eukaryotic Cells, Cambridge, University Press.

5. Skulachev, V.P. (1972) Energy Transformation in Biomembranes, Nauka, Moscow.

6. Kitasato, H. (1968) J.Gen.Physiol., 52, 60-87.

7. Pitman, M.G. (1970) Plant Physiol., 45, 787-790.

8. Hager, A., Menzel, H., Krauss, A. (1971) Planta, 100, 47-75.

9. Polevoy, V.V., Fedoseenko, A.A., Malo, A. (1973) Fiziol.Rast. (Moscow), 20, 499-503.

10. Marrè, E., Lado, P., Rasi Caldogno, F., Colombo, R. (1973)
    Plant Sci.Lett., I, 179-184; ibid.: I, 185-191.

11. Cleland, R. (1973) Proc.Nat.Acad.Sci.USA, 70, 3092-3093; (1975)
    Planta, 127, 233-242.

12. Rayle, D.L. (1973) Planta, 114, 63-73.

13. Polevoy, V.V. (1969) Physiology and Biochemistry of Auxin and
    Gibberellin Action, Doct. Thesis, Leningrad University.

14. Polevoy, V.V. (1972) Trudi Petergof.Biol.Instit., 21, 191-207.

15. Polevoy, V.V. (1972) In: Abstracts of Internat.Conf. on Natural
    Plant Growth Substances, Liblice, Czechoslovakia, 83.

16. Polevoy, V.V., Salamatova, T.S., Malo, A. (1973) Ukrain.Bot.
    Zhurnal, 30, 292-299.

17. Polevoy, V.V., Salamatova, T.S., Maksimov, G.B., Popova, R.K.
    (1974) In: The Growth and Hormonal Regulation of Plant Life,
    Irkutsk, 47-65.

18. Polevoy, V.V. (1975) XII Internat.Bot.Congr.,Abstracts II, 308.

19. Malo, A., Polevoy, V.V., Proshina, R.A. (1974) Biolog.Nauki
    (Moscow), II, 79-85.

20. Malo, A.,and Polevoy, V.V. (1974) In: The Growth and Hormonal
    Regulation of Plant Life, Irkutsk, 151-167.

21. Polevoy, V.V. and Salamatova, T.S. (1975) Fiziol.Rast.(Moscow),
    22, 519-525.

22. Archakov, A.I. (1975) Microsome Oxidation, Nauka, Moscow.

23. Rehm, W.S. (1972) In: Metabolic Pathways, vol. 6, Acad.Press,
    New York a. London, 187-241.

24. Davies, P.J. (1973) Bot. Rev., 39, 139-171.

25. Evans, M.L. (1974) Annual Rev. Plant Physiol., 25, 195-223.

26. Hardin, J.W., Cherry, J.H., Morré, D.J., Lembi, C.A. (1972)
    Proc.Nat.Acad.Sci.USA, 69, 3146-3150.

27. Licholat, T.V., Pospelov, V.A., Morozova, T.M., Salganik, R.I.
    (1974) Fiziol. Rast.(Moscow), 21, 939-945.

28. Polevoy, V.V. and Salamatova, T.S. (1975) In: Biology of Plant
    Development, Ed. by M.Kh. Chailakhian, Nauka, Moscow, 111-125.

29. Dgapharov, A.I., Kols, O.R., Sumarukov, G.V. (1968) In: Physico-
    Chemical Basis of Autoregulation in Cells, Nauka, Moscow,
    112-114.

Regulation of Cell Membrane Activities in Plants
E. Marrè and O. Ciferri eds.
© *1977, Elsevier/North-Holland Biomedical Press, Amsterdam*

FUSICOCCIN: STRUCTURE - ACTIVITY RELATIONSHIPS

Alessandro Ballio
Institute of Biological Chemistry
University of Rome
Italy

Fusicoccin is the major phytotoxic compound isolated from cultures of *Fusicoccum amygdali* Del., the fungus causing canker of peach and almond trees[1,2]. When introduced into an almond shoot, it produces symptoms on leaves closely resembling those that follow an infection of *F. amygdali*[2]. It has been recently demonstrated[3] that fusicoccin is also produced, in amounts high enough to account for toxicity symptoms on leaves, both in tissues of unripe peaches artificially infected by the pathogen and in cankers formed on naturally infected almond shoots; this represents good evidence for a definite role of fusicoccin in pathogenesis.

Preliminary investigations on the mechanism of action of fusicoccin[2,4,5] led to the discovery that this substance, initially isolated as a phytotoxin, affects several important physiological processes in higher plants, such as leaf transpiration through opening of stomata[6-9], cell enlargement of different tissues[10,11], germination of dormant seeds[12], etc. These properties have prompted investigations directed to establish the structural features necessary for the display of biological activity by fusicoccin.

The structure of fusicoccin (I) has been elucidated in 1968 by Ballio *et al.*[13], and soon after confirmed by an independent investigation of Barrow *et al.*[14]. The complexity of this highly asymmetric and polyfunctional molecule makes the total

Fusicoccin (I)

synthesis of analogues quite a difficult task. However, a number of substances
structurally related to fusicoccin were isolated from culture filtrates of
F. amygdali and several others were prepared by chemical modification of fusicoccin itself. Fourteen minor metabolites of the fungus were obtained and their
structures determined[15]: ten of them differ from the major metabolite only in the
number and/or the position of the *O*-acetyl groups, whereas four other compounds
(II-V), besides additional minor modifications have a different oxygenation
pattern in the aglycone moiety. In particular compound (III), which carries an
extra hydroxy group on C-3, has a strict chemical similarity with the cotylenins
(VI), a family of plant growth promoting substances produced by a *Cladosporium* sp.,
recently described by Sassa [16]. Nearly twenty other related substances have been
prepared, chiefly in the course of the structural investigations[13,15], by chemical
modification of fusicoccin: examples are provided by dihydrofusicoccin (which
carries a *t*-pentyl instead of a *t*-pentenyl unit on the glucose moiety), de-*t*-
pentenylfusicoccin (which lacks the C_5 unit on the glucose moiety), deacetyl-9-
*epi*fusicoccin (which has an inverted configuration at C-9), the aglycones of most

	R_1	R_2	R_3	R_4
(II)	$-C(CH_3)_2CH=CH_2$	$-OH$	$-CH_2OCH_3$	$-H$
(III)	$-C(CH_3)_2CH=CH_2$	$-OH$	$-CH_2OCH_3$	$-OH$
(IV)	$-C(CH_3)_2CH=CH_2$	$-OH$	$-H$	$-CH_2OH$
(V)	$-H$		$-H$	$-CH_2OH$	$-H$

Some minor metabolites of *F. amygdali*

Cotylenin E (VI)   Isomer of fusicoccin aglycone (VII)

of the natural fusicoccins and of deacetyl-9-*epi*fusicoccin, and by compound VII
(an isomer of the fusicoccin aglycone)[17].

Reports on the phytotoxic activity of the majority of these substances have
already been published[18,19], and with the exception of dihydrofusicoccin, all
modified fusicoccins have a strongly reduced phytotoxicity on tomato cuttings.

More recently the activity of several natural and chemically modified
fusicoccins has been compared with that of the parent compound in the following
tests: opening of stomata of broad beans, according to Ballio *et al.*[3]; cell
enlargement and proton extrusion in pea stem segments, according to Marrè *et al.*[20];
cell enlargement in isolated squash cotyledons, according to Marrè *et al.*[21];
germination of dormant lettuce and radish seeds, according to Lado *et al.*[12]. Some
analogues have also been tested in the binding assay to membrane preparations of
corn coleoptiles developed by Dohrman *et al.*[22]

Results so far obtained, to be published in detail elsewhere, can be summarized
as follows.

Opening of stomata: 3'-Monodeacetylfusicoccin, dihydrofusicoccin, (II),
cotylenin A, C and E, cotylenol (the aglycone common to all cotylenins) and the
aglycone of (II), are more or less as active as fusicoccin. De-*t*-pentenyl-
fusicoccin, deacetylfusicoccin and its epimer at C-9, (IV), (V), the aglycone of
deacetyl-9-*epi*fusicoccin and (VII) are very weakly active, whereas (III) and the
aglycones of fusicoccin and of (V) show moderate activity. Besides considerations
common to those pertaining to other tests, it appears that the effect on opening
of stomata is, at least in part, dependent on the overall polarity of the molecule.

This is particularly evident on comparing the activities of the different aglycones and on abserving the increased potency of the aglycone of (V) over that of (V) itself.

Cell enlargement and proton extrusion in pea stem segments: Dihydrofusicoccin, de-*t*-pentenylfusicoccin, deacetylfusicoccin, (II), (III), (IV), (V), cotylenins A, C and E, the aglycones of fusicoccin, of (II) and of the cotylenins, are more or less as active as fusicoccin. Deacetyl-9-*epi*fusicoccin, the corresponding aglycone and (VII) are inactive. As expected[20,21,23], proton extrusion was consistently found to parallel stimulation of cell enlargement with all analogues tested. In this test the glucose moiety and oxygen atoms on C-12 and C19 appear to be features not necessary for activity. The positive results with (IV) and (V) are noteworthy as these compounds are moderately or very weakly active in all other tests.

Cell enlargement in isolated squash cotyledons: Dihydrofusicoccin, de-*t*-pentenylfusicoccin, deacetylfusicoccin, (III), cotylenins A, C and E are more or less as active as fusicoccin; cotylenol and (II) are about 50% as active as fusicoccin; (IV) and the fusicoccin aglycone are moderately active, and deacetyl-9-*epi*fusicoccin, the corresponding aglycone and (V) are inactive. Therefore, structural requirements for the enhancement of cell enlargement in cotyledons are greater than in pea stem segments, namely the glucose moiety, the *O*-methyl group on C-16 and the stereochemistry of C-3 appear to be of importance.

Germination of dormant lettuce and radish seeds: The response to the different compounds of lettuce seeds shortly irradiated with far-red light compares well with that of radish seeds treated with abscisic acid, except in the case of the aglycones of fusicoccin and of (II) which display a more rapid effect on the second material. Dihydrofusicoccin, de-*t*-pentenylfusicoccin, deacetylfusicoccin, (III), cotylenins A, C and E and cotylenol are more or less as active as fusicoccin. (II), the aglycone of fusicoccin and that of (II) are less effective, (IV) and (V) are weakly active and deacetyl-9-*epi*fusicoccin, the corresponding aglycone and (VII) are inactive. Qualitatively, the results of this test parallel those obtained in the assay on cell enlargement of cotyledons.

Binding to membrane preparations of corn coleoptiles: Tritiated dihydrofusicoccin was added to membrane preparations together with different concentrations of unlabelled fusicoccin, or one of its analogues, and the pellettable radioactivity was measured after proper incubation. From experimental results the molar concentration producing 50% inhibition of the binding of labelled dihydrofusicoccin was calculated for each analogue. Figures so far obtained with a limited number of compounds are: fusicoccin $3.5 \cdot 10^{-9}$, deacetylfusicoccin $1.0 \cdot 10^{-8}$, (II) $1.5 \cdot 10^{-8}$, fusicoccin aglycone $2.0 \cdot 10^{-6}$ and cotylenin A $2.5 \cdot 10^{-9}$. The specificity of the binding is supported by the complete absence of competition of (VII), an analogue inactive in all physiological tests, as well as of the chemically unrelated abscisic acid which counteracts the effects of fusicoccin in several processes[8,9,12,24,25].

Results so far obtained from studies of structure-activity relationships in the fusicoccin family of compounds are far from being conclusive. Nonetheless they yield some information about the importance of certain features of these complex substances, as for instance the correct configuration on C-9 and the conformational flexibility of the 8-membered ring; in fact, deacetyl-9-*epi*fusicoccin and (VII) (the latter has a nearly "frozen" conformation) have been found inactive in all physiological tests chosen for the present investigation. Furthermore, informations have been obtained about sites of the fusicoccin molecule unnecessary for activity, thus permitting a rational approach to the design of derivatives adequate for the synthesis of stationary phases to use for the isolation of fusicoccin binding molecules by affinity chromatography.

The different response of different tissues observed with certain analogues is probably imputable to factors affecting, in broad terms, the interaction with the receptor, rather than to the existence of different receptors. In fact, data so far produced by Marrè's group and by other laboratories with a number of plant materials, favour the hypothesis[23] that all the physiological effects of fusicoccin depend on the activation of a single mechanism which operates at the cell membrane level and involves the energy-dependent proton extrusion associated with cation uptake.

ACKNOWLEDGEMENTS

Experimental results reported in this paper have been obtained through the cooperation of four laboratories; Centro di Studio del C.N.R. per la Biologia Cellulare e Molecolare delle Piante, Istituto di Scienze Botaniche dell'Università, Milan (P. Lado and M.I. De Michelis); Centro di Studio del C.N.R. per le Tossine e i Parassiti Sistemici dei Vegetali, Istituto di Patologia Vegetale dell'Università, Bari (A. Graniti); Istituto di Chimica Organica dell'Università, Naples (G. Randazzo); Istituto di Chimica Biologica dell'Università, Rome (A. Ballio and R. Federico).

A.B. thanks Dr. T. Sassa, Department of Agricultural Chemistry, Yamagata University, Japan, for generous samples of cotylenins and cotylenol, as well as the Italian National Research Council (C.N.R.) for financial support.

REFERENCES

1. Ballio, A., Chain, E.B., De Leo, P., Erlanger, B.F., Mauri, M. and Tonolo, A. (1964) Nature 203, 297.

2. Graniti, A. (1964) The role of toxins in the pathogenesis of infections by *Fusicoccum amygdali* Del. on Almond and Peach. In host-parasite relations in Plant Pathology. Z. Kiraly and G. Ubriszy, eds., Budapest, 211-217.

3. Ballio, A., D'Alessio, V., Randazzo, G., Bottalico, A., Graniti, A., Sparapano, L., Bosnar, B., Casinovi, C.G., and Gribanovski-Sassu, O. (1976) Physiol. Plant Pathol. 8, 163-169.

4. Ballio, A., Graniti, A., Pocchiari, F. and Silano, V. (1968) Life Sci. 7, 751-760.

5. Bottalico, A. (1969) Phytopath. medit. 8, 41-46.

6. Turner, N.C. and Graniti, A. (1969) Nature 223, 1070-1071.

7. Graniti, A. and Turner, N.C. (1970) Phytopath. medit. 9, 160-167.

8. Tucker, D.J. and Mansfield, T.A. (1971) Planta, 98, 157-163.

9. Squire, G.R. and Mansfield, T.A. (1972) Planta, 105, 71-78.

10. Ballio, A., Pocchiari, F., Russi, S. and Silano, V. (1971) Physiol. Plant Path. 1, 95-103.

11. Marrè, E., Lado, P., Rasi-Caldogno, F. and Colombo, R. (1971) Rend. Accad. Naz. Lincei 50, 45-49; subsequent literature quoted in Lado, P., De Michelis, M.I., Cerana, R. and Marrè, E. (1976) Plant Sci. Lett. 6, 5-20.

12. Lado, P., Rasi-Caldogno, F., and Colombo, R. (1974) Physiol. Plant. 31, 149-152.

13. Ballio, A., Brufani, M., Casinovi, C.G., Cerrini, S., Fedeli, W., Pellicciari, R., Santurbano, B. and Vaciago, A. (1968) Experientia 24, 631-635.

14. Barrow, K.D., Barton, D.H.R., Chain, E.B., Ohnsorge, U.F.W. and Thomas, R. (1968) Chem. Comm. 1197-1198.

15. Ballio, A., Casinovi, C.G., Randazzo, G. and Rossi, C. (1970) Experientia 26, 349-351; subsequent literature quoted in Ballio, A., Casinovi, C.G., Grandolini, G., Marta, M. and Randazzo, G. (1975) Gazz. Chim. Ital. 105, 1325-1328.

16. Sassa, T., Tojyo, T. and Munakata, K. (1970) Nature 227, 379; subsequent literature quoted in Sassa, T. and Takahama, A. (1975) Agr. Biol. Chem. 39, 2213-2215.

17. Casinovi, C.G., Santurbano, B., Conti, G., Malorni, A. and Randazzo, G. (1974) Gazz. Chim. Ital. 104, 679-691.

18. Ballio, A., Bottalico, A., Framondino, M., Graniti, A. and Randazzo, G. (1971) Phytopath. medit. 10, 26-32.

19. Ballio, A., Bottalico, A., Framondino, M., Graniti, A. and Randazzo, G. (1973) Phytopath. medit. 12, 22-29.

20. Marrè, E., Lado, P., Rasi-Caldogno, F. and Colombo, R. (1973) Plant Sci. Lett. 1, 179-184.

21. Marrè, E., Colombo, R., Lado, P. and Rasi-Caldogno, F. (1974) Plant Sci. Lett. 2, 139-150.

22. Dohrman, U., Hertel, R., Pesci, P., Cocucci, S., Marrè, E., Randazzo, G. and

Ballio, A., unpublished data.

23. Marrè, E., Lado, P., Rasi-Caldogno, F., Colombo, R., Cocucci, M. and De Michelis, M.I. (1975) Physiol. Vég. 13, 797-811.

24. Pilet, P.E. (1975) Plant Sci. Lett. 5, 137-140.

25. Halloin, J.M. (1976) Plant Physiology 57, 454-455.

Regulation of Cell Membrane Activities in Plants
E. Marrè and O. Ciferri eds.
© *1977, Elsevier/North-Holland Biomedical Press, Amsterdam*

EFFECT OF RED LIGHT ON THE BINDING OF NAA
ON MAIZE COLEOPTILE MICROSOMES

Guy Normand, Francis Schuber, Pierre Benveniste et Denis Beauvais
Laboratoire de Biochimie Végétale - E.R.A. n°487 du C.N.R.S.
Institut de Botanique
28, rue Goethe - 67087 Strasbourg cédex
France

INTRODUCTION

Time course of cell elongation in response to auxin suggests
that rapid events are implied : proton secretion (1), transmembrane
potential changes (2), glucan synthetase activation (3), etc....
Also auxin polarized transport plays a role during the elongation
process (4). All these events suggest on interaction between auxin
and membrane components. Recently, Hertel et al. (5) demonstrated a
specific interaction of hormones with membranes receptors. This interaction was shown to be specific, reversible and saturable. In our
laboratory we confirmed these findings and the receptor was shown to
be associated with an endoplasmic reticulum enriched fraction (6).
Several authors have observed an interference of light with auxin
action. In coleoptiles, Muir et Chen-Chang (7) observed an inhibitory effect of P_{FR} on elongation. An analogous effect was observed by
Kondo et al. (8) in mesocotyls, an organ in which growth is strongly
promoted by IAA. It was also observed that light decreases polarized
auxin transport in either coleoptiles (9) and mesocotyls (8). In
this latter case auxin immobilization at the level of the node was
considered to be responsive of mesocotyl elongation inhibition. Finally, at the level of transmembrane potential, Tanada (10) showed
that mung bean roots attachment to negatively charged glass promoted
by P_{FR} could be reversed by auxin. In this context it is interesting
that phytochrome has been shown to be bound to membranes (11) and
specifically to endoplasmic reticulum (12).

These facts suggest the possibility of an interaction between
phytochrome and auxin at the membrane level. At the molecular level,
such an interaction could be viewed as a cooperative effect of phytochrome on auxin binding (K effect) or translocation rate (V effect).

Among other possibilities, phytochrome could also act on the number
of available receptor sites. Using microsomes derived from maize
coleoptiles and mesocotyls, we have studied the influence of phytochrome on NAA binding *in vitro*. Under our experimental conditions
no effect could be detected so far.

MATERIALS AND METHODS

$[1-^{14}C]$-NAA (44 mCi/mmole) was a Radiochemical Centre product
(Amersham, U.K.). NAA was purchased from Sigma Chemicals Co (St-
Louis, Mo., U.S.A.).

Maize seeds (*Zea mays* L., variety INRA 170) are allowed to germi-
nats in darkness and at 25°C. For this purpose, plastic boxes containing six layers of wet filter paper are used. After five days,
coleoptiles or mesocotyls are collected.

Light treatment : When given light tratments, the coleoptiles or
mesocotyls   were halved. The control was kept in complete darkness. The coleoptiles or mesocotyls were given 10 min red light
(660 nm) followed or not by 10 min far red light (730 nm) at 25°C
in wetted plastic boxes.

Preparation of microsomes : Coleoptiles or mesocotyls (70 g)
were ground in a mortar for 1 min in the presence of a medium (70 ml)
containing : mannitol 0.5 M, EDTA 1 mM, BSA 0.5 %, tris-HCl 0.1 M
(pH 7.5). The homogenate was filtered through four layers of cheese-
cloth and the filtrate was centrifuged 10 min at 6 000 x g. The
supernatant obtained was made 10 mM in $MgCl_2$ and centrifuged 60 min
at 106 000 x g. Under these conditions it has been shown (13) that
pelletability of phytochrome was maximum. Microsomal pellet was
resuspended with an Elvehjem-Potter in a medium (30 ml) containing :
sucrose 0.5 M, sodium citrate 10 mM, $MgSO_4$ 5 mM, pH 6.0. Appropriate
aliquots of the microsomal suspensions were kept for protein measurement by the Lowry procedure.

Binding assays : The binding assay was carried out at 4°C according to the isotope dilution technique. To each centrifuge tube was
added 1 ml of microsomal suspension, 20 µl ^{14}C -NAA solution
(12 300 dpm) in ethanol containing from 0-18 600 ng non radioactive
NAA. This isotopic dilution corresponds to an NAA concentration, in
the assay tube, going from $1.14 \cdot 10^{-7}$M to 2.10^{-4}M. The content of the
tubes is carefully mixed and the mixture was allowed to equilibrate
for 5 min. The tubes were then centrifuged 30 min at 36 000 x g
(MSE centrifuge). All the preceeding steps besides specific light

treament were carried out in darkness or under green safe light.
The supernatant was discarded and the bottom of the tuber were cut
off and transferred into scintillation vials containing 1 ml EtOH
(5). After 4 hrs, the scintillation cocktail (Bray) was added and
the vials were counted.

 Calculation of dissociation constants and site numbers : The NAA-
receptor dissociation constants (Kd) and the number of sites (lm)
were calculated according to the Scatchard plots :

$$\frac{1}{L} = \frac{lm}{Kd} - \frac{1}{Kd}$$

where 1 = specifically bound ligand concentration
     L = free ligand concentration
1 was estimated from the counts retained in the pellets knowing
that the amount of NAA bound at the highest isotopic dilution
$(2 \cdot 10^{-4} M)$ corresponds to the unspecific binding. From lm was calculated the number (N) of binding sites per mole of microsomal protein
assuming for these latter an average molecular weight of 50 000.
The equation was analysed using a least square method.

RESULTS

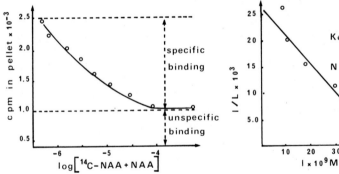
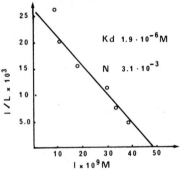

Fig. 1. Binding of NAA to maize
coleoptiles microsomal fractions
using isotope dilution technique
Each point corresponds to an
average of three determinations.

Fig.2. Scatchard representation
of the values derived from Fig.1.

On Fig. 1 and 2 are represented typical binding assays of NAA with its membrane receptor obtained from maize coleoptiles.

The coleoptiles, mesocotyls and the node region of maize coleoptiles were irradiated with red light or red light followed with far red light. For each experiment a dark control was also performed. The NAA receptor dissociation constant (Kd) and the concentration of receptor sites (N) were determined (see methods). The results obtained were shown in table I.

Table I

EFFECT OF RED LIGHT ON PARAMETERS RELATIVE TO THE BINDING
OF NAA WITH ITS MEMBRANE RECEPTOR

	Darkness		Red		Red + Far red	
	Kd (10^{-6}M)	N 10^{-3}	Kd (10^{-6}M)	N 10^{-3}	Kd (10^{-6}M)	N 10^{-3}
Coleoptiles[a]	2.0 ± 0.1	3.1 ± 0.5	2.1 ± 0.1	3.2 ± 0.5	2.2 ± 0.1	2.1 ± 0.5
Mesocotyls[b]	0.75	0.25	0.65	0.20	-	-
Mesocotyl[c] + node	3.1	2.1	4.0	1.8	-	-

a, b, c, Average of respectively 7, 2 and 3 experiments.

Table I shows that red light has no detectable effect on either the dissociation constant or the binding site number in the case of coleoptiles and mesocotyls. In the case of the node region, a slight increase of the dissociation constant following red light irradiation has been observed. However more experiments have to be performed in order to show whether this effect is significant or not.

DISCUSSION

In vivo studies with intact coleoptiles or mesocotyls have indicated an antagonistic effect of phytochrome on growth promoted by auxin. Our results which indicate no effect of red light *in vivo* irradiation on binding parameters suggest that this antagonism does not originate from a direct effect of phytochrome on the affinity constant of the auxin receptor for its ligand and cannot result from an effect of phytochrome on receptor concentration (spare receptor or receptor sheding). The methodology used in this study (Scatchard plots) warrants the validity of this conclusion.

However this study which constitutes the first approach of the interaction phytochrome-auxin at the molecular level bears its basic limitations, it does not for example allow to show if phytochrome acts on auxin translocation. This effect might be of importance for example at the nodal region in the regulation of mesocotyl elongation according to the data of Kondo (8). On another hand, the topological relationship between the hormone and phytochrome might be raised. Phytochrome has been claimed to be bound to the plama membrane (14), to etioplasts (15), to mitochondria (16) and to ribonucleoproteic fractions (17). An effect of phytochrome on auxin might be indirect and could necessitate mediators which were lost in *in vitro* experiments. Recent work of Weintraub and Lawson (18) on wheat coleoptiles indicated that growth response of coleoptile segments to light depends on the initial position of the excised segment in the intact coleoptile. Red light stimulates the growth of the apical cells but inhibits that of the cells of the basal regions. In our experiments we took the complete coleoptile and it is possible therefore that these opposing effects averaged out. However this argument does not apply to the node and mesocotyl regions.

REFERENCES

1. Rayle,D.(1973) Planta 114,63-73.
2. Marré,E., Lado,P., Ferroni,A. and Ballarin Denti,A.(1974) Plant Science Letters 3,257-265.
3. Ray,P.M.(1973) Plant Physiol. 51,601-608.
4. Hertel,R. and Flory,R.(1968) Planta 87,123-144.
5. Hertel,R., Thomson,K.S. and Russo,V.E.A.(1972) Planta 107, 325-340.
6. Normand,G., Hartmann,M.A., Schuber,F. and Benveniste,P.(1975) Physiol. Veg. 13,743-761.
7. Muir,R.M. and Chen-Chang,K.(1974) Plant Physiol. 54,286-288.
8. Kondo N., Fujii,T. and Yamaki,T.(1969) In Development, Growth and Differenciation, vol 11, n° 1.
9. Sherwin,J.E. and Furuya,M.(1973) Plant Physiol. 51,295-298.
10. Tanada,T.(1973) Plant Physiol. 51,150-153.

11. Marmé,D., Boisard,J. and Briggs,W.R.(1973) Proc. Natl. Acad. Sci. USA 70,3861-3865.
12. Williamson,F.A., Morré,D.J. and Jaffe,M.J.(1975) Plant Physiol. 56,738-743.
13. Grombein,S., Rüdiger,W., Pratt,L. and Marmé,D.(1975) Plant Science Letters 5,275-280.
14. Yu,R.S.T.(1975) J.Exp. Bot.26,808-822.
15. Evans,A. and Smith,H.(1976) Proc. Natl. Acad. Sci. USA 73, 138-142.
16. Manabe,K. and Furuya,M.(1974) Plant Physiol. 53,343-347.
17. Quail,P.H.(1975) Planta 123,223-234.
18. Lawson,V.R. and Weintraub,R.L.(1975) Plant Physiol. 56,44-50.

ACKNOWLEDGEMENTS

We thank Professor Hertel (Biologisches Institut, Freiburg, GFR) who kindly showed the NAA binding assay.

Regulation of Cell Membrane Activities in Plants
E. Marrè and O. Ciferri eds.
© *1977, Elsevier/North-Holland Biomedical Press, Amsterdam*

ACTION OF INDOLEACETIC ACID ON MEMBRANE STRUCTURE AND TRANSPORT

U. Zimmermann and E. Steudle
Institute of Biophysical Chemistry, ICH/2
Nuclear Research Centre Jülich, FRG

SUMMARY

Experiments are reported concerning the effect of indoleacetic acid (IAA) on the molecular structure of bilayer membranes and on the membrane permeability and conductance of cell membranes of Valonia utricularis. In bilayers the location of IAA within the membrane can be determined by low frequency (0.1 to 100 Hz) dispersion measurements of the impedance. From such measurements the individual values of the capacitance and the conductance of the hydrocarbon and polar-head regions can be derived. At low KCl concentration in the bulk solutions (1 mM) in the presence of IAA the capacitance of the polar-head region is markedly reduced, whereas the conductance of the hydrocarbon layer is unaffected. With increasing KCl concentration the conductance of the hydrocarbon layer increases dramatically. Conversely, the effect of IAA on the capacitance of the polar-head groups decreases until a value at high KCl concentration is reached which is close to that measured in the absence of IAA. The results suggest that at low KCl concentration IAA is located in the polar-head region and that the IAA molecules are relocated completely from the polar head region to the hydrocarbon region with increasing KCl concentration. It is postulated that salt-induced relocation of IAA within a cell membrane is an important factor for the action of IAA (and other hormones) on the ionic relations of plant cells.

In Valonia cells IAA interacts with the pressure - dependent K^+-fluxes. In the absence of IAA, K^+-influx decreases with turgor pressure, whereas the efflux increases with both pressure and cell volume. This finding indicates a relation between K^+-efflux and the elastic modulus of the cell wall, ε, which is also increased with pressure and volume, but is not changed by IAA in Valonia. In the presence of IAA these relations remain qualitatively unchanged, but the absolute values of the fluxes are varied; that is, K^+-influx is enhanced and K^+-efflux is reduced. This is in agreement with the finding that the maximum in membrane resistance observed in Valonia in dependence on turgor pressure is shifted to higher values when IAA is present.

The results suggest that the elastic modulus couples the ion fluxes to water flow. On the basis of this phenomenon it is postulated that effects of IAA (and other hormones) on ion transport in plant cells will affect indirectly the water permeability. This has to be taken into account when effects of hormones on the water permeability are considered.

INTRODUCTION

Our knowledge concerning the molecular mechanisms, which are involved in the hormonal regulation of plant growth, is still rudimentary. Yet, there is a good body of experimental evidence that ion transport in plant cells is under hormonal control[1, 2]. However, the manner in which hormones interact with membrane transport has to be established; there are several mechanisms under consideration such as direct participation in the ion pumps, direct effects on the membrane structure (permeability and conductance) or induction of enzymes that are linked with the ion transport. In order to shed light on this subject matter, it is of considerable importance to study the effect of hormones on unicellular and membrane systems because of the complexity of higher plants. From a physical point of view the action of a hormone on the membrane processes should be independent of whether the hormone is naturally present or whether it is artificially introduced into the system. This is evidently true only if we consider the primary processes induced or triggered by the hormone in the membrane; that is to say adsorption, penetration and location, and the subsequent changes in the membrane permeability and electrical conductance caused by the hormone.

In this communication we study the effect of indoleacetic acid (IAA) on the molecular structure of an artificial bilayer and on the cell membranes of <u>Valonia utricularis</u>. Bilayers were used to get insight into the penetration and location of this hormone in the hydrophobic and hydrophilic layers of the membrane. In bilayers techniques are available to locate the hormone by determining the capacitance and conductance dispersion in the low frequency range[3]. Giant algal cells such as <u>V. utricularis</u> have the considerable advantage that primary effects induced by IAA in the membrane can be distinguished from secondary effects on the cell wall, because the cell wall elasticity can be determined by measuring the elastic modulus of the cell wall directly[4].

SALT INDUCED RELOCATION OF IAA IN BILAYER MEMBRANES

Lüttge et al.[5] reported that the effectiveness of IAA in stimulating K^+ uptake in old <u>Mnium</u> gametophytes depends on the KCl concentration in the external bulk solution. At 0.2 mM KCl IAA promoted the K^+ uptake whereas at higher KCl concentrations (10 mM) IAA had no significant effect on the K^+ transport. On the basis of this result Zimmermann suggested that "unspecific" salt concentrations in the bulk solution lead to a displacement of specific substances such as hormones or inhibitors within the membrane; this, in turn, could result in a triggering of diverse reactions which are involved in metabolic and growth processes. If such a relationship between the location of hormones, or inhibitors, in the membrane layers and the salt concentration exists, then under conditions where net ion fluxes and/or water flow occur the action of such substances depends not only on the salt concentration in the bulk solution, but also on the concentration polarization near the membrane surface. Concentration profiles near the membrane arise due to unstirred layers and transport number effects.

Recently, experimental evidence was obtained to support this hypothesis. Zimmermann et al.[6] studied the effect of IAA on the capacitance and conductance of the hydrocarbon layer and of the polar-head groups of a bilayer membrane in the presence of varying salt concentration in the external solutions facing the bilayer membrane. The individual values for the capacitance and conductance of both the hydrophobic and the hydrophilic layers in a bilayer membrane can be obtained from the very low frequency range (0.1 to 100 Hz) dispersion of the electrical impedance[3].

From the experimentally measured frequency variation of the overall capacitance and conductance of the bilayer membrane, it is possible to determine the individual values of the capacitance and conductance of both layers by analysing the data in terms of a parallel arrangement of the hydrocarbon layer and the two layers of the polar-head groups. The capacitance and the conductance of the hydrocarbon layer, C_H and G_H, respectively, are fixed by the very low frequency (0.1 to 0.5 Hz) values of the overall capacitance and conductance of the bilayer membrane. From this it is possible to obtain the values of the capacitance and the conductance of the polar-head groups, C_p and G_p, respectively, by fitting the experimental data to the theoretically derived curves for the overall equivalent parallel combination of the capacitance and conductance for a bimolecular lipid membrane. This procedure enables us to determine changes in the molecular structures of the polar-head groups and the hydrocarbon layer induced by IAA and KCl[6]. As reported by Coster and Smith[3], C_H and C_p are nearly independent of the KCl concentration in the range of 1 mM to 1 M. On the other hand, G_H does depend on the KCl concentration in the bulk solution (see also Fig. 1), whereas G_p is only weakly influenced by KCl concentration. IAA added to the external solutions on both sides of the bilayer membrane[+] at a concentration of about 10^{-6} M (p_H of the solution 5.4) has a pronounced effect on the conductance of the hydrocarbon layer, G_H, and on the capacitance of the polar-head region, C_p. The effect of IAA on both parameters depends strongly on the KCl concentration in the bulk solutions as indicated in figure 1 and 2. For comparison, the values of G_H and C_p as a function of increasing KCl concentration in the absence of IAA are also presented. It is obvious that at low KCl concentrations IAA has a significant effect on the capacitance of the polar-head region whereas the conductance of the hydrocarbon layer is only slightly influenced. With increasing KCl concentration the effect

---

[+] The membranes were made by the common technique of painting a film of n-tetradecane containing egg phosphatidylcholine over a hole in a polycarbonate septum, which divided a plexiglass cell into two compartments.

of IAA on G_H becomes more pronounced. Conversely, its effect on the capacitance of the polar-head groups is decreased. At high KCl concentrations the value for C_P was close to the value measured in the absence of this hormone, while G_H was increased about 10 fold. On the other hand, as the KCl concentration increases, only a small decrease in G_P was observed. C_H was practically not affected by IAA.

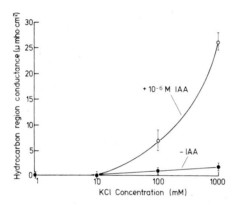
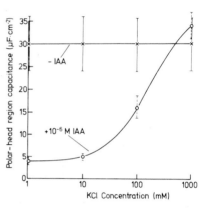

Fig. 1. The effect of IAA on the electrical conductance of the hydrocarbon region of a phosphatidylcholine bimolecular lipid membrane as a function of the external KCl concentration (logarithmic scale). The vertical bars indicate standard deviations of 15 measurements per point (taken from Zimmermann et al.[6]). Note, that at high KCl concentrations the conductance is increased up to 10 fold in the presence of IAA compared to the control experiments.

Fig. 2. The effect of IAA on the capacitance of the polar-head region of bilayers (same experiments as in figure 1) as a function of increasing KCl concentrations. Note, that at low concentrations IAA has a dramatic effect on the capacitance, whereas at high KCl concentrations the value of the capacitance becomes close to that measured in the absence of IAA (taken from Zimmermann et al.[6]).

In the light of the results presented here we can conclude that the effect of IAA on the structural parameter of the polar-head and hydrocarbon layers is highly sensitive to the salt concentration in the bulk solutions (that is, if it is allowed to generalize the KCl effect). It turns out that at low salt concentrations IAA is predominately located in the polar-head regions and is probably arranged as a monolayer parallel to the polar-head layer[6]. Adsorption of IAA in the polar-head region would lead to a decrease in the capacitance term of this region without affecting the capacitance and conductance of the hydrocarbon layer. It is evident that the polar-head conductance is not strongly influenced by IAA adsorption since the conductance of the polar-head region is normally very high.

With increasing salt concentration we would need to postulate that the IAA molecules are completely displaced from the polar-head region into the internal hydrocarbon layer. This relocation of IAA molecules would result in an increase of the
conductance of the hydrocarbon layer and to a corresponding increase of the capacitance value of the polar-head region. It is not surprising that the capacitance
value that is reached at higher salt concentration is nearly the same as in the
absence of IAA. The facilitated penetration of IAA across the polar-head layer
into the membrane interior with increasing salt concentration can be readily explained by assuming that the number of hydrogen bonds between the polar-head groups
is decreased in response to the increasing salt concentration.

If we extrapolate the data to the action of IAA on transport processes in biological membranes, then we may postulate that the interaction of IAA with specific sites in the membrane should depend strongly on the salt concentration at the
membrane surface. It should be noted again that the concentration profiles near
the membrane surface can be changed not only by varying the salt concentration in
the external solution, but also by concentration polarization in the presence of
net ion fluxes and water flow. On the basis of the fluid mosaic model introduced
by Singer and Nicolson[7], specific protein sites for the binding of IAA[8] can be
either buried in the hydrocarbon layer or located in the polar head region. Due
to this variance in the location of sites, salt-induced effects of IAA should vary
from species to species and this may be the reason for the manifold phenomena of
IAA effects reported in the literature.

Finally, it should be pointed out that the range of salt concentration where
the effect of KCl on the hormone action was observed with bilayers is abnormally
high from a physiological point of view. However, for biological membranes containing fixed charge regions of opposite sign, small changes in the salt concentration could be amplified by pronounced variations in the number of hydrogen bonds
between the polar-heads, the proteins, and the polar-heads and the proteins.

THE ACTION OF IAA ON CELL MEMBRANES OF VALONIA CELLS

A clear cut indication of IAA effects on the electrical conductance and fluxes
of a cell membrane was recently reported for cells of Valonia utricularis by Zimmermann et al.[9]. The results of these experiments revealed a tight connection between the action of IAA, the ion fluxes, and the wall elasticity. In order to
provide the knowledge for interpreting IAA effects that are observed in more complex systems such as higher plants, it seems worthwhile to consider these relationships in more detail. For Valonia cells in the absence of IAA a pronounced
biphasic course of the membrane resistance in dependence on turgor pressure is
observed; with increasing pressure the membrane resistance increases up to a
maximum value and then decreases again. In these experiments the pressure was

236

measured directly using the pressure probe developed by us[4]. The maximum in membrane resistance which occurs at a certain pressure is associated with a drop in the membrane potential. The pressure at the maximum resistance depends on the cell volume; it shifts to higher pressures with decreasing cell volume (Fig. 3).

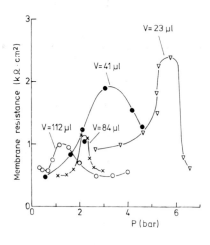

Fig. 3. Membrane resistance of Valonia utricularis in dependence on turgor pressure, P, shown for 4 different cells. The pressure where the maximum in membrane resistance occurs increases with decreasing cell volume (for experimental details see (9)).

Simultaneously, the absolute value of the maximum resistance is increased. This finding can be explained by assuming that turgor pressure has a different and opposite effect on the ion fluxes and that at least the magnitude of one of these fluxes is dependent on cell volume. This assumption was confirmed by measuring the long-term uptake and extrusion of K-42 at different pressures for cells of varying volume[10]. The long-term K^+-influx was strongly pressure-dependent in the pressure range between 0 and approximately 2 bar (Fig. 4); the influx decreases from 55 pM cm^{-2} s^{-1} at 0.5 bar to 17 pM cm^{-2} s^{-1} at 2 bar. This result is consistent with the finding reported by Gutknecht[11] using a perfusion technique on cells of Valonia ventricosa. Above 2 bar the K^+-influx becomes nearly independent of pressure; however, the K^+-efflux increases over the entire pressure range of 0 to 5 bar (Fig. 4). The magnitude of the efflux at each pressure was dependent on the cell volume, but a volume dependence of the influx could not be detected in the limits of accuracy[10].

If we assume that the electrical resistance of the cell membrane is mainly determined by the K^+-fluxes, then the biphasic course of the membrane resistance with pressure can be explained on the basis of these results. It turns out that the pressure at which the maximum in the membrane resistance is observed, probably corresponds to the steady state of the potassium-fluxes, i. e. where the net-flux is zero (see inset of Fig. 4). The occurence of the maximum resistance at higher pressures in small cells means that the "threshhold" for potassium-

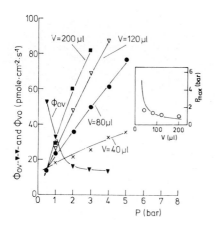

Fig. 4. Long-term K^+-fluxes of _Valonia utricularis_ as a function of turgor pressure, $\overline{P}$. Vacuolar influx, ϕ_{ov}, is decreased with increasing pressure and becomes nearly independent of pressure above 2 bar, whereas the efflux, ϕ_{vo}, is pressure-dependent over the entire pressure range. Note the pronounced effect of K^+-efflux on cell volume. From the intersections between $\phi_{ov}(P)$- and $\phi_{vo}(P)$-curves different steady state pressures are obtained for different cell volumes. If these values are plotted against cell volume, as shown in the inset, then the steady state values (0) coincide with the curve obtained from the relationship between turgor pressure at maximum resistance, P_{max} (see Fig. 3), and cell volume, V^9.

extrusion from the cells is shifted to higher pressures. Since KCl is the main osmotically active solute in _Valonia_ cells, the consequence is that in smaller cells higher turgor pressures can be built up in the cell interior than in larger cells. High turgor pressure is required for plant growth. This is well established by several experimental observations where it was found that turgor must exceed a certain critical level to induce extension growth[12, 13]. If there is any relationship between turgor pressure and growth, then we can assume that IAA should have an influence on the pressure dependence of the potassium fluxes and the membrane resistance.

This could be verfied experimentally by adding IAA to the external seawater and measuring the electrical membrane resistance as a function of pressure (Fig. 5). As indicated in this figure the pressure, where the maximum in resistance occurs in large cells is shifted to higher pressures. Concomitantly, the height of the maximum increases markedly. The effect of IAA is independent of IAA concentration in the range of 10^{-5} to 10^{-7} M. In preliminary experiments where the effect of IAA on the labelled K^+-fluxes were investigated, it was shown that IAA enhances the K^+ uptake over the entire pressure range although the course of the influx as pressure increased was nearly unaffected. On the other hand, the magnitude of the efflux was reduced at any pressure in the presence of IAA. Effects of turgor pressure and auxin supply as reported here seem to be common for plant cells. Similar changes in membrane permeability (conductance) and potential after application of auxin or changes in the osmotic pressure of the environment have been observed in bean roots by Jenkinson and Scott[14].

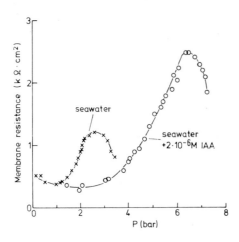

Fig. 5. Effect of IAA on the pressure-resistance characteristic of a cell of Valonia utricularis. When IAA is added the maximum in the membrane resistance is shifted to higher pressure values. Simultaneously, the height of the maximum resistance increases (for experimental details, see (9)).

The effect of IAA on the membrane permeability and resistance must be related to a direct effect of IAA on the membrane structure. An indirect effect by cell wall loosening can be excluded. This was demonstrated by measuring the elastic modulus of the cell wall, ε. This elastic modulus by which the cell wall elasticity can be described is defind by the following equation introduced by Philip[15]:

$$\varepsilon = V \frac{dP}{dV} \qquad (1)$$

Since small pressure changes, dP, and the corresponding changes in the cell volume, dV, can be monitored by our pressure probe[4], the elastic modulus and its variation with pressure and cell volume can be determined (Fig. 6). From figure 6 it is evident that ε increases not only with increasing pressure, but also with increasing volume. It should be pointed out that neither the magnitude of ε nor its dependence on pressure and volume is changed when IAA is added to the seawater, at least not during the time lapse, up to 4 hours, in our experiments.

On the other hand, the volume-dependence of ε suggests that the volume-dependence of the membrane resistance and K^+-fluxes (Figs. 3 and 4) can be related to the variance of the elastic properties of the cell wall with respect to volume.

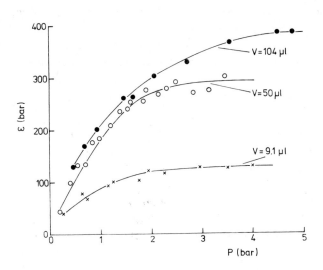

Fig. 6. Variation of the elastic modulus of the cell wall, ε, with turgor pressure, P, for 3 different cells of <u>Valonia utricularis.</u> Note, the significant dependence of the modulus on both pressure and cell volume, V.

This interdependence links the IAA effect to the elastic properties of the cell wall, although IAA does not directly affect the magnitude of the elastic modulus of the cell wall. In addition, as shown in the following consideration, the elastic modulus of the cell wall couples the potassium fluxes to the water flow; both transport processes are interdependent. On the basis of these relationships (see also scheme in Fig. 7) IAA should also affect the apparent water permeability of <u>Valonia</u> cells. Despite the fact that such measurements are not available at the present we will illustrate our point theoretically in order to shed light on experiments performed with higher plants, where effect of hormones on the water permeability were reported (e. g. (16)).

A <u>Valonia</u> cell is subjected to salt or water stress either by changing the osmolarity of the external seawater or by changing the turgor pressure using the pressure probe[4]. In response to the osmotic stress turgorregulation is initiated. Two phases can be distinguished. Immediately after the onset of the regulation a rapid phase of swelling and shrinking due to water flow is observed. This phase is completed after several minutes when a quasistationary state is reached, where the net water flow is vanishing, but a net ion flux is still present. This water transport phase is subsequently followed by a much slower phase of turgorregulation, which is governed by salt transport (mainly KCl), until the original turgor pressure is restored. For the sake of simplicity we restrict our considerations to the salt transport phase. We assume that the vacuole contains only KCl, that the reflection coefficients for KCl and NaCl (which is the main solute in the seawater) are 1, and that KCl is actively transported from the seawater into the cell. (The latter assumption implies that we restrict our attention only to osmoregulation in the low pressure range.) The assumptions do not change the basic sense of the following considerations. For a more rigorous mathematical treatment the reader is referred to Steudle et al.[10].

With these presuppositions we can theoretically divide the changes in the turgor pressure, P, and the osmotic pressure difference, $\Delta\Pi$, during the phase of active salt uptake, J_s, into two steps. The quasistationary state is infinitesimally disturbed when the internal osmotic pressure is changed by active ion uptake. The turgor pressure and the osmotic pressure difference are unbalanced, and, in turn, a rapid water flow is induced and a new quasistationary state is immediately established. This process continues until the steady state of both the water flow and ion fluxes is restored. The water flow observed can be described by the phenomenological equations of the thermodynamics of irreversible processes:

$$J_W = L_p [P - \Pi^i + \Pi^o] \tag{2}$$

where L_p is the hydraulic conductivity, that is the thermodynamic defined water permeability and Π^i and Π^o denote the osmotic pressure of the cell sap and the seawater, respectively. With the conditions outlined above in mind the change in turgor pressure is equal to the change in the osmotic pressure difference:

$$dP = d\Pi^i \tag{3}$$

The change in the osmotic pressure depends on both the concentration of the solution taken up, which is equal to the ratio of $\frac{J_s}{J_w}$ and the swelling of the cell, which is related to the turgor pressure by the Philip equation (Eq. (1)):

$$d\Pi^i = (\frac{J_s}{J_w} \cdot RT - \Pi^i) \frac{dV}{V} \tag{4}$$

Combining Equations (3) and (4) and substituting dP by Eq. (1) yields:

$$J_s = \frac{\varepsilon + \Pi^i}{RT} J_w \tag{5}$$

From Eq. (5) it is immediately evident that changes in the ion fluxes induced by a hormone such as IAA can cause a change in the water flow, since both fluxes are linked by the elastic properties of the cell wall (see also Fig. 4).

Another interesting feature of the coupling phenomenon described by Eq. (5) can be obtained by calculating the resistance to water flow, $\frac{dP}{dJ_w}$, from Eqs. (5) and (2) (see also Steudle et al.[10]):

$$\frac{dP}{dJ_w} = \frac{1}{L_p} - \frac{J_s \cdot RT}{J_w^2} \qquad (6)$$

The resistance to water flow, $\frac{dP}{J_w}$ is the parameter measured experimentally both in single cells and in higher plants. Eq. (6) states that this resistance comprises two terms, the thermodynamic one ($\frac{1}{L_p}$) and a second one which depends on the net ion flux and the water flow. It is obvious from Eq. (6) that the contribution of this term can be high if $J_s \cdot RT/J_w^2$ is of the same magnitude as the "thermodynamic" resistance, $1/L_p$. Thus, the calculation of L_p from changes of J_w in response to changes in turgor can lead to an apparent increase in L_p, if the second term is neglected and J_s is inwardly directed (i. e., if J_s has a negative sign). Such an increase in L_p was found in the low pressure range (on approaching the plasmolytic point) on Valonia, Nitella and Chara cells[17, 18, 19].

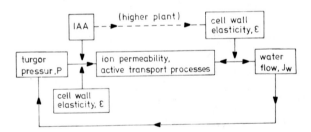

Fig. 7. Proposed scheme for the action of IAA on ion fluxes and water flow in Valonia (for further explanatio see text).

In addition, changes in the active ion flux component in the presence of hormones can bring about changes in the resistance to water flow, even though the hydraulic conductivity is not altered by the hormone. Effects of hormones on the hydraulic conductivity, L_p, as reported for higher plant tissues in the literature[16], require a more cautious interpretation when considering the relationships established experimentally for a single cell. In higher plant cells the situation may become much more complex because the hormone additionally alters the magnitude of the elastic modulus. This is indicated schematically in figure 7, where the interactions between IAA the ion fluxes and the water flow are represented for <u>Valonia</u> cells.

ACKNOWLEDGEMENTS

We are grateful to H. Jaeckel for expert technical assistance and to Miss V. Hudson, University of North Carolina at Chapel Hill, USA, for helping to prepare the manuscript. Part of this work was done, while U. Z. was on leave at the University of New South Wales, Sydney, Australia. This work was supported by a grant from the Deutsche Forschungsgemeinschaft, Sonderforschungsbereich 160.

REFERENCES

1. Ray, P. M. (1974) in Recent Advances in Phytochemistry (Runeckles, V. C., Sondheimer, E. and Walton, J. C., eds.), Vol. 7, 93 - 122, Academic Press, New York

2. Steveninck, R. F. M. van (1976) Encyclopedia of Plant Physiology, New Series, Vol. 2, Part B (Lüttge, U. and Pitman, M. G., eds.) 307 - 342, Springer-Verlag, Berlin

3. Coster, H. G. L. and Smith, J. R. (1974) Biochim. Biophys. Acta 373, 151 - 164

4. Zimmermann, U. and Steudle, E. (1974) in Membrane Transport in Plants (Zimmermann, U. and Dainty, J., eds.) 64 - 71, Springer-Verlag, Berlin

5. Lüttge, U., Higinbotham, N. and Pallaghy, C. K. (1972) Z. Naturforsch. 27b, 1239 - 1242

6. Zimmermann, U., Ashcroft, R., Coster, H. G. L. and Smith, J. R. (1976) Biochim. Biophys. Acta, in press

7. Singer, S. J. and Nicolson, G. L. (1972) Science 175, 720 - 731

8. Hertel, R. (1974) in Membrane Transport in Plants (Zimmermann, U. and Dainty, J. eds.) 457 - 461, Springer-Verlag, Berlin

9. Zimmermann, U., Steudle, E. and Lelkes, P. I. (1976) Plant Physiol., in press

10. Steudle, E., Zimmermann, U. and Lelkes, P. I. (1976) International Workshop on "Transmembrane Ionic Exchange in Plants", Rouen 1976, in press

11. Gutknecht, J. (1968) Science 160, 68 - 70

12. Cleland, R. (1971) Annu. Rev. Plant Physiol. 22, 197 - 222

13. Green, P. B., Erickson, R. O. and Buggy, J. (1971) Plant Physiol. 47, 423 - 430

14. Jenkinson, I. S. and Scott, B. I. H. (1961) Aust. J. Biol. Sci. 14, 231 - 243

15. Philip, J. R. (1958) Plant Physiol. 33, 264 - 271

16. Collins, J. C. (1974) in Membrane Transport in Plants (Zimmermann, U. and Dainty, J., eds.) 441 - 443, Springer-Verlag, Berlin

17. Zimmermann, U. and Steudle, E. (1974) J. Membr. Biol. 16, 331 - 352

18. Steudle, E. and Zimmermann, U. (1974) Biochim. Biophys. Acta 332, 399 - 412

19. Zimmermann, U. and Steudle, E. (1975) Aust. J. Plant Physiol. 2, 1 - 12

Regulation of Cell Membrane Activities in Plants
E. Marrè and O. Ciferri eds.
© *1977, Elsevier/North-Holland Biomedical Press, Amsterdam*

AUXIN ACTION ON K^+- H^+ - EXCHANGE AND GROWTH, $^{14}CO_2$ - FIXATION AND MALATE ACCUMULATION IN AVENA COLEOPTILE SEGMENTS

Hans-Peter Haschke and Ulrich Lüttge

Institut für Botanik
Fachbereich Biologie
Technische Hochschule
D-6100 Darmstadt
Germany

INTRODUCTION

Elongation growth of plant cells depends primarily on the extensibility of the cell wall. Indoleacetic acid (IAA) as well as solutions of low pH have been shown repeatedly to increase the elastic and particularly the plastic properties of the cell wall[1-8] and thus to induce elongation. On the basis of these and other facts HAGER and coworkers[9] proposed a hypothesis in which IAA is thought to stimulate a membrane-bound H^+-excretion mechanism which leads to an acidification of the cell wall. This is supported by the finding that IAA and other agents, which stimulate extension growth, cause a rapid excretion of protons to the external solution[10-14].

The paper presented here deals with some ionic and metabolic relations that are involved in this mechanism.

MATERIAL AND METHODS

Coleoptile segments (10 mm long) were obtained from etiolated seedlings of Avena sativa (cv. "Flemings Krone"). For the experiments coleoptile segments were incubated in various experimental solutions which always contained 0.1 mM $CaSO_4$ and which were slightly buffered. K^+-uptake was determined by flame photometry. Malate was estimated enzymatically.

Labelled products of $^{14}CO_2$-dark fixation were separated by ion exchange resins and TLC, identified by autoradiography and co-chromatography of known compounds and assayed in a scintillation counter. More details of materials and methods used are described elsewhere[15,16].

RESULTS

Indoleacetic acid is known to stimulate K^+-uptake by various plant tissues [18-20]. In Avena coleoptiles K^+-uptake electrochemically balances auxin induced H^+-excretion; both growth and H^+-excretion are stimulated by K^+ in the external medium [21-24].

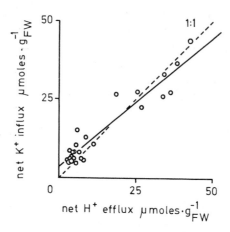

Fig. 1.   Correlation between IAA-induced net-K^+-influx and net-H^+-efflux in Avena coleoptile segments. Individual data were obtained by varying external K^+-concentration and duration of incubation.

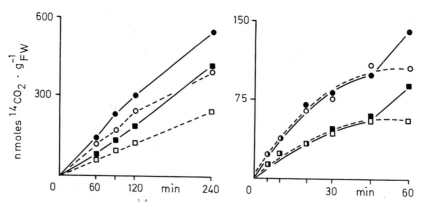

Fig. 2.   Fixation (o,•) of $^{14}CO_2$ by coleoptile segments and its incorporation into malate (□,■). Experimental solutions contained 1 mM K-phosphate buffer (pH 6.5), 0.1 mM $CaSO_4$ and 0.5 mM $NaHCO_3$ labelled with $H^{14}CO_3^-$ (10 mCi/mmole) with (•,■) or without (o,□) 10 μM IAA.

Figure 1 summarizes the results we obtained with various tissue batches in a series of experiments. It shows a stoichiometric correlation between auxin-enhanced net-K$^+$-influx and net-H$^+$-efflux. The linear regression line through the experimental points in good approximation indicates a 1 : 1 exchange of H$^+$ and K$^+$.

Figure 2A shows the stimulating effect of 10 μM IAA on fixation of $^{14}CO_2$ by coleoptile segments and its incorporation into malate. The additional CO_2-fixation observed in the presence of auxin is almost entirely due to malate synthesis. The earliest effect of IAA in these experiments was observed after 60 minutes (Fig. 2B). Pulse-chase experiments showed that most of the malate synthesized by $^{14}CO_2$ dark fixation in the presence of IAA is not subjected to metabolic turnover. Hence malate is accumulated in the tissue[16], probably in the vacuoles of the cells, reaching amounts of up to 30 μmoles per g fr. wt. within 16 hours[15]. Other metabolic pathways leading to amino acid synthesis or gluconeogenesis are not affected by IAA[16].

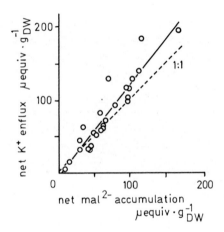

Fig. 3. Correlation between auxin-enhanced net-K$^+$-influx and malate accumulation in Avena coleoptile segments.

Auxin-induced malate accumulation is stoichiometrically correlated with net-K$^+$-uptake (Fig. 3.) and hence with IAA-dependent K$^+$- H$^+$- exchange.

DISCUSSION

The data reported above are summarized in the following model (Fig. 4):

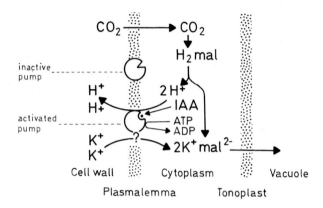

Salient features of this model can be stated as follows:

1. As suggested already by HAGER et al[9] indoleacetic acid stimulates a mechanism at the plasmalemma which extrudes protons by use of metabolic energy. This leads to an acidification of the cell wall and an increase of its extensibility enabling the cell to undergo elongation.

2. Growth regulator enhanced H^+-extrusion is electrochemically balanced and stimulated by K^+-ions, which are taken up in stoichiometric amounts. Coupling between K^+-uptake and H^+-extrusion could be electrochemical as suggested by PITMAN and coworkers[24,25], chemical by means of a real exchange pump as postulated by MARRÈ and coworkers[26,27] or even different for individual growth promoters[14].

The present discussion does not aim to make a contribution to this particular detail.

3. Continuous withdrawal of protons from the cytoplasm by an H^+-extrusion pump would lead to an increase of intracellular pH. Changes of cytoplasmic pH obviously must be prevented or at least minimized to maintain balance between metabolic pathways. A system to stabilize intracellular pH, which is quite common in plants is based on synthesis and breakdown of malic acid[28,29]. Phosphoenolepyruvate carboxylase which catalyses malic acid synthesis by CO_2-dark-fixation is stimulated in vitro by even small increases of pH above 6.5[30].

Thus auxin stimulated $^{14}CO_2$ dark fixation appears to be a consequence of auxin-induced proton extrusion leading to a pH-mediated increase of CO_2-incorporation into malic acid[15, 16, 31-33]. This conclusion is corroborated by the differential lag-phases that have been observed between the application of the hormone and the onset of the individual processes:

H^+ - extrusion	14 min	ref. 23
K^+ - (Rb+) - uptake	15 - 20 min	ref. 33
$^{14}CO_2$ - fixation	45 - 60 min	refs. 16, 33, and present paper

Additionally, auxin in vitro does not influence activity of PEP-carboxylases from Avena coleoptiles (ref. 34, and HAGER, pers. comm.), indicating that IAA-stimulated CO_2-fixation is caused by an intermediary step and not by a direct action of the hormone on the metabolic reactions.

Continuous synthesis of malic acid plays a multiple role in this mechanism:

i.) According to the "pH-stat"-theory cited above, it provides protons for stabilizing intracellular pH during H^+-extrusion.

ii.) Malate-anions, remaining in the cytoplasm, electrochemically balance K^+ ions taken up. Probably both are transported together to the vacuoles. This may be concluded from the fact that malate which derives from $^{14}CO_2$-fixation after application of auxin is withdrawn from metabolic turnover[16].

iii.) Finally, accumulation of K^+- and malate^{2-}- ions in the vacuole may contribute to stabilization of the osmotic potential of the cell sap and hence to turgor pressure in spite of extensive water influx during elongation growth[35].

SUMMARY

Indoleacetic acid stimulates a K^+ - H^+ - exchange mechanism at the plasma-lemma with concomitant synthesis of malic acid with a stoichiometric 1 : 1 - relationship.

A basic model is presented concerning the different aspects of this mechanism, underlying auxin-induced elongation growth.

REFERENCES

1. Strugger, S.: Ber. Dt. Bot. Ges. 50, 77-92, 1932
2. Bonner, J.: Protoplasma (Wien) 21, 406-423, 1934
3. Hager, A.: Untersuchungen über einen durch H^+-Ionen induzierbaren Zellstreckungsmechanismus. Habil.-Schr., Naturw. Fak. Univ. München, 1962
4. Cleland, R.: Physiol. Plant. 11, 599-609, 1958

5. Menzel, H.: Die pH-abhängige Veränderung mechanischer und chemischer Eigenschaften der Zellwand und ihr Zusammenhang mit der Wuchsstoffwirkung. Dissertation Naturw. Fak. Univ. München, 1966

6. Cleland, R.: Ann. N. Y. Acad. Sci. 144, 3-18, 1967

7. Rayle, D. L., R. Cleland: Plant Physiol. 46, 250-253, 1970

8. Yamamoto, R., Y. Masuda: Plant & Cell Physiol. 11, 947-956, 1970

9. Hager, A., H. Menzel, A. Krauss: Planta 100, 47-75, 1971

10. Lado, P., F. Rasi-Caldogno, R. Colombo, E. Marrè: Rend. Accad. Naz. Lincei 53, 583-588, 1972

11. Marrè, E., P. Lado, F. Rasi-Caldogno, R. Colombo; Rend. Accad. Naz. Lincei 53, 453-459, 1972

12. Cleland, R.: P. N. A. S. 70, 3092-3093, 1973

13. Rayle, D. L.: Planta 114, 63-73, 1973

14. Cleland, R. E.: Planta 128, 201-206, 1976

15. Haschke, H.-P., U. Lüttge: Plant Physiol. 56, 696-698, 1975

16.   --   ,   --   : Pl. Sci. Lett. (in press)

17. Higinbotham, N., M. J. Pratt, R. J. Foster: Plant Physiol. 37, 203-221, 1962

18. Ilan, I.: Nature 194, 203-204, 1962

19. Lüttge, U., N. Higinbotham, C. K. Pallaghy: Z. Naturforsch. 27b, 1239-1242, 1972

20. Haschke, H.-P., U. Lüttge: Z. Naturforsch. 28c, 555-558, 1973

21. Marrè, E., P. Lado, F. Rasi-Caldogno, R. Colombo, M. J. de Michelis: Pl. Sci. Lett. 3, 365- 379, 1974

22. Haschke, H.-P., U. Lüttge: Z. Pflanzenphysiol. 76, 450-455, 1975

23. Cleland, R.: Plant Physiol. 57, suppl. 1976

24. Pitman, M. G., N. Schaefer, R. A. Wildes: Pl. Sci. Lett. 4, 323-329, 1975

25.   --   ,   --   ,   --   : Planta 126, 61-73, 1975

26. Marrè, E., R. Colombo, P. Lado, F. Rasi-Caldogno: Pl. Sci. Lett. 2, 139-150, 1974

27. Cocucci, M., E. Marrè, A. Ballarin-Denti, A. Scacchi: Pl. Sci. Lett. 6, 143-156, 1976

28. Davies, D. D.: Symp. Soc. Exp. Biol. 27, 513-529, 1973

29. Smith, F. A., J. Raven: in U. Lüttge and M. G. Pitman (eds.): Encycl. Plant Physiol. New Series, vol. II, part A, 317-346, Springer, 1976

30. Kluge, M., C. B. Osmond: Z. Pflanzenphysiol. 66, 97-105, 1972

31. Yamaki, T.: Sci. Pap. Coll. Gen. Educ. Univ. Tokyo pp. 127-154, 1954

32. Johnson, K. D., D. L. Rayle: Plant Physiol. 57, suppl. 1976

33. Stout, R. K. D. Johnson, D. L. Rayle: Plant Physiol. 57, suppl. 1976

34. Bown, A., J. Dymock, B. Hill: Plant Physiol. 57, suppl. 1976

35. Burström, H. G.: Ann. Agr. Coll. Sweden 10, 1-30, 1942

PHYSIOLOGICAL ASPECTS OF THE REGULATION OF MEMBRANE ACTIVITIES

Regulation of Cell Membrane Activities in Plants
E. Marrè and O. Ciferri eds.
© 1977, Elsevier/North-Holland Biomedical Press, Amsterdam

HORMONAL CONTROL OF APICAL DOMINANCE IN *CICER ARIETINUM* SEEDLINGS:
A TENTATIVE UTILISATION OF THE MODELS OF HORMONAL REGULATION OF CELL MEMBRANE
PROPERTIES

By Jean GUERN, Madeleine USCIATI and Annie SANSONETTI
Service des Hormones végétales, Université Pierre et Marie Curie
T 53-3E, 4 Place Jussieu, 75230 PARIS Cedex 05

The axillary bud of the second leaf of 8 days old etiolated seedlings reacted by a strong stimulation of growth to the direct application of 2 µl of a 10^{-3}M benzyladenin solution. A fusicoccin treatment (2 µl of a 10^{-4}M solution) induced a transient release of the treated axillary buds from apical dominance, a fraction only of the treated buds continuing further growth. Evidences are presented showing that the growth of the basal axis under the first leaf of the axillary bud was stimulated at first when the release from apical dominance was induced. Decapitated axillary buds still reacted to benzyladenin or fusicoccin treatment. The basal axis of axillary buds when isolated and incubated on mineral solution showed an important proton extrusion in response to a fusicoccin treatment. These results suggest a possible involvment of the regulation of ion exchanges and water status in the control of the inhibition of axillary bud growth.

Despite various reviews of this problem appearing regularly (1,2,3,4), up to now, there is no definite answer to the question of the mechanisms by which an apical bud inhibits the growth of the axillary buds on the same shoot. With respect to the hormonal regulation of this process the antagonistic influences of cytokinins and auxins are well documented but the approach of the action of these hormones is mainly phenomenological. This situation probably explains the lack of hypotheses and data dealing with the cell regulatory mechanisms in which the preceeding hormones are involved to regulate the axillary bud growth.

A search for such mechanisms at the cell or molecular levels is tremendously impaired by the by nature high level of complexity of the biological system used. The various components of this system (apical bud, cells of the shoot axis, cells of the axillary buds, etc...) are potential receptors of the hormonal actions and no precise answer can be given to the question of the identity and location of cells being the primary receptors of the hormonal action.

During the time that some attention has been continuously given to the problem

of the mechanisms of growth correlations, important progress has been made with
respect to the knowledge of the mechanisms by which biological responses, at a
rather low level of integration, are initiated by hormones applied to simplified
and well defined biological materials. One can wonder if the models of cell regulation by plant hormones issued from such investigations have an operational
value to investigate much complex situations such as growth correlations.

The present paper deals with such a tentative approach on *Cicer arietinum* etiolated seedlings where the characteristics of the reactivation of growth of the
axillary buds, specially by cytokinin application, have been extensively studied
in our laboratory (3,5,6).

MATERIAL AND METHODS

The cultivation of plant material, treatment by hormones solutions and measurement of the growth of axillary buds have been previously described (5,6).
Briefly, an aqueous solution of the growth regulator studied was applied as a
2 µl drop on the axillary bud of the second leaf of an 8 days old etiolated
seedling, control plants received a 2 µl drop of distilled water. In some cases,
the axillary buds were treated after being decapitated (removal of the meristematic area and young leaves leaving the basel part of the bud as an axis, 1mm
long with a diameter 0.3 mm). Extension growth of axillary buds was measured
directly under a binocular microscope.
To test the eventual action of different growth regulators on proton extrusion
by the cells of the basal axis of axillary buds, the following procedure has
been adopted. Basal axis of axillary buds of the second leaf have been isolated
and preincubated for about two hours on a $Ca\ Cl_2$ (0,5 mM) and $MgCl_2$ (0,5 mM)
solution ; 20 to 30 sections (about 20 mg fresh weight) were then incubated on
200 µl of a $Ca\ Cl_2$, $Mg\ Cl_2$ solution with or without KCl and various growth
regulators. pH changes were followed with an Heito pH meter PSD 111 equiped
with a special contact microelectrode BRM 5.

RESULTS AND DISCUSSION

1 - Diversity of the positive effectors of the axillary bud growth
Fig. 1 shows that the axillary buds of *Cicer arietinum* seedlings revealed a
strong positive response to the treatment by 6-benzylaminopurin (BAP) as already
reported (3,4,6,7,8,9). The growth of inhibited axillary buds was also promoted
by fusicoccin (FC). This effect of fuxicoccin has been obtained regularly, the
response measured after 6 days being lower than that obtained for BAP. However,

253

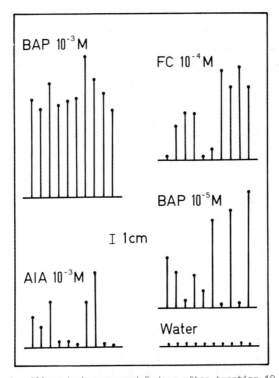

Fig. 1 - Length of axillary buds measured 6 days after treating 10 plants
populations by various growth regulators
Groups of 10 etiolated *Cicer arietinum* seedlings were treated by
6-benzylaminopurine (BAP 10^{-3}M or 10^{-5}M), fusicoccin (FC 10^{-4}M) or
auxin (AIA 10^{-3}M). The axillary bud of the second leaf received at
zero time 2 µl of one growth regulator solution, control buds receiving 2 µl of water. Lengths of treated buds were measured after 6 days.

for short times following the treatment (1 to 2 days) the intensity of the
response to FC treatment is of the same magnitude than that corresponding to
the BAP treatment ; the length of the axillary bud being 0.7 to 1.0 cm at that
time, compared to 0.3 cm initial length. A fraction only of the FC population
continued further growth from this initial activated stage and this was at the
origin of the frequent variability of the response measured after 6 days.
The response to an AIA treatment was much more erratic, repeatedly a fraction
only of the treated population reacted, often to a limited extent. Such a stimulation of the growth of the axillary bud by a direct application of auxin, has
already been discribed (9). The variability in the intensity of the response

between individuals of the treated population was also observed when axillary
buds received 2 µl of a 10^{-5}M BAP solution (100 times less than the dose giving
the maximum response). The response to an FC treatment was dose dependant (fig.2)
but FC solutions at concentrations higher than 10^{-3}M or frequently repeated
treatments with 10^{-5}M solutions revealed to give toxicity symptoms ; the same
was observed for AIA solutions.

Positive effects of FC and BAP on cell enlargement, proton extrusion and transmembrane potential in auxin insensitive materials such as *Cucurbita maxima*
cotyledons have been described previously (10,11,12). The promoting action of
these two growth regulators on seed germination (13) and stomatal opening (14)
are also documented. This suggests a common action of these regulators at the
membrane level, modifying ion and water exchanges.

The question arised of an eventual involvement of the same regulatory mechanisms
at the membrane level in the control of the axillary bud growth either by FC or
BAP. To answer this question it was necessary first to identify which part of
the biological system could be considered as the receptor of the growth regulator action.

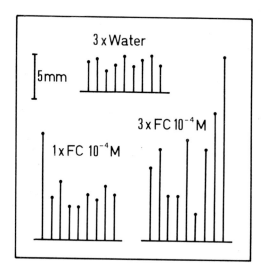

Fig. 2 - A typical stimulation of the growth of axillary buds of *Cicer arietinum*
seedlings treated by fusicoccin

Axillary buds of the second leaf of 8 days old etiolated plants received either a 2 µl drop of a 10^{-4}M FC solution (1xFC) or 3 drops (3 µl,
3 µl, 4 µl) at 5 hours intervals of the same solution (3xFC) or 3 drops,
as above, of distilled water (3xwater). Lengths of axillary buds were
measured 6 days after the treatment.

2 - Reaction of the basal part of the axillary bud to BAP and FC applications

Fig. 3 shows that the basal part of the axillary bud of the second leaf of etiolated 8 days old plants was represented by an axis (b) about one third of the
total length (l) of the bud. The growth stimulation of the axillary bud induced
by a 6-benzyladenin treatment affected markedly the basal axis which was responsible for the largest part of the total increase in length of the bud at
least over a 1 to 2 days period.

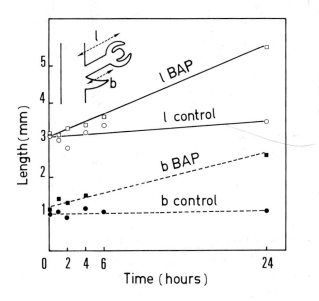

Fig. 3 - Stimulation of the growth of the basal part of axillary buds receiving
a direct application of BAP
Treated plants received a 2 µl drop of a 10^{-3}M BAP solution applied
directly on the axillary bud of the second leaf. Control plants received a 2 µl drop of water. The evolutions with time of total length (l)
and length of the basal axis (b) of treated and control buds were
followed.

When the axillary bud was decapitated, leaving only its basal axis (b.o.) attached to the epicotyl, a direct treatment of intact buds or operated buds by a
BAP solution or a release of apical dominance by decapitation of the apex of the
plants, stimulated markedly the elongation of the basal axis b.o. (fig. 4).
This stimulation was quite comparable to that corresponding to the same organ

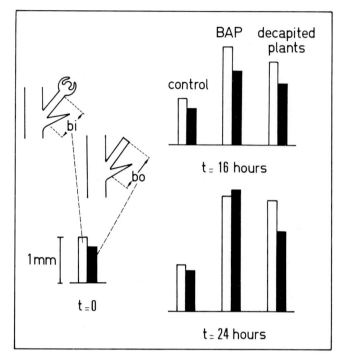

Fig. 4 - <u>Independance of the growth stimulation induced by a BAP treatment</u>
<u>towards the apical part of the axillary bud</u>
Axillary buds of the second leaf of 8 days old etiolated seedlings
were decapitated and the plants so obtained then treated either by
deposition of a 2 µl drop of water (control) or 10^{-3}M BAP solution
(BAP) on the operated base (bo) of the axillary bud or by decapitating
the apex of the seedling epicotyls (decapitated plants). The same
treatments were applied to seedlings the axillary buds of which were
not decapitated. The evolution with time of the length of the basal
axis of decapitated axillary buds (bo - black bars) or non-decapitated
buds (bi-white bars) was followed.

on intact axillary buds (b.i.) at least during the 24 hours following the cytokinin treatment. For longer periods the length of b.i. axis was higher than that
of b.o. axis. Fusicoccin stimulated also the elongation of the basal axis of the
axillary bud.

These results showed that the early events induced by benzylaminopurin and fusicoccin treatments were not accounted for by modifications of the properties

of the apical part of the axillary bud at least for periods of growth following the treatment of 1 to 2 days.

It was likely to think that the growth regulators used were acting either directly on the cells of the basal axis or on the epicotyl tissues regulating the intensity of exchanges between the shoot and the axillary bud.

The last hypothesis (i.e. a control of axillary bud growth by regulating the intensity of exchanges between the bud and the shoot axis) has been already presented and discussed (15,3,5,6,7).

Fig. 5 shows that if it is assumed that Xi is the concentration of the limiting factor of growth within the reactive cells at the base of the bud this concentration could be regulated by positive or negative modulators (M_2^{+-}) acting for example on the exchange properties of the plama membrane and consequently on

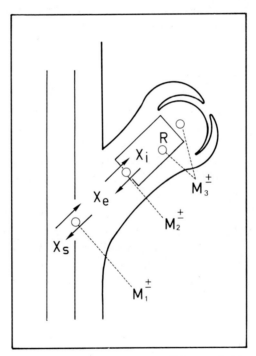

Fig. 5 - Schematic representation of various possible regulations of the intracellular concentration (Xi) of the limiting factor X of the growth of cells at the base of the bud

Xe, Xs : extracellular and sap concentrations of X. R : receptor molecule for X - M_1^{+-}, M_2^{+-}, M_3^{+-} : positive and negative modulators of the lateral transferts from transport conduits (M_1), of the transmembrane exchanges of the reactive cells (M_2), of the receptor properties (M_3).

the quantitative relationships between the extracellular concentration of X (Xe) and Xi. A modification of the intensity of the lateral transfer of X from the transport conduits due to positive or negative action of modulators M_1 could also account for a modification of Xe and Xi. Finally, if R is the receptor of X inside the reactive cells, for a same value of Xi, a modification of the properties of R induced by modulators $M_3^{\pm}$ could be at the origin of different intensities of response. This scheme accounts for the lack of specificity of positive effectors of the growth of axillary buds. It clearly shows that for a unique limiting factor, the concentration of X, there could exist several positive effectors M_1^+, M_2^+, M_3^+ of the axillary bud growth. Consequently, the stimulation of growth registered following the treatment of the bud by a growth regulator is not a sufficient argument to identify this growth regulator to X, that is to say to postulate that the inhibition of the axillary bud is linked to a shortage in the supply of the natural growth regulator from the shoot to the axillary bud.

As discussed above, it is likely to think that a basis of the diversity of the positive effectors of the axillary bud growth could lie in the fact that these effectors have in common the property to modulate the same cellular process, the intensity of which is limiting the growth of axillary buds.

Regulatory actions common to cytokinins, auxins and fusicoccins have already been demonstrated (10, 11, 12). These three growth regulators appear to act by modifying membrane properties resulting in modifications of the intensity of ions exchange, metabolism of organic acids, transmembrane potential values. Consequently, it seems logical to hypothesize that one, at least, of the common consequences of the action of these three regulators plays a central role in the control of the growth of the axillary bud, X being either a cation or water. The idea that the intracellular concentration (Xi) of a cation could be the limiting factor of the axillary bud growth is in good agrement with the results obtained by others with a different approach on a different biological material (16).

3 - Evidence for a modification of membrane properties of the cells at the base of the axillary bud

It was interesting to check that the fusicoccin effect on the growth of axillary buds could be associated with one of its well known action on membrane exchange properties : the stimulation of proton extrusion.

Excised bases of axillary buds were incubated as described in Material and Methods in a salt medium with different growth regulators and the pH changes followed with time. Fig. 6 shows that fusicoccin induced a rather important proton extrusion while BAP and AIA did not give a significant pH modification

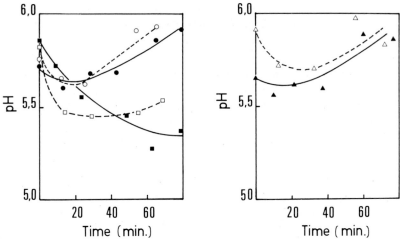

Fig. 6 - Proton extrusion by excised basal axis of axillary buds incubated in various growth regulators solutions

25 excised basal axis of axillary buds (1mm length - 0.3 mm diameter- were incubated into 200 µl of each growth regulator solution and the pH changes were followed with time as described in Material and Methods.

○——○ : Ca Cl$_2$, MgCl$_2$ solution -   ●——● :   CaCl2, MgCl2 solution + 10 mM KCl -   □——□ :   CaCl2, MgCl$_2$ solution + 2x10^{-5}M FC -   ■——■ :   CaCl2, MgCl2, KCl solution +2x10^{-5}M FC - △——△ :   CaCl2, MgCl2, KCl solution + 2x10^{-5}M BAP - ▲——▲ : CaCl2, MgCl2, KCl solution + 2x10^{-5}M AIA.

of the medium compared to control conditions. Unfortunately, it was not possible up to now, due to the very small amount of tissue available, to measure the eventual increase in fresh weight caused by the various treatments. These results clearly show that the cells of the basal axis of the axillary buds were directly sensitive at least to the fusicoccin action. This suggested that modifications of the membrane properties of these cells could be at the origin of the release from growth inhibition.

Rather surprisingly BAP which has a strong growth stimulating activity, much higher than that given by fusicoccin, did not conversely stimulate significantly proton extrusion, this last results being in good agreement with previously described comparative data (10, 11). Preliminary results clearly show that the BAP induced growth stimulation of the basal part of axillary bud was accounted for, both by cell division and cell enlargement. Further investigations are necessary to compare the cellular basis of BAP and FC stimulated elongation of the basal axis. The scheme from fig. 5 suggest that FC and BAP could either act as different modulators, each one having a unique property, or regulate a common process superimposed to a specific action, at least for one of them. It is obvious from fig. 5 that a large part of the growth response could depend upon topographical characteristics of this biological material. When a 2 µl drop is applied on the axillary bud, growth regulator molecules can penetrate either directly into the axillary bud or into the cells of the epicotyl and second leaf. Consequently, furthur investigations are necessary to check that the membrane modifications induced by FC on isolated basal parts of axillary buds are also expressed, for example by transmembrane potential modifications, by the same cells of intact treated axillary buds attached to the plant.

LITERATURE CITED

1 - CHAMPAGNAT P., 1965 - In : Encycl. Plant Physiol., XV (1), 1106-1164.

2 - PHILLIPS, I.D.J., 1969 - In : Physiology of Plant Growth and Development, Wilkins ed., Mc Graw Hill, London, 165-202.

3 - GUERN J. and USCIATI M., 1972 - In : Hormonal Regulation in Plant Growth and Development, Kaldewey and Vardar ed., Verlag-Chemie, Weinhem, 383-400.

4 - PHILIPPS I.D.J., 1975 - Ann. Rev. Plant Physiol., 26, 341-367.

5 - USCIATI M., CODACCIONI M. and GUERN J., 1972 - J. exp. Bot., 23, 1009-1020.

6 - USCIATI M., CODACCIONI M., MAZLIAK P. and GUERN J., 1974 - Plant Science Letters, 2, 295-301.

7 - GUERN J. and USCIATI M., 1976 - In : Etudes de Biologie végétale, R. JACQUES éd., in press.

8 - SACHS T. and THIMANN K.V., 1964 - Nature (Lond.), 201, 939-940.

9 - SACHS T. and THIMANN K.V., 1967 - Amer. J. Bot., 54, 136-144.

10 - MARRE E., COLOMBO R., LADO P. and RASI-CALDOGNO F., 1974 - Plant Science Letters, 2, 139-150.

11 - MARRE E., LADO P., FERRONI A. and BALLARIN-DENTI A., 1974 - Plant Science Letters, 2, 257-265.

12 - MARRE E., LADO P., RASI-CALDOGNO F., COLOMBO R., COCUCCI M. and DEMICHELIS M.I., 1975 - Physiol. Vég., 13 (4), 797-811.

13 - LADO P., RASI-CALDOGNO F. and COLOMBO R., 1974 - Physiol. Plant., 31, 149-152.

14 - SQUIRE G.R. and MANSFIELD T.A., 1972 - Planta, 105, 71-78.

15 - PANIGRAHI B.P. and AUDUS L.J., 1966 - Ann. Bot., N.S., 30, 457-473.

16 - DESBIEZ M.O. and THELLIER M., 1975 - Plant Science Letter, 4, 315-321.

Regulation of Cell Membrane Activities in Plants
E. Marrè and O. Ciferri eds.
© 1977, Elsevier/North-Holland Biomedical Press, Amsterdam

CELL DIVISION AND ION TRANSPORT AS TESTS FOR THE DISCRIMINATION BETWEEN THE ACTIONS OF 2,4-D AND FUSICOCCIN

Franco Rollo, Erik Nielsen and Rino Cella*
Istituto di Microbiologia e Fisiologia Vegetale, Università di Pavia
*Laboratorio di Genetica Biochimica ed Evoluzionistica del CNR
Via S. Epifanio 14, 27100 Pavia, Italy

SUMMARY

Addition of 2,4-D restores cell division in sycamore cell suspension cultures made resting by transferring them into medium without auxin. Addition of fusicoccin at concentrations between 10^{-8} and 10^{-5}M cannot substitute for the effect of auxin. However fusicoccin clearly influences cell membrane activity in this material, as shown by the increase of K^+ uptake and proton secretion. On the other hand no significant effect of 2,4-D on ion transport in sycamore cell suspension cultures was detected. These results suggest that the characteristic effect of auxin on cell multiplication is not linked to the activation of a K^+/H^+ exchange mechanism at cell membrane level.

INTRODUCTION

Fusicoccin (FC), a fungal toxin[1], induces in a number of experimental plant materials responses similar, though amplified, to those induced by plant hormones, i.e. extension growth and stimulation of cell membrane transport[2-4]. Therefore FC has been proposed as a useful tool to understand at least in part, the mechanism of action of plant growth hormones. In particular a correlation has been established between cell enlargement, induced either by auxin or FC, and responses at cell membrane level such as proton extrusion, cations uptake and hyperpolarization of transmembrane electric potential[5-9]. On the other hand auxins are also known to be required in order to support cell division in several plant tissue cultures[10]. However no data are available to demonstrate whether an auxin effect at cell membrane level can act as a trigger for the processes leading to cell division. Sycamore cell suspension cultures need 2,4-D in order to grow. It has been demonstrated that when exponentially growing sycamore cells are transferred to a

Abbreviations: FC, fusicoccin; 2,4-D, 2,4 dichlorophenoxy-acetic acid.

medium lacking 2,4-D, they stop dividing after a few generations[11]. Addition of 2,4-D to such a culture restores normal cell division. In the present investigation we have compared the ability of 2,4-D and FC to promote cell division, proton extrusion and K^+ uptake in 2,4-D-starved sycamore cells.

MATERIALS AND METHODS

Culture conditions: Acer pseudoplatanus L. cells, strain A M, were grown in a modified liquid Heller's medium(4×10^{-6}M 2,4-D)at 25° C on a rotatory shaker (120 r.p.m.)[12,13]. Under these conditions the doubling time was about 48 hours. To obtain a starving culture, exponentially growing cells were harvested, washed three times with Heller's medium lacking 2,4-D, and resuspended in the same medium at a concentration of 8×10^4 -10^5 cells/ml. The residual growth proceeded for about 2 generations. The resting culture was then distributed in 60 ml aliquots into 250 ml Erlenmayer flasks to which sterile FC or 2,4-D were added.

Cell count: 5 ml aliquots were withdrawn from cultures, mixed with 5 ml of 20% CrO_3 aqueous solution and incubated for 10' at 70°C. Macerated cell suspensions were passed 3 times through a syringe needle (n° 21) in order to disaggregate cell clumps. The cell number was evaluated by microscopic examination in a Nageotte cell (every value was calculated from the average of 50 or 70 microscopic field counts).

H^+ extrusion: Resting cells were harvested by filtration on nylon cloth (25 mesh), washed with 1 mM EDTA and a salt solution containing 10^{-2}M KCl, 10^{-3}M $MgSO_4$, 7.5×10^{-3}M $NaNO_3$ and 4×10^{-4}M $CaCl_2$. Cells were then weighed and resuspended in the same salt solution at a concentration of 0.5 g /10 ml. Proton or base extrusion was evaluated at constant pH values (6.20). Equal amounts of 10^{-3}N NaOH or H_2SO_4 were added to the samples at frequent intervals to maintain the pH of the medium within 0.02 units of the selected value.

K^+ uptake: 0.2 g cell samples were mixed with 5 ml of 10^{-2}M KCl solution containing ^{86}Rb as tracer for K^+. After 10' 2,4-D or FC was added at the indicated concentrations and the suspension was incubated for 30' at 25°C. Cells were collected by vacuum filtration on glass fibre filters, washed 3 times with 10^{-2}M KCl, resuspended in 4 ml of H_2O + 10 ml Insta-Gel Packard scintillation cocktail, and finally counted in a Packard Tri Carb 2425 counter.

RESULTS

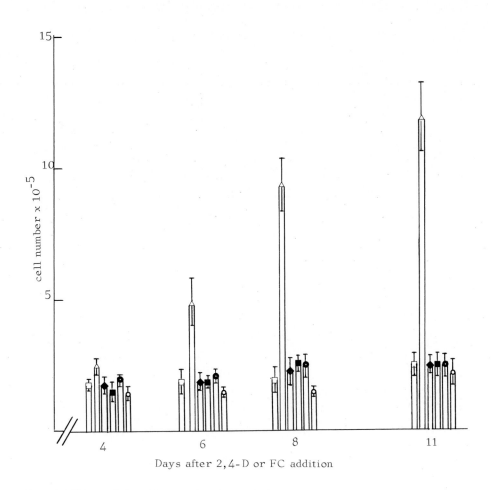

Fig. 1. Effect of the addition of 2,4-D and FC on resumption of cell division of starving sycamore cells. No addition ◻ ; 4×10^{-6}M 2,4-D △; FC: 10^{-8}M ◆; 10^{-7}M ■ ; 10^{-6}M ● ; 10^{-5}M ○ .

Figure 1 shows the effect of the addition of 2,4-D or FC to resting sycamore cells. 2,4-D at a concentration of 4×10^{-6}M restores cell division whereas FC, at concentrations ranging from 10^{-8} to 10^{-5}M, fails to produce a similar effect. On the other hand K^+ uptake in 2,4-D-starved sycamore cells is not enhanced by the addition of 2,4-D, while FC induces a stimulation which is linear for concentrations ranging from 10^{-8} to 10^{-4}M (Fig. 2). Similarly when 4×10^{-6}M or 4×10^{-5} M 2,4-D is added to resting sycamore cells no acidification of the medium is detected, while under the same conditions, FC promotes H^+ extrusion as shown in figure 3.

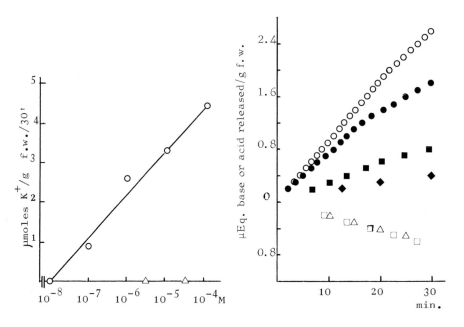

Fig. 2. Effect of 2,4-D and FC on K^+ uptake. FC ⊙–⊙; 2,4-D △–△. Controls have been subtracted.

Fig. 3. Effect of 2,4-D and FC on time course of proton or base excretion. No addition □ ; 4×10^{-6}M 2,4-D △ ; FC: 10^{-4}M ○ ; 10^{-5}M ● ; 3×10^{-6}M ■ ; 10^{-6}M ◆.

DISCUSSION

When FC is added to 2,4-D-starved sycamore cells, a sharp stimulation of cell membrane activities occurs as shown by the rapid acidification of the medium and the increase of K^+ uptake. However FC-induced activation of these membrane processes is not accompanied by any detectable stimulation of cell division. On the other hand, 2,4-D restores cell division, while it has no appreciable effect on either H^+ extrusion or K^+ uptake. Thus, in our material, the activation of H^+ and K^+ transport does not seem to be required for the promotion of cell division. These data suggest that the division-promoting activity of 2,4-D (and probably other auxins) has to be sought in cellular processes differing from membrane ion transport.

ACKNOWLEDGEMENT

We are very grateful to Prof. E. Marrè for helpful discussions and for the generous gift of FC.

REFERENCES

1. Ballio, A., Brufani, M., Casinovi, C.G., Cerrini, S., Fedeli, W., Pellicciari, R., Santurbano, B. and Viciago, A. (1968) Experientia, 24, 631.

2. Lado, P., Rasi Caldogno, F., Pennacchioni, A. and Marrè, E. (1972) Planta, 110, 331.

3. Lado, P., Pennacchioni, A., Rasi Caldogno, F., Russi, S. and Silano, V. (1972) Physiol. Plant Pathol., 2, 75.

4. Marrè, E., Lado, P., Rasi Caldogno, F., Colombo, R., Cocucci, M., and De Michelis, M.I. (1975) Physiol. Vég. 13, 797.

5. Marrè, E., Lado, P., Rasi Caldogno, F. and Colombo, R. (1973) Plant Sci. Lett., 1, 179.

6. Marrè, E., Lado, P., Rasi Caldogno, F. and Colombo, R. (1973) Plant Sci. Lett., 1, 185.

7. Marrè, E., Lado, P., Rasi Caldogno, F., Colombo, R. and De Michelis, M. I. (1974) Plant Sci. Lett., 3, 365.

8. Lado, P., Rasi Caldogno, F., Colombo, R., De Michelis, M.I. and Marrè, E. (1976) Plant Sci. Lett., 7, 199.

9. Marrè, E., Lado, P., Ferroni, A. and Ballarin Denti, A. (1974) Plant Sci. Lett., 2, 257.

10. Yeoman, M.M. (1973) In: Plant Tissue and Cell Culture, Ed. H.E. Street, Blackwell, Oxford, pp. 31-58.

11. Leguay, J.J. and Guern, J. (1975) Plant Physiol., 56, 356.

12. Stewart, C.R. and Street, H.E. (1969) J. Exp. Bot., 20, 556.

13. Bayliss, M.W. and Gould, A.R. (1974) J. Exp. Bot. 25, 772.

Regulation of Cell Membrane Activities in Plants
E. Marrè and O. Ciferri eds.
© *1977, Elsevier/North-Holland Biomedical Press, Amsterdam*

CHANGES IN MEMBRANE PROPERTIES AND TRANSCELLULAR ION MOVEMENTS IN DEVELOPING PLANT CELLS

Manfred H. Weisenseel
Botanisches Institut der Universität Erlangen-Nürnberg
Schloßgarten 4, 8520 Erlangen
Germany

ABSTRACT

To test the hypothesis that changes in the plasma membrane and selfelectrophoresis are the key mechanisms by which cells become firmly polarized, membrane potentials and endogenous electrical currents were measured in developing fucoid eggs and in lily pollen. The results of several investigations, including our own, show that the plasma membranes of Fucus eggs, Pelvetia eggs and lily pollen hyperpolarize by roughly 50 mV after activation. An electrical current of about 60 pA in Pelvetia eggs and about 200 pA in lily pollen traverses the cells before visible growth begins. These currents indicate qualitatively different membrane areas which develop during early stages of cell differentiation. Currents of about 100-300 pA enter the growing end and leave the non-growing end of Pelvetia embryos and lily pollen. These currents probably contain a significant amount of Ca^{++}-ions. Red light, a strong environmental signal for plant cell development, was found to depolarize the plasma membrane of Nitella internodes by 10-30 mV. This depolarization is probably mediated by phytochrome and it involves Ca^{++}-ions.

INTRODUCTION

Evidence is accumulating that cell differentiation takes a centripetal course, the cell periphery changing first and the cell nucleus changing last[1]. One special problem of cell differentiation is the mechanism which establishes a firm polarity in an unpolar cell. For this mechanism we assume that localized changes in the plasma membrane and some sort of selfelectrophoresis are the key events[2,3,1].

Plant cells are favorable objects to study the mechanism and structures involved in establishing a cell polarity. There are plant cells like the eggs of the brown algae Fucus and Pelvetia which are practically unpolar at the beginning of their free existence. Plant cells in general cling tightly to their polarity once it is established.

Various investigations, including our own, show that the perm-
selectivity and electrical potential of the plasma membrane of developing cells change. They show further that differentiating cells
drive large electrical currents through themselves. (Electrical
current is defined here as a flow of positive ions.) To increase
the support for our hypothesis we are currently investigating the
membrane potential of growing lily pollen and the effect of red
light on the membrane potential of Nitella internodes.

MATERIALS AND METHODS

For the measurements of the electrical membrane potentials we
used KCl-filled micropipettes with tip diameters of about 1 μm, tip
resistances of below 60 megohms and tip potentials, measured in the
culture medium, of less than 10 mV. The micropipettes were connected to the recording devices by Ag-AgCl-wires. The recording
apparatus consisted of a high impedance electrometer (Keithley's
610 C) or a preamplifier (Bioelectric's NF 1) connected to a storage oscilloscope (Tectronix's 5031) and to a chart recorder (Sie-
mens, Sargent).

The measurements of transcellular currents were made extracellu-
larly with a highly sensitive vibrating electrode which detects the
electrical fields near developing cells. "The sensor ....is a sphe-
rical, 30 μm platinum-black electrode at its tip which measures
voltages with respect to a coaxial reference electrode... . The
probe is vibrated at about 200 cycles/s in a horizontal plane between two extracellular points 30 μm apart. Vibration between these
points converts any steady voltage difference between them into a
sinusoidal output measurable with the aid of a lock-in amplifier
tuned to the vibration frequency. Since the electrical field will
be nearly constant over this small distance, it is approximately
equal to the voltage difference divided by this distance. The current density in the direction of vibration and at the center of
vibration is then given by this field multiplied by the medium's
conductivity"[4].

For autoradiography of lily pollen we pulse-labelled growing
pollen tubes with Ca-45, washed them free of extracellular Ca-45
and stranded them on a 10 μm thin Nuclepore filter. Filter and pollen were quickly frozen in liquid nitrogen and exposed on nuclear
emulsion over dry ice for several days[5].

Fucus serratus and Pelvetia fastigiata eggs were obtained from cut fronds and grown in artificial sea water (483 mM Na^+, 10 mM K^+, 10 mM Ca^{++}, 55 mM Mg^{++}, 564.5 mM Cl^-, 28 mM SO_4^{--}, 2.5 mM HCO_3^-, 10 mM Tris-HCl, pH = 8.0)[6],[7]. Pollen for the membrane potential measurements was taken from fresh flowers of Lilium longiflorum and sown into a medium of 290 mM mannitol, 1.3 mM $Ca(NO_3)_2$, 2.0 mM $CaCl_2$, 1.0 mM KNO_3, 3.9 mM H_3BO_3 (pH = 3.9). For current measurements the medium was slightly modified to give a higher pH and it contained 290 mM mannitol, 1.3 mM $Ca(OH)_2$, 2.0 mM $CaCl_2$, 1.0 mM KNO_3, 3.9 mM H_3BO_3 (pH = 5.4)[8]. Nitella spec. was grown in natural pond water and was put into artificial pond water of 1.0 mM NaCl, 0.1 mM KCl, 5.0 mM $CaCl_2$ (pH = 7.0) just before the experiment.

RESULTS AND DISCUSSION

Changes in membrane potential during early development

The average membrane potential of unfertilized eggs of the brown alga Fucus serratus is -20 to -33 mV[9],[7]. In eggs of Pelvetia fastigiata it is about -20 mV[6]. After fertilization the plasma membrane of the zygote hyperpolarizes by 40-60 mV before any visible growth. During this hyperpolarization the plasma membrane increases its perm-selectivity for K^+ substantially[9],[6]. In Pelvetia the intracellular K^+-concentration doubles, increasing from about 170 mM to 340 mM[10].

Our preliminary results with lily pollen show a similar increase in membrane potential before visible growth of the pollen tube begins (Fig.1, Leo Matschkal and M.H.Weisenseel, work in progress). The plasma membrane hyperpolarizes by roughly 50 mV and, as with the seaweed eggs, remains at the higher electrical potential during further development.

Similar hyperpolarization of the membrane occurs in developing eggs of Echinoderms[11],[12],[13], Tunicates[14] and Vertebrates[15],[16],[17]. What might be the developmental significance of such a hyperpolarization? A hyperpolarization of, for example, 50 mV amounts to an increase of 50000 Vcm^{-1} in the electrical field across the plasma membrane. Such an increase could cause changes in the function of membrane associated enzymes and could therby give signals to the genetic machinery. Or, if membrane patches on opposite sides of the cell hyperpolarize to different heights, this difference could become a driving force for a transcellular current.

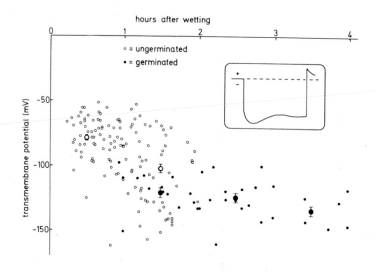

Fig. 1. Membrane potentials of developing lily pollen. Data from
several experiments are combined in this figure. Every measurement
represents the highest recorded potential during a 10 min puncture
of a different grain of developing pollen and is corrected for the
average increase in the micropipette's tip potential of +20 mV.
(This correction is probably not final.). The large symbols represent the average potential values at hourly intervals $\pm$ the standard error of the mean. The inset shows a representative recording
of a single measurement.

Evidence for localized membrane changes

An important question for our hypothesis of cell polarization is
whether there are changes in the plasma membrane which are confined
to certain areas. Localized changes can be the origin of spatial
gradients in developing cells. Recent experiments with Pelvetia
eggs show that the membrane of the prospective rhizoid pole is more
permeable to Ca^{++}-ions than is that of the prospective thallus pole.
Robinson and Jaffe[18] took a nickel screen, plugged each hole with
an egg, applied unilateral light to polarize these eggs, and measured the influx and efflux of radioactive Ca^{++} at the dark (future
rhizoid) and light (future thallus) ends of the eggs. They found
that at six hours after fertilization, Ca^{++} enters the dark end
five times faster than the light end. The efflux of Ca^{++} is four
times higher at the light end. This indicates a Ca^{++}-current of
2 pA through the developing egg. During the next few hours this
Ca^{++}-current decreases and at the time of germination the influx
and efflux of Ca^{++} is about equal at both ends. Another current of
about 60 pA, of which Ca^{++} is only a part, startes traversing the

cell[19,20,21].

In the close vicinity of ungerminated lily pollen we have discovered an electrical field which indicates a strong current entering the prospective site of germination and leaving the opposite
end of the grain[8]. This current starts in most cases at about one
hour after sowing the pollen grains into the medium. The inward
current shows a somewhat greater density than the outward current
which suggests that the site of current entry is smaller than the
site of current exit. The absolute current density entering the
grain is about 4 μAcm^{-2} at the plasma membrane. From this we estimate the total current traversing a single pollen grain before
germination to be about 200 pA. Such a transcellular current implies that membrane patches with different qualities develop within a single pollen grain.

Origin of membrane changes

What can cause changes in the properties of the plasma membrane?
In plant cells, which depend heavily on environmental inputs for
their development, localized changes can be caused by environmental
vectors like gravity, ion or hormone gradients, or light gradients.
Light is probably one of the most important stimuli. To give just
one example, red light absorbed by the pigment phytochrome causes
a change in the growth direction of the protonemata of the moss
Funaria[22]. Might phytochrome have changed some membrane properties?

To tackle this question we have started to investigate the possible effect of phytochrome on the membrane in a larger cell than
the Funaria protonema, namely the Nitella internode. Our preliminary results with Nitella show some interesting features (Hermann
Ruppert and M.H.Weisenseel, work in progress). Red light, as compared to white light, always causes a depolarization of the membrane. This depolarization has an average lag phase of 1.5 s (0.4-
3.2 s in 16 measurements). The height of depolarization increases
significantly with the Ca^{++}-concentration of the medium. The depolarization reaches 10-30 mV (maximum 57 mV) at 5 mM Ca^{++}. The
effect of Ca^{++}-ions seems highly specific because it cannot be
mimicked by adding Na^+, K^+, Mg^{++}, La^{+++} or by increasing the osmotic
potential of the medium with mannitol. The red light induced depolarization has features which indicate the involvement of phytochrome (Fig. 2 and 3). Far-red light alone has no effect on the
resting membrane potential; far-red light applied together with
red light significantly reduces the height of depolarization and
significantly increases the rate of repolarization.

We would like to interpret these findings to mean that phytochrome causes an influx of Ca^{++} through the membrane. Ca^{++} then triggers some functional changes in the membrane (such as an increase in Cl^--permeability). The increase in cytoplasmic Ca^{++} could be of influence for development and for oriented growth by creating an electrical field in the cytoplasma[21].

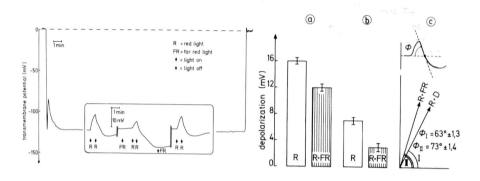

Fig. 2. Representative example of the effects of red light and far-red light on the membrane potential of a Nitella internode. The internode was surrounded by artificial pond water of 5 mM Ca^{++}. After being punctured the whole internode was treated with 20 min of far-red light and 30 min of darkness before the first irradiation with red light. The light regimes are indicated by the lettered arrows. Red light intensity 3.4 Wm^{-2}, far-red light intensity 4.4 Wm^{-2}. In the intervals the internode was kept in darkness. The micropipette was filled with 0.1 M KCl.

Fig. 3. Effect of red light alone and of red light plus far-red light applied simultaneously on the amount of depolarization and the rapidity of repolarization of the membrane potential in Nitella internodes. Red light (R) was applied for 60 s each, far-red light from several minutes before to several minutes after the red light irradiation. The artificial pond water contained 5 mM Ca^{++}. The average resting membrane potential was about -150 mV. a) Red light intensity 8.4 Wm^{-2}, far-red light intensity 4.4 Wm^{-2}. Measurements from 26 different internodes were compiled. b) Red light intensity 3.4 Wm^{-2}, far-red light intensity 4.4 Wm^{-2}. Measurements from 16 different internodes were compiled. c) Rapidity of repolarization after red light for the results shown in a). The larger the angle Ø, which is defined above, the more rapid is the repolarization. All errors given are the standard errors of the mean.

Amplification of membrane changes

How can small differences in the plasma membrane be amplified? How can such differences lead to stable polarity? The mechanism might be selfelectrophoresis, i.e. an electrical field within the cell could cause the transport of ionized components to opposite ends of the cell.

There is some older evidence indicating the existence of electrical fields within the cytoplasm of cells. By puncturing the tips of growing Neurospora hyphae or fish melanophores with two microelectrodes some distance apart Slayman and Slayman[23] and Kinosita[24] measured electrical fields of 0.1-0.6 Vcm^{-1}.

The origin of such fields could be an electrical current traversing the cell. Jaffe et al. found one in Pelvetia embryos[19,20,21,25,4] and we discovered one in lily pollen[8]. In both cases the current always enters the growing end (the rhizoid in Pelvetia, the tube in pollen) and leaves the non-growing end. The total steady current through a growing pollen is about 100-300 pA and reaches the considerable density of 60 μAcm^{-2} inside the tube. In Pelvetia the current consists of a steady component of roughly 100 pA and of larger spontaneous spikes of 100 s duration. The current through growing pollen consists mainly of a K^+-influx and a H^+-extrusion[26]. In both cell types there is some evidence that part of the inward current is Ca^{++}. The involvement of Ca^{++}-ions is deduced from autoradiographs which show an accumulation of radioactive Ca^{++} within the growing tip of pollen tubes[5] and growing Pelvetia embryos (Laurinda A. Jaffe and John Gilkey, pers. comm.).

In addition to establishing an electrical field within the cytoplasm, a transcellular current might concentrate or remove certain ion species from the cytoplasm. Changes of the ion composition and concentration might be signals to activate or repress enzymes and the genetic machinery.

ACKNOWLEDGEMENTS

I would like to thank the Deutsche Forschungsgemeinschaft for financial support of our research, Mr. A. Dorn for technical assistance and Dr. M.E. Feinleib for reading the manuscript.

REFERENCES
1. Jaffe, L.F. (1969) Dev.Biol.Suppl. 3, 83-111
2. Bentrup, F.W. (1968) Berichte der Deutschen Botanischen Gesellschaft 81, 311-314
3. Lund, E.J. (1947) Bioelectric Fields and Growth. The University of Texas Press, Austin
4. Nuccitelli, R. and Jaffe, L.F. (1975) J.Cell Biol. 64, 636-643
5. Jaffe, L.A., Weisenseel, M.H. and Jaffe, L.F. (1975) J.Cell Biol. 67, 488-492

274

6. Weisenseel, M.H. and Jaffe, L.F. (1972) Dev.Biol. 27, 555-574
7. Weisenseel, M.H. and Jaffe, L.F. (1974) Exptl.Cell Res. 89, 55-62
8. Weisenseel, M.H., Nuccitelli, R. and Jaffe, L.F. (1975) J.Cell Biol. 66, 556-567
9. Bentrup, F.W. (1970) Planta (Berl.) 94, 319-332
10. Allen, R.D., Jacobsen, L., Joaquin, J. and Jaffe, L.F. (1972) Dev.Biol. 27, 538-545
11. Tyler, A., Monroy, A., Kao, C.Y. and Grundfest, H. (1956) Biol. Bull. 111, 153-177
12. Tupper, J., Saunders, J.W. and Edwards, C. (1970) J.Cell Biol. 46, 187-191
13. Steinhardt, R.A., Lundin, L. and Mazia, D. (1971) Proc.Nat. Acad.Sci.USA 68, 2426-2430
14. Takahashi, K., Miyazaki, S. and Kidokoro, Y. (1971) Science 171, 415-417
15. Bennett, M.V.L. and Trinkaus, J.P. (1970) J.Cell Biol. 44, 592-610
16. Ito, S. and Hori, N. (1966) J.Gen.Physiol. 49, 1019-1027
17. Palmer, J.F. and Slack, C. (1970) J.Embryol.Exp.Morphol. 24, 535-553
18. Robinson, K.R. and Jaffe, L.F. (1975) Science 187, 70-72
19. Jaffe, L.F. (1966) Proc.Nat.Acad.Sci. USA 56, 1102-1109
20. Jaffe, L.F. (1968) Advan.Morphog. 7, 295-328
21. Jaffe, L.F., Robinson, K.R. and Nuccitelli, R. (1974) Membrane Transport in Plants (eds. U. Zimmermann and J. Dainty Springer Verlag Berlin) 226-233
22. Etzold, H. (1965) Planta 64, 254-280
23. Slayman, C.L. and Slayman, C.W. (1962) Science 136, 876-877
24. Kinosita, H. (1953) Annot.Zool.Japanases 26, 115-127
25. Nuccitelli, R. and Jaffe, L.F. (1974) Proc.Nat.Acad.Sci.USA 71, 4855-4859
26. Weisenseel, M.H. and Jaffe, L.F. (1976) Planta, accepted

Regulation of Cell Membrane Activities in Plants
E. Marrè and O. Ciferri eds.
© 1977, Elsevier/North-Holland Biomedical Press, Amsterdam

AUXIN INDUCED PROTON RELEASE, CELL WALL STRUCTURE AND

ELONGATION GROWTH: A HYPOTHESIS

A G Darvill, C J Smith and M A Hall
Department of Botany and Microbiology
University College of Wales, Aberystwyth,
Dyfed SY23 3DA, United Kingdom

INTRODUCTION

Since the first discovery of auxin induced proton release and its relationship to extension growth (1) much work has sought to determine how an alteration in the pH of the cell wall might lead to changes in its properties (2). The work described below using coleoptiles of Zea mays and cell 'sheets' derived from this tissue seeks to provide possible answers to this question.

MATERIALS AND METHODS

Coleoptiles of Zea mays L. were grown and harvested as described by Smith (3). Cell sheet material, 1-3 cells thick was prepared by maceration in 5mM phosphate-citrate buffer, pH 7.0 (for further details see 4). Details of methods for measurement of growth of coleoptiles and cell sheets and of analysis of cell wall structure are given in 3 and 4.

RESULTS AND DISCUSSION

Figure 1 shows the effect of IAA upon growth of maize coleoptiles and of cell 'sheets' derived from these. Figure 2 shows how the same samples respond in terms of acidification of the medium in which they are incubated. The same amount of growth is made in both cases although the initial rate is higher with the cell sheets. Equally the rate of drop in pH is faster with the sheets than with whole coleoptiles and the final H$^+$ concentration lower. Cell sheets showed approximately 10% increase in growth over a 30 min incubation period when incubated at pH 4.5. No further growth was observed up to 8 h. 'Peeled' coleoptiles (4) gave intermediate results. The inclusion of 0.2M mannitol in the medium inhibits elongation markedly (Fig. 1) but has little effect on acidification (Fig. 2).

Changes in the composition of the cell wall and the nature of any carbohydrate released was measured after an 8 h incubation period using cell sheets. The results are shown in Table 1. With the exception of the changes in t-glucose, 4-glucose and 4,6-glucose which are derived from starch in the preparation, the most striking changes are the release of free arabinose and galactose, 5-arabinose and free glucuronic and galacturonic acids. With the exception of the arabinose these changes were only observed where the pH of the medium was held at 4.5 or where the sheets were incubated in 10^{-5}M IAA. The composition of the cell wall after treatment reflected the release of the sugars described above. Thus, on a mole percent basis of total carbohydrate both with 10^{-5}M IAA and pH 4.5

276

there is a general reduction in terminal and 5-linked arabinose, in terminal and 4-linked galactose, in 4-linked glucuronic acid, in terminal and 4-linked galacturonic acid and in 3,4-linked xylose in the cell wall.

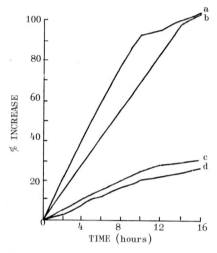

Fig. 1. Growth kinetics of intact coleoptiles and 'cell sheets'. Cell sheets + 10^{-4}M IAA (a), whole coleoptiles + 10^{-4}M IAA (b), cell sheets + 10^{-4}M IAA + mannitol (c), whole coleoptiles + 10^{-4}M IAA + mannitol (d).

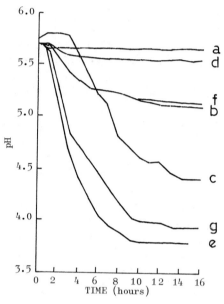

Fig. 2. Acidification by coleoptiles and cell sheets. Whole coleoptiles (a), whole coleoptiles + 10^{-4}M IAA (b), peeled coleoptiles + 10^{-4}M IAA (c), cell sheets (d), cell sheets + 10^{-4}M IAA (e), whole coleoptiles + 10^{-4}M IAA + mannitol (f), cell sheets + 10^{-4}M IAA + mannitol (g).

One further interesting observation was made, namely a release of 3-linked glucose residues. The total amount of the latter residues was greater in all treatments (taking cell wall and supernatant together) after 8 h incubation although again this was more marked in the 10^{-5}M IAA and pH 4.5 treatments. The proposed explanation for this effect is described below in the section on the pulse chase experiments.

Maize coleoptiles were pulsed with ^{14}C-(U)-glucose for 1.5 h, chased with ^{12}C glucose for a further 2 h prior to incubation in ^{12}C glucose in either 10^{-4}M IAA, 10^{-4}M IAA + 0.2M mannitol, 0.2M mannitol or with no other additions. The results of this experiment are shown in Figure 3a-f. Two main effects were observed, namely; (1) continued incorporation or no change in radioactivity during the chase (t-mannose, t-glucose, t-galactose, 5-arabinose, 4-glucose, 4-xylose (all treatments), 4-glucuronic acid, 3,4-xylose, t-arabinose (f), t-arabinose (p) (glucose, glucose + mannitol)); (2) incorporation followed by decorporation

(4-glucuronic acid, 3,4-xylose, t-arabinose (f), t-arabinose (p) (auxin, auxin + mannitol), 3-glucose (all treatments).

TABLE 1

METHYLATION ANALYSIS OF SUGARS RELEASED FROM MAIZE COLEOPTILE CELL WALL
MATERIAL AFTER TREATMENT AT DIFFERENT pH VALUES AND VARIOUS
CONCENTRATIONS OF AUXIN

SUGAR LINKAGE	COMPOSITION OF INITIAL CELL WALL (MOLE %)	10^{-5}M IAA			pH 4.5			pH 7.0		
		cw[1]	sup[2]	%[3]	cw[1]	sup[2]	%[3]	cw[1]	sup[2]	%[3]
t-ara (f)	14.45	8.13	5.4	29.78	10.65	3.05	18.68	11.48	1.96	10.28
t-ara (p)	2.45	1.01	1.63	8.99	1.53	1.24	6.78	2.21	0.91	4.78
2-ara	0.07	0.07	0	0	0.06	0	0	0.08	0	0
3-ara	0.69	0.42	0	0	0.43	0	0	0.53	0	0
5-ara	1.97	0.34	1.13	6.23	0.23	1.00	5.47	1.24	0.64	3.36
2,5 ara	Tr	Tr	0	0	Tr	0	0	Tr	0	0
Arabinose	Tr	Tr	0	0	Tr	0	0	Tr	0	0
TOTAL	19.72	9.97	8.16	45.00	12.99	5.29	28.94	15.54	3.51	18.42
t-xyl	0.23	0.18	0	0	0.14	0	0	0.29	0	0
4-xyl	5.17	3.98	0	0	4.23	0	0	5.37	0	0
3,4-xyl	13.73	9.79	0.83	5.62	11.00	0.54	3.39	13.01	0	0
TOTAL	19.13	13.95	0.83	5.62	15.37	0.54	3.39	18.67	0	0
t-man	0.34	0.29	0	0	0.14	0	0	0.34	0	0
4-man	1.69	1.17	0	0	1.20	0	0	1.62	0	0
TOTAL	2.03	2.00	0	0	1.34	0	0	1.96	0	0
t-gal	1.26	0.46	0.84	22.40	0.02	0.93	28.62	1.03	0	0
4-gal	2.32	1.84	0	0	1.57	0	0	2.68	0	0
3,4-gal	0.03	0.08	0	0	0.02	0	0	0.05	0	0
4,6-gal	0.84	0.53	0	0	0.53	0	0	1.31	0	0
TOTAL	4.45	2.91	0.84	22.40	2.32	0.93	28.62	5.07	0	0
t-glu	0.66	0.54	0.98	17.57	0.66	10.30	19.84	0.41	1.24	2.66
3-glu	6.30	6.24	5.98	11.70	5.89	6.42	12.37	6.01	2.32	4.98
4-glu	34.66	15.31	10.14	19.84	13.25	11.75	22.63	24.32	8.42	18.08
3,4-glu	1.37	0.34	1.08	2.11	0.20	1.03	1.98	1.21	0	0
4,6-glu	2.45	1.89	0.62	1.21	2.00	0.42	0.81	2.31	0.34	0.73
TOTAL	45.47	24.32	26.80	52.62	22.00	29.92	57.63	34.26	12.32	26.45
t-gal UA	5.56	3.21	2.39	23.39	3.00	2.02	22.70	5.20	0	0
4-gal UA	0.86	0.64	0	0	0.72	0.09	1.01	0.71	0	0
t-glu UA	0	0	1.96	19.18	0	1.94	21.80	0	0	0
4-glu UA	2.92	2.02	0	0	1.13	0	0	2.67	0	0
TOTAL	9.34	5.87	4.35	42.56	4.85	4.05	45.51	8.58	0	0

1 = mole % of cell wall sugar remaining after treatment

2 = mole % of cell wall sugar released during treatment

3 = % of sugar released during treatment of each total sugar

As regards the changes in 3-glucose these appear to be due to changes in callose ($\beta 1 \rightarrow 3$ glucan) in the tissue. If coleoptiles were stained with resorcinol blue a narrow band of cells (c. 0.5 mm) at each end of the sections stained blue, indicating the presence of definitive callose. In the absence of

278

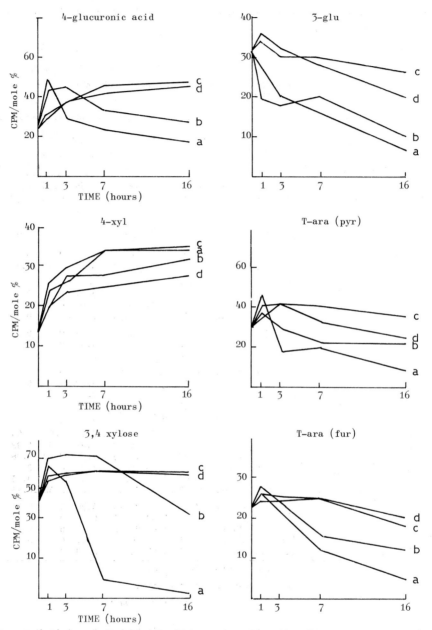

Fig. 3.   Turnover of glycosidic linkages in cell walls of maize coleoptiles.
10^{-4}M IAA (a), 10^{-4}M IAA + 0.2M mannitol (b), 0.2M mannitol (c), glucose alone (d).

auxin this colouration developed first at that end of the coleoptile which had been adjacent to the apex, subsequently appearing at the basal end. Sections incubated in auxin developed callose at both ends very rapidly. This finding is consistent with results showing IAA-induced increases in $\beta 1 \to 3$ glucan synthetase activity (5) and may to some extent explain the observations of Masuda (6) on this subject. However, the restriction of the effect to the extremities of the coleoptile suggests that it is unlikely to play an important role in the mediation of extension growth.

The results from the pulse-chase studies agreed well in the main with the studies on carbohydrate release from the cell sheets, such minor discrepancies as exist (for example in the case of terminal galactose) are probably explicable in terms of limitations imposed by the techniques relating to detection of radio-activity.

In associated work an attempt has been made to construct a model of the distribution and interconnection of the polysaccharide and protein components of the cell wall matrix. This structure is illustrated in Figure 4. While other structures are possible and the nature of the interconnections still in doubt in some cases, nevertheless we feel that the figure represents a composite picture which is entirely consistent with our structural studies (the linkage composition of the polysaccharides of the wall is shown in Table 1. 4-glucose includes both starch and cellulose).

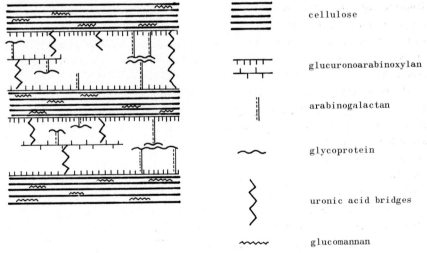

cellulose

glucuronoarabinoxylan

arabinogalactan

glycoprotein

uronic acid bridges

glucomannan

Fig. 4.   Proposed model of maize coleoptile cell wall

The principal component is a glucuronoarabinoxylan which we have extracted by treatment with an endopolygalacturonase obtained from <u>Sclerotium rolfsii</u> and further purified by gel filtration. The main features of the structure are a

1→4 linked xylan backbone with frequent side chains attached at the 3 position, comprising either a) single arabinose units; b) arabinose 5→1 galactose 4→ ; c) xylose 4→1 glucuronic acid 4→1 galacturonic acid 4(?); d) single 4-0-methyl glucuronic acid units. We believe that the galactose in the side chains is connected to the serine in the wall protein which latter appears to have arabinose side chains attached to the hydroxyproline residues, as is commonly found (7). The individual glucuronoarabinoxylan chains may hydrogen bond to the cellulose microfibrils in the wall (a property which we have shown they possess _in vitro_), and seem to be interconnected via uronic acid 'bridges'. The other minor component of the wall, a glucomannan, does not appear to be covalently attached to the remainder of the matrix.

The results of both the sugar release and pulse chase studies are consistent with an auxin/H^+ induced breakage of the linkages connecting glucuronoarabinoxylan both with other similar molecules (changes in uronic acid linkages) and with the protein (4-galactose linkages). The changes in arabinose linkages are consistent with a loss of arabinose side chains from the glucuronoarabinoxylan (and possibly protein).

These observations suggest that the auxin induced increase in plasticity can be explained by a) a decreased interaction between components of the matrix; b) a decrease in viscosity occasioned by loss of side chains from a highly branched molecule. They further suggest that the changes can be accounted for by the induction by auxin of a 'proton pump' since similar changes are observed as a result of incubation at low pH. The hypothesis does not exclude the implication of H^+ effects on the physical properties of the matrix, by, for example, protonation of the uronic acid residues, or alteration in the degree of hydrogen bonding.

The observation that, in general, mannitol inhibits turnover of the linkages which we feel to be significant might be construed as showing that the effects are growth-dependent rather than growth-inducing. We feel that this is unlikely for several reasons. Firstly, although turnover is inhibited it is still greater in the case of mannitol treated sections than in controls (glucose alone) which nevertheless show similar growth. Secondly, that where terminal linkages are involved, that is where cleavage results in the release of a monomer not connected to another part of the matrix (e.g. t-arabinose), then the inhibitory effect of mannitol is quite small. Furthermore, the results are consistent with the concept that linkages once broken may reform at the same position in the absence of turgor-driven extension (8).

Since none of the linkages which break in response to auxin/H^+ treatment are acid labile at the concentrations obtaining, it seems likely that the breakage is enzyme mediated. Whether the enzymes involved are capable both of hydrolysis and resynthesis of the appropriate linkages or whether separate enzyme systems are

involved remains obscure. In either case it seems that the reformation of linkages is accomplished sufficiently effectively at low pH. These findings would
agree with other work on enzymes which may be involved (9). Although for technical reasons it was impossible to obtain satisfactory results over very short
intervals (e.g. 10 min after auxin addition), nevertheless, examination of Fig. 3
shows that turnover of the critical linkages is very rapid (e.g. of the 3,4 xylose
linkages which become labelled during the pulse and subsequent chase more than 80%
had turned over after 7 h in IAA). The results are not at variance with the finding that continual cell extension involves polysaccharide synthesis although it
seems probable that net synthesis is induced at a later stage than the changes
reported here. This would be in agreement with other work on polysaccharide
synthetases and may represent a process which is growth-dependent rather than
growth-causative.

The hypothesis discussed here is represented diagramatically in Figure 5.

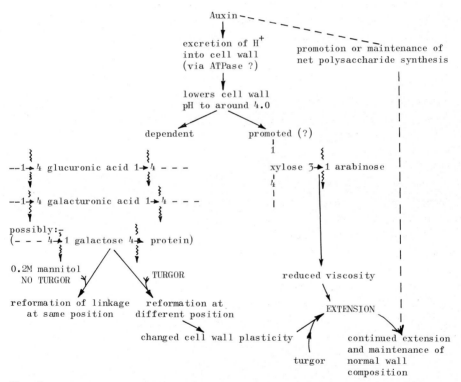

Fig. 5.

Proposed hypothesis for auxin-induced extension growth.

REFERENCES

1. Hager, A., Menzel, H. and Kraus, A. (1971) Planta 100, 47-75.

2. Evans, M.L. (1974) Ann. Rev. Plant Physiol., 25, 195-223.

3. Smith, C.J. (1976) PhD Thesis, University of Wales.

4. Darvill, A.G. (1976) PhD thesis, University of Wales.

5. Smith, J. (1976) PhD Thesis, University of Wales.

6. Masuda, Y. and Wada, S. (1967) Bot. Mag. Tokyo, 80, 100-102.

7. Lamport, D.T.A. and Miller, D.H. (1971) Plant Physiol., 48, 454-456.

8. Cleland, R. (1971) Ann. Rev. Plant Physiol., 22, 197-222.

9. Johnson, K.D., Daniels, D., Dowler, M.J. and Rayle, D.L. (1974) Plant Physiol., 53, 224-228.

ACKNOWLEDGEMENTS

   We are grateful to the Science Research Council for providing research studentships for two of us (A.G.D. & C.J.S.) and for the provision of GCMS facilities. Thanks are also due to Professor P F Wareing FRS for encouragement and discussion and to Professor P Albersheim for advice on analytical techniques.

Regulation of Cell Membrane Activities in Plants
E. Marrè and O. Ciferri eds.
© *1977, Elsevier/North-Holland Biomedical Press, Amsterdam*

ON THE POSSIBLE ROLE OF HYDROGEN IONS IN
AUXIN-INDUCED GROWTH

David Penny
Botany and Zoology Dep't
Massey University
Palmerston North
New Zealand

SUMMARY

Several lines of evidence have been presented that do not appear to be consistent with the hypothesis that acidification of the cell wall is the main response to auxin. Another test is whether the same compounds are required for both growth and for acidification of the external medium. The results presented do not show any evidence of a requirement for exogenous chemicals (other than water) for auxin-induced growth, at least in short term experiments, but the requirements for acidification of the medium are not yet known. In addition it is difficult to explain by one reaction the observed kinetics of transient changes in the growth rate which includes a latent period, rapid acceleration, and an oscillation in the growth rate.

There is evidence that enzymes in the cell wall are important in cell expansion and such enzymes would have a definite dependence on pH with a pH optimum. There is still no good direct evidence that the auxin-induced acidification is necessary for rapid cell expansion, but it is still a reasonable hypothesis that it is one part of a two component system linked to give negative feedback control.

INTRODUCTION

The possible involvement of hydrogen ions in the expansion of plant cells has attracted considerable attention. There is no direct evidence that auxin controls external pH with the same kinetics as for auxin-induced growth. A drop in cell wall pH cannot be detected before growth increases[1], and low pH was found to cause a decrease in diameter rather than the increase that was found with auxin[2]. On the other hand, auxin effects on external pH do occur and perhaps more significantly other hormones that stimulate growth will also lower the external pH.

One of the main problems in improving our understanding is a lack of knowledge of the mechanisms involved. This has been complicated in   growth studies by referring to a "proton pump" or "H^+ excretion" etc. with the implication that the drop in external pH depends on the movement of protons across the plasma membrane. We have pointed out previously[1] that there are in principle, many ways in which the external pH could decrease. Some of these possible mechanisms are shown in Fig. 1 in the style used by Thomas[4]. They are only examples of possible mechanisms and for example K^+ is shown as the counterion for exchange mechanisms but other cations are possible and have been suggested in animal systems[4]. And again there

284

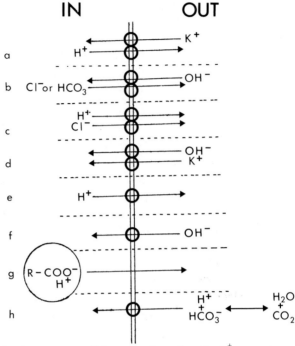

Fig. 1. Diagram showing some possible mechanisms for the H^+ or proton pump in plant cell membranes. The mechanisms are essentially those from ref.1.
a. K^+ - H^+ exchange (cations other than K^+ are possible); b. OH^- exchange with another anion; c. a coupled HCl pump; d. a coupled KOH pump (or other cation); e. simple uncoupled H pump; f. simple uncoupled OH^- pump; g. mechanisms involving changes in transport or metabolism of carboxyl groups in the cell wall; h. one of a series of possible mechanisms involving bicarbonate.

are many possibilities involving CO_2 and bicarbonate which will depend on whether the tissue is or is not photosynthetic, and whether it is light or dark grown etc. Effects of CO_2 on acidification have been reported[5,6] and Sloane and Sadava[7] have implicated increased CO_2 levels in the response.

One of the conclusions from Fig.1 is that some mechanisms require other ions to be in the external medium outside the plasma membrane. The interest in our laboratory has not been in the mechanism of acidification but rather in developing methods of simultaneously measuring growth along with other parameters and at short term intervals. We have already reported that lupins responded better to auxin in distilled water than in a buffer solution[8] and this lead us to suggest that there was nothing in the external solution that was necessary for cell expansion, at least over the short term. In this paper some additional experiments of this type are considered and the possible role of hydrogen ions is discussed.

MATERIALS AND METHODS

Seedlings of *Lupinus angustifolius* cv. New Zealand Bitter Blue were grown under continuous light for 4 days before 25 mm segments were cut for growth measurements. Three segments were run simultaneously and the elongation of each was measured by a displacement transducer. Experiments were run at $25°$ and supplementary light was provided by a 40 W incandescent bulb. All experiments were run at least three times with three segments. Full details of the growth conditions and recording techniques have been described[9].

Intact seedlings were used in experiments where a weight was used to increase the tension on a hypocotyl. The "Ministron" described previously[9] was modified for this purpose and the details will be published in a later publication.

CO_2 free air was obtained by passing 2.5 1/min of air through a 35x2 cm column of non-deliquescent soda lime ("Carbosorb") 10-16 mesh. There was no detectable CO_2 left when it was measured with an infra red gas analyser that should readily detect 2-3 ppm of CO_2.

Deionised water was made by passing glass distilled water through a column of 'Amberlite' monobed resin MB-1, analytical grade. The specific conductivity was used as a measure of the total dissolved ions and was measured with a Radiometer CDM 2e Conductivity Meter. Results are expressed as μ Siemen (μ ohm^{-1}).

RESULTS AND DISCUSSION

Exogenous compounds: It has not yet been possible to detect any exogenous compound other than water that is required for auxin-induced growth in short term experiments. A series of responses to 30 μM indolyl-3-acetic acid (IAA) are shown in Fig. 2. The first graph(2A) serves as a base line and is the response to IAA in 5 mM sodium phosphate buffer pH 6.0 and with a conductivity of about 500 μSiemens. There is little difference in short term experiments if a potassium salt is used.

The next graph (2B) gives a response to IAA in distilled water. The water has a pH of about 5.0 and a specific conductivity of 3.8 μSiemens. The pH falls to about 4.0 and the conductivity increases to 6.0 when 30 μM IAA is dissolved in it. The fall in the pH is not significant in affecting growth because these pH's do not induce growth in this tissue unless the cuticle is damaged, and because the pH can be adjusted to 6.0 with NH_4OH and the response is not changed. NH_4OH was chosen because the pH could be raised without introducing K^+ or Na^+. Organic ions could also be tried.

Deionised water has even fewer dissolved ions and consequently a lower conductivity, but the normal response to IAA was still obtained (2C). The conductivity with and without IAA was 3.4 and 1.2 μSiemens and the pH's were 4.1 and 5.2. This conductivity in the absence of IAA is equivalent to about 10 μM KCl and since this is for the total ions, the concentration of any individual ion will be much less. The conclusion is that the ions in the medium can be reduced 500 times from 5 mM to about 10 μM with no consistent effect on the growth rate.

Two obvious problems that remain are that the concentration of ions in the cell

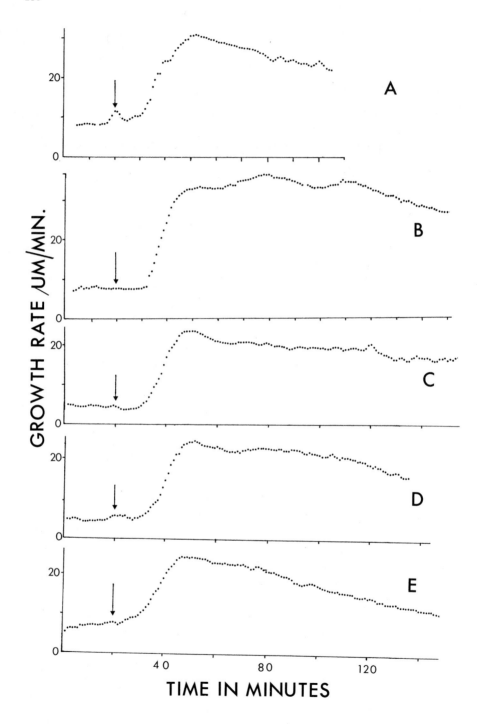

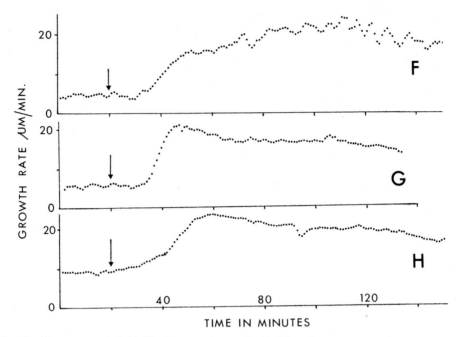

Fig. 2. Responses of individual segments to 30 μM IAA. Segments were chosen at random excluding those at the extremes of the range. Segments were treated in: A, aerated 5 mM sodium phosphate buffer; B, aerated distilled water; C, aerated deionised water; D, distilled water plus CO_2 free air; E, deionised water plus CO_2 free air; F, aerated 5 mM buffer with segment predipped in ether for 10 secs; G, aerated deionised water with ether dipped segment; H, deionised water plus CO_2 free air with either dipped segment.

wall is not known and that dissolved CO_2 and bicarbonate in equilibrium with CO_2 in the air will still be present in the solution.

The solutions used in the growth studies are aerated and in some additional experiments CO_2 was removed. The growth responses to IAA in both distilled and deionised water that had been aerated by CO_2 free air is shown in 2D and 2E. Rather surprisingly[10], there is still no significant drop in the response to IAA although the growth is slightly less whether or not auxin is present. It is not yet possible to say that external CO_2 or HCO_3^- is unnecessary until the level of, say bicarbonate, has been measured in the solution. There will be a small amount of CO_2 released from the 3 segments (300 mg of tissue) and leakage has not yet been eliminated.

The cuticle is a significant barrier to movement of compounds into the segment (unpublished experiments). Most methods of reducing the effect of the cuticle also reduce the response of lupin hypocotyls to auxins. Dipping the intact seedlings in ether for 2-10 sec reduces the resistance of the cuticle to diffusion (unpublished experiments). The half time for efflux of tritiated water is reduced from about 20 min for nontreated segments to about 6½ min for segments dipped in ether for 10 sec. With a 10 sec ether dip there may be some reduction in auxin-induced growth even in 5 mM phosphate buffer pH 6.0 (Fig 2F). There is a slight reduction about 20 min for nontreated segments to about 6½ min for segments dipped in ether for 10 sec. With a 10 sec ether dip there may be some reduction in auxin-induced growth even in 5 mM phosphate buffer pH 6.0 (Fig 2F). There is a slight reduction in growth rate caused by the ether treatment but auxin gives relatively the same effect (Fig. 2 G&H) as non ether treated. It should be noted that in all cases the segments were in the treatment solution for 3 or more hours before auxin was added.

The tentative conclusion is that there is no exogenous compound that is necessary for auxin-induced growth. However this does not answer the question about any relationships between cell expansion and a low pH in the cell wall. It does however, give a new way of testing whether auxin-induced acidification of the cell is necessary for auxin-induced growth. The question is whether conditions can be found that will allow growth without acidification. Does acidification stop in the absence of an exogenous supply of K^+, while growth continues?

It is quite possible that, for example, K^+ diffuses passively across the plasma membrane and is then used in a K^+-H^+ exchange system. There would be no net movement of K^+. There is leakage from the free space of the tissue and the concentration of mobile ions in the cell wall is likely to be low in deionised water. The solutions bathing the segments were changed to deionised water after they had been in 5 mM buffer for 5 hours. After flushing the system, the conductivity of the new solution was measured and did show an increase which could be detected for up to an hour.

The equilibrium concentration in the wall would be expected to be low but above the maximum concentration in the water of about 10 μM. If the concentrations in the wall are going to be effective it would require a mechanism similar to System I in roots[11]. There are several reports of the effect of auxins on uptake mechanisms[12] but most are not short term effects and unfortunately most measurements have been made at concentrations characteristic of System II. It is not yet clear whether mecanisms similar to System I occur in hypocotyls and stems, and if they do whether they result in acidification of the cell wall. However the general approach of determining what ions must be in the external medium before acidification can occur, will be a powerful test on the necessity for acidification in auxin-induced growth. There have already been studies along these lines[13,14,15] but short term growth should be measured simultaneously.

KINETICS

In a review still in press[12], we pointed out that direct auxin activation of a "proton pump" would not explain both the latent period and the rapid increase in the growth rate to a maximum value. This is because pH is a logarithmic scale and

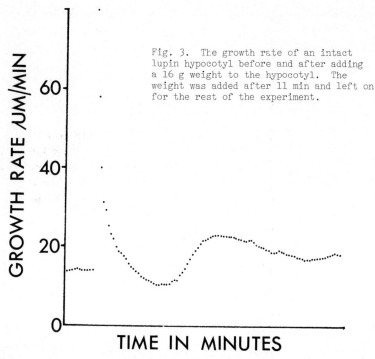

Fig. 3. The growth rate of an intact
lupin hypocotyl before and after adding
a 16 g weight to the hypocotyl. The
weight was added after 11 min and left on
for the rest of the experiment.

if a mechanism will lower the pH from 6.0 to 5.0 in say, 10 min then it will take
100 min to lower it from 5.0 to 4.0. And a system that may be inhibited at lower
pH's[16] will only make it more difficult to explain the kinetics by a single reaction.

We have recently referred to several conditions that give a marked oscillation
in the growth rate[17], one of which is shown in more detail in Fig. 3. Here the
growth rate is recorded before and after a weight is applied to an intact seedling.
A part of the high peak is due to mechanical properties of the walls but the oscillation is still there even if the weight is applied for only one minute and then
removed. Other conditions that show two measurable responses of the growth rate
have been reported[18].

It is not certain yet whether the two responses are independent or not, but two
linked responses could give a damped oscillation. Any involvement of acidification
of the wall in growth could be one of the responses and could maintain the pH of
the wall at a suitable range for enzyme action.

ENZYMES IN THE CELL WALL

There has always been interest in the possible involvement of cell wall enzymes
in cell expansion but until recently there has been little good evidence for the
involvement of any particular enzyme. One line of work has focused on B,1-3 glucanase in the wall and nojirimycin has been used since it is an inhibitor of the
enzyme. Nojirimycin inhibits both auxin-induced growth and the auxin-induced loss

of noncellulosic glucose from the cell wall[19]. A B,1-3 glucanase is readily detected in lupin hypocotyl cell walls[20] and nojirimycin inhibits growth of this tissue
(unpublished experiments). Evidence has recently been published that a B,1-3 glucan occurs in new plant cell walls[21].

Evidence of this type supports the hypothesis that extracellular enzymes may be
important in cell expansion and consequently the pH of the cell wall will alter the
activity of such enzymes. The observation that strong buffers in the neutral or
alkaline range reduce auxin-induced growth is not strong evidence that auxin acts
by a "proton pump", because the buffers could be maintaining the pH of the wall
outside the active range of the enzymes involved[12]. From our knowledge of enzymes,
it would be surprising to find growth to be independent of pH. Although we might
expect it to be advantageous to the plant to be able to maintain the pH of the wall
in a desirable range, this does not tell us whther changing the pH is a normal
method of controlling the growth rate.

But equally it does mean that changes in cell wall pH could be significant in
controlling growth. But of course lowering the pH artificially may activate an enzyme that is usually regulated in a different manner. It does seem unlikely that
acidification of the wall is the sole response to auxin that stimulates growth[12,12].
The present results do not exclude the possibility that it is one of two responses
that are linked in a feeback model. One way of testing this is to determine whether growth and acidification have the same requirements in the external medium.

REFERENCES

1 Penny,P.,Dunlop,J.,Perley,J.E.and Penny,D.(1975) Plant Sc. Lett. 4,35-40
2 Perley,J.E.,Penny,D.andPenny,P.(1975) Pl. Sc. Lett. 4,133-136
3 Marre,E.,Colombo,P.,Lado,P.and Rasi-Caldgno,F.(1974) Pl. Sc. Lett. 2,139-150
4 Thomas,R.C. (1976) Nature 262,54-55
5 Lucas,W.J.and Smith,F.A. (1973) J. Exptl. Bot. 24,1-14
6 Boron,W.F.and De Weer,P. (1976) Nature 259, 240-241
7 Sloane,M.and Sadava,D. (1975) Nature 257, 68
8 Penny,P.,Penny,D.,Marshall,D.and Heyes,J.K.(1972) J. Exptl. Bot. 23,23-36
9 Penny,D.,Penny,P.and Marshall,D.C. (1974) Can. J. Bot. 52,959-969
10 Bown,A.W.,Dymock,I.J. and Aung,T. (1974) Pl.Physiol. 54,15-18
11 Epstein,E. (1971)Mineral Nutrition of Plants: Principles and Perspectives, Wiley
   New York.
12 Penny,P.and Penny,D. (1976) in Plant Hormones and Related Compounds (Letham
   D.S.,Higgins,T.J. and Goodwin,P.B.,eds.),A.S.P. Biological and Medical,Amsterdam.
   (in press)
13 Bayer,M.H.and Sonka,J. (1976) Z. Pflanzenphysiol. 78,271-278
14 Haschke,H-P.and Luttge,U. (1975) Z. Pflanzenphysiol. 76,450-455
15 Lado,P.,De Michelis,M.I.,Cerana,R.and Marre,E.(1976) Pl.Sc. Lett. 6,5-20
16 Cleland,R.E. (1975) Planta 127,233-242
17 Marshall,D.C. and Penny,D. (1976) Aust. J. Plant Physiol. 3,237-246
18 Vanderhoef,L.N. and Stahl,C.A.(1975) Proc. Natl. Acad. Sci. U.S. 72,1822-1825
19 Nevins,D.J. (1975) Plant Cell Physiol. 16,347-356
20 Monro,J.A. (1974) Carbohydrate Fractionation and Elongation of Lupin Hypocotyl
   Cell Walls Ph.D. thesis, Massey University, Palmerston North
21 Fulcher,R.G.,McCully,M.E.,Setterfield,G.and Sutherland,J. (1976) Can. J.Bot.
   54,539-542

Regulation of Cell Membrane Activities in Plants
E. Marrè and O. Ciferri eds.
© *1977, Elsevier/North-Holland Biomedical Press, Amsterdam*

MODEL OF A SWITCHING "ON" AND "OFF" PUMP-AND-LEAK, AS A RELAY AND

AMPLIFICATION-MECHANISM IN THE CONTROL OF MORPHOGENESIS

Michel Thellier[*] and Marie-Odile Desbiez[**]
* L.A.T.P.E.I.C., Faculté des Sciences, I.S.H.N., 76130 Mont-St-Aignan, France
**Laboratoire de Phytomorphogénèse, 4 à 6 rue Ledru, 63000 Clermont-Ferrand, France

SUMMARY

In multicellular organisms, some cells would work as a relay and amplification-mechanism, especially for the control of morphogenesis. Such cells would be characterised by an active ionic pump liable to be switched "on" or "off", and by a membrane with a non (or poorly) reversible possibility of transition between two states of passive permeability ("small leak" or "big leak"). Some signals from the surrounding medium would switch the pump, and some others the leak. The reaction of the cells to those signals would thus be a modification of their ionic content, which in turn would shift their metabolism (and particularly their biosynthesis of some specific proteins) from one direction to another one. Such a model also explains the memorisation of some signals, which was experimentally encountered.

INTRODUCTION

In the course of its development, any multicellular organism undergoes a certain number of "shifts" in its morphogenetic behaviour: from the resting egg to the developing embryo, from the quiescent to the growing bud or from the vegetative to the flowering state in a plant, at the metamorphoses for some animals, etc.

We assume that a same type of "relay and amplification-mechanism" could apply to the control of many of those processes. We try here to build an interpretative model. It is based primarily upon experimental data obtained on the "mechanically induced precedence between axillary buds", which we have particularly studied. But it might be more general, as will be seen thanks to other data found in the literature.

REMINDER OF OUR EXPERIMENTAL DATA

The experimental data, obtained on plantlets of *Bidens pilosus L.* (var. *radiatus*), have already been described in detail[1-9]. Thus they will only be briefly mentioned again here.

Consider a plantlet of *Bidens* with its apical bud and no other developed leave than the two cotyledons. As long as the apex functions, the axillary bud remains quiescent (fig. 1a). If the apex is cut off, the cotyledonary buds will start growing; but, most often, one of them only will continue to grow and finally replace the eliminated terminal bud, while the other one will become quiescent again. This is the precedence effect (fig. 1b).

In homogeneous conditions, the two cotyledonary buds have the same chance (50 %

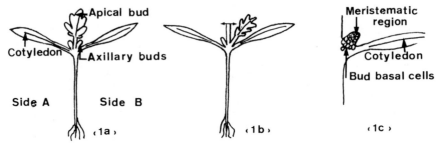

Fig. 1. *Schematic representation of the Bidens plantlet. (a): normal plant, with the arbitrary orientation A and B. (b): decapitated plantlet; in this case, bud B has taken precedence on bud A. (c): schematic histology of the axillary bud.*

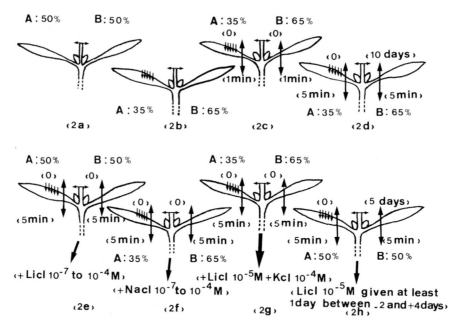

Fig. 2. *Summary of the experimental data. The percentages of dominating axillary buds on sides A and B are given for the different experimental situations.*

each) to take precedence on the other one (fig. 2a). But a few prickings in one of the cotyledons (refered to as cotyledon A) are enough to give a significant advantage to the axillary bud of cotyledon B: the relative chances now become about 35 % and 65 % (fig. 2b). Removing both cotyledons, less than 1 minute after the prickings, will not modify the final percentage of dominating buds on sides A and

B (fig. 2b and 2c): this means that the pricking-signal is rapidly transmitted from the wounded cotyledon to the axillary buds. But, when it has reached the buds, the signal can be memorised there for a long time, as can be seen by pricking a cotyledon before decapitation of the plant: if the prickings are made up to 10 days before decapitation, the percentages will still be 35 % and 65 % (fig. 2d). Li^+, even at very low concentration, suppresses the effect of the prickings, whereas Na^+ does nothing (fig. 2e and 2f); the effect of Li^+ can be reversed by K^+ (fig. 2g); and Li^+ acts on the memorisation mechanism, as it remains efficient even if given between prickings and decapitation (fig. 2h). The direct effect of K^+ is complicated: it allows or suppresses the effect of the prickings, depending on the concentration, the photoperiodicity and the presence or absence of the root system. Last, the total growth of the axillary buds (bud A + bud B) is not much modified by these ionic treatments.

From the morphological point of view, the axillary bud can be divided into two parts (fig. 1c): an apical region with small meristemetic cells, and a basal region with somewhat bigger cells (although never exceeding much 10 μm in diameter). The growth of the bud always begins by enlargement of the basal cells during the first days which follow the decapitation of the plant: during this period, both axillary buds will grow more or less; and even if, at a given moment, one bud seems to dominate on the other one, the precedence remains reversible. Later on the meristematic cells begin to divide, and the bud differentiates in such a way that it joins the conductive tissues of the axis: the precedence is now generally apparent and irreversible.

PROPOSED INTERPRETATIVE MODEL

Statement of the model

The basal cells of the axillary buds would act as a regulatory module for the growth of the bud. They would have the following characteristics (fig. 3 and 4):
- They would possess an active ion pump and a passive leak. For the sake of simplicity, and by analogy with what is known from other plant systems[10], we shall suppose that the active ionic pumping works on H^+-extrusion, and that K^+ is the main compensating ion.
- The pump could be switched "on" or "off"; and the leak could be either "small" or "big", as corresponding to a non-reversible (or, at least, poorly reversible) transition of the cell-membrane. This finally corresponds to 4 possible states for those cells (fig. 3a-d).
- In the basal cells, the synthesis of the various permeases (and possibly of some other specific proteins) would be dependent on the K^+-content of the cell (K^i), with a simoïdal response curve (fig. 4). As a first approximation, we can consider a critical value (K^i_c) of the K^+-content of the basal cells: when $K^i < K^i_c$

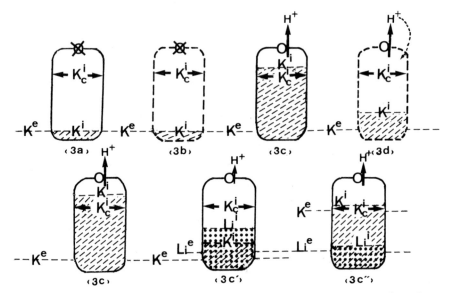

Fig. 3. Model-representation of the ionic status of the basal cells. K^e, Li^e: ionic concentrations in the surrounding domain. K^i, Li^i: ionic concentrations in the basal cells. K^i_c: critical concentration (see fig. 4).

(fig. 3a, b, d) those cells would possess little or no permeases for the various organic and inorganic nutrients; when $K^i > K^i_c$ (fig. 3c), they would synthetize permeases.

Interpretation of the experimental data

Prior to the plant decapitation, the whole trophic capacity of the plant is diverted towards the terminal bud. The pump of the basal cells of the axillary buds is probably switched off, perhaps only by lack of available energy; hence, whatever is the leak (fig. 3a-b), we shall have $K^i < K^i_c$ and those cells will have no permeases: lacking both substrates and permeases, they will naturally remain quiescent. No growth can occur in the axillary buds, in such conditions.

Following plant decapitation, the whole trophic capacity of the plant becomes available to the buds. And we shall admit that, either for purely trophic or for hormonal reasons, plant decapitation also switches on the active pump of the basal cells of the axillary buds.

The enlargement of the basal cells (which is the first manifestation of growth of the buds) can then be explained in the following way.

Let τ_A and τ_B be the proportions of the basal cells with a "small leak" (fig. 3c), respectively in buds A and B. The proportions of the basal cells with a "big leak" (fig. 3d) will then be $(1 - \tau_A)$ and $(1 - \tau_B)$.

The basal cells which have a "big leak" (hence $K^i < K^i_c$) will not synthetise permeases. We shall say that their overall "conductance" (to the absorption of the various organic or inorganic substrates from the surrounding tissues) will remain low: let us call $\bar{1}$ such a low "conductance".

Conversely, the basal cells which have a "small leak" (hence $K^i > K_c^i$) will synthetise permeases: they will thus get a high "conductance" to the absorption of nutrients, which we shall call 1.

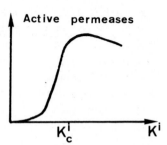

Fig. 4. *Model-representation of the permease-activities in the basal cells of an axillary bud, in terms of the K^+ content of the cells.*

Respectively for buds A and B, the resultant conductances can then be expressed in a normalised way, as

$$\begin{cases} L_A = \tau_A 1 + (1 - \tau_A)\bar{1} = \tau_A(1 - \bar{1}) + \bar{1} \\ L_B = \tau_B 1 + (1 - \tau_B)\bar{1} = \tau_B(1 - \bar{1}) + \bar{1} \end{cases} \quad (1)$$

And, $\bar{1}$ being defined as small when compared to 1, this simplifies to

$$\begin{cases} L_A \simeq \tau_A 1 + \bar{1} \\ L_B \simeq \tau_B 1 + \bar{1} \end{cases} \quad (2)$$

If we call T the overall trophic supply coming from the rest of the plant, it will be shared, between the two sets of basal cells A and B, as

$$\begin{cases} T_A = TL_A/(L_A + L_B) \\ T_B = TL_B/(L_A + L_B) \end{cases} \quad (3)$$

with

$$T = T_A + T_B \quad (4)$$

The respective enlargements of both sets of basal cells A and B will be proportional to T_A and T_B: in fact, T_A and T_B can be considered as measurements, in arbitrary units, of the enlargements of the sets of basal cells.

At this early stage, when the growth of a bud practically corresponds to the enlargement of its basal cells only, T_A and T_B will thus measure the growth of buds A and B respectively.

Hence, following the model, during this first period of the growth, both axillary buds will grow; the bud with the biggest value of τ will take an advantage on the other one; but this apparent precedence will remain reversible; the total growth (bud A + bud B is independant of the absolute values of τ_A and τ_B (equation 4), as far as T can be considered independant of τ_A and τ_B.

In a later stage of the growth, the basal cells will also synthetise diffusible activators (or eliminate inhibitors) for the rest of the bud. This will tend to activate the meristems and to join the bud to the conductive tissues of the axis. The bud with the highest value of τ will have a better chance to be the first one to achieve those processes. And when this is done, its precedence will be irreversibly gained.

In homogeneous conditions, when no prickings are made on the cotyledons, τ_A might be different from τ_B in each plant; but, statistically, there is approximately the same number of cases when τ_A is bigger than τ_B and when it is smaller. And this corresponds to 50 % plants where bud A will finally dominate, and 50 % where bud B will dominate.

The signal initiated by pricking a cotyledon would shift some of the basal cells of both buds from the little-permeable state (small leak) to the highly permeable one (big leak). This would tend to diminish τ_A and τ_B, but, naturally, τ_A more than τ_B when cotyledon A is the pricked one: and this would give a statistical advantage to bud B.

The memorisation of the pricking signal is easily understood: the shift of a basal cell (from the non-permeable to the permeable state) might occur whether the pump is switched on (fig. 3d) or off (fig. 3b).

Li^+, if present in the plant, will tend to accumulate in the basal cells when they start pumping. There, it probably inhibits the pump; and, anyway, by pure electrochemical effect, it diminishes the K^+-content of the cell (K^i) (fig. 3c'). Whatever their leak, all the basal cells are thus in a situation where $K^i < K_c^i$, which will correspond to conductance-values remaining fairly close to $\bar{1}$. Li^+ would thus restore the symmetry between buds A and B, even if $\tau_A \neq \tau_B$.

A greater K^+-supply, in the presence of Li^+, would tend to diminish the interference of the Li^+-ions, thanks to a pure electrochemical effect. It could thus reintroduce the dissymmetry between buds A and B, when $\tau_A \neq \tau_B$ (fig. 3c'').

POSSIBLE GENERALISATION OF THE MODEL

Two sets of experimental data, on plant or animal systems, might lead to an extension of the model.

First, some organs (the hypophysis, salivary glands, digestive system) which modulate the secretion-velocity of definite proteins, also tend to accumulate Li^+ much more than the neighbouring organs do[11].

Secondly, many different mechanisms of morphogenesis are found to react to ionic (Li^+ and K^+) treatments in a manner very similar to the one which has been described here for the cotyledonary buds. This is the case for the regeneration of some invertebrates[12-15], sea-urchin embryology[16-19], and the induction of a plant to flower[20].

One might thus wonder if similar types of control-models could not be considered for all those processes where definite shifts in the protein-metabolism are encountered: the adjustement of the ionic content of some target cells (thanks to a "pump-and-leak" ionic mechanism) would then appear as a very general relay and amplification-system of intercellular control.

POSSIBLE EXPERIMENTAL APPROACHES TO TEST THE MODEL

The experimental test of the model will not be easy. Especially, the electro-physiological study of the basal cells of the buds is impossible for two main reasons. First, those cells are small and deeply buried in the rest of the plant tissues: it would be very difficult, if not impossible, to know the exact position of a microelectrode. Besides, we have seen that, given to the cotyledons, prickings could modify the behaviour of the basal cells of the axillary buds. Hence one would never know if a direct pricking of those basal cells with a microelectrode would not also perturb them!

At present, the best way of approach would be to study the distribution of the different ions in the basal cells, under the different experimental conditions. This can be done with the classical methods for most of the ions, and with the help of a (n,α) nuclear reaction for lithium[11].

Besides, if the phenomenon is really general, we should be able to find some other living samples where it would, both, exist, and be more easily accessible to experiment (especially by biochemical means). The recovery of some "shock-effects" in cell-suspension cultures (as studied by Guern's group and ourselves[21-22]), or the processes of ageing in leaves (as studied by Carlier[23]), might be relevant for that purpose.

DISCUSSION

Clear enough, our model is still hypothetical and might have to be modified in the future. For instance, we have taken into account only two possible values of the "conductance" (1 and $\bar{1}$), whereas the consideraration of intermediate values would probably be more realistic. Besides, an "overall conductance" has not a precise biological signification; and one should rather consider a conductance term for each possible substrate. However, as it is, the model will at least help to classify and discuss the amount of experimental data already available.

Furthermore, beside the well-known rapid mechanism of control (as encountered in nerves, for instance), our model thus comes to postulate the existence of a long-term mechanism of memorisation and transfer of informations. The former is found in highly differentiated cells; it undergoes rapidly-reversible transitions of the cell-membrane; and the signal is the very electric depolarisation of the cell, without any metabolic biosynthesis accompanying it. Conversely, the latter would be found in much more commonplace cells; the transition of the membrane would be non-reversible (or, perhaps, only slowly reversible); and the main consequence of the signal would be the biosynthesis of new substances (especially proteins). But both mechanisms would, at least, have in common the existence of cells with an ionic pump and two conformational states of leak. They might thus represent two different evolutionary states of the same fundamental process.

298

ACKNOWLEDGEMENTS

This work was supported by grants of the C.N.R.S. (L.A. 45, L.A. 203, and R.C.P. 285) and of the D.G.R.S.T. (74 7 0194). We are very thankful to Professors Champagnat and Marré for scientific discussions.

REFERENCES

1. Desbiez, M.O. (1975) State Doctorate Thesis, Clermont-Ferrand, France.
2. Desbiez, M.O. (1971) Biol. Plant. (Praha) Vol. 13, 375-382.
3. Desbiez,M.O. (1973) Z. Pflanzenphysiol. Vol. 69, 174-180.
4. Champagnat, M., Desbiez, M.O. and Delaunay, M. (1973) Ann. Sci. Nat., Bot., Paris, 12ème série, Vol. 14, 71-86.
5. Desbiez, M.O. and Chamel, A. (1974) C.R. Acad. Sci., Paris, série D, Vol. 279, 1433-1436.
6. Desbiez, M.O. and Thellier, M. (1975) Plant Science Letters Vol. 4, 315-321.
7. Desbiez, M.O. (1976) Physiol. Plant. Vol. 36, 11-15.
8. Champagnat, P. and Desbiez, M.O. Physiol. Veg. (under the press).
9. Desbiez, M.O. and Thellier, M. C.R. Intern. Workshop on the Transmembrane Ionic Exchanges in Plants (under the press).
10. Marré, E., Lado, P., Rasi-Caldogno, F., Colombo, R., Cocucci, M. and de Michelis, M.I. (1975) Physiol. Veg. Vol. 13, 797-811.
11. Thellier, M., Stelz, T. and Wissocq, J.C. (1976) Biochim. Biophys. Acta (under the press)
12. Bondi, C. (1957) Arch. Zool. Ital., Napoli, Vol. 42, 105-122.
13. Kanatani, H. (1958) J. Fac. Sci. Tokyo Univ. (Zool.) Vol. 8, 245-251.
14. Stephan-Dubois, F. and Morniroli, C. (1967) Bull. Soc. Zool. Fr. Vol. 92, 335-344.
15. Yasugi, S. (1974) Development, Growth and Differ. Vol. 16, 171-180.
16. Herbst, C. (1892) Z. Wiss. Zool. Vol. 55, 446-518.
17. Lallier, R. (1960) Exptl. Cell Res. Vol. 21, 556-563.
18. Lallier, R. (1963) Exptl. Cell Res. Vol. 29, 119-127.
19. Lallier, R. (1973) C.R. Acad. Sci., Paris, série D, Vol. 276, 3473-3476.
20. Kandeler, R. (1970) Planta Vol. 90, 203-207.
21. Dorée, M., Leguay, J.J. and Terrine, C. (1972) Physiol. Veg. Vol. 10, 115-131.
22. Thoiron, A., Thoiron, B., le Guiel, J., Guern, J. and Thellier, M. (1974) in: Membrane transport in Plants, Zimmermann and Dainty Eds., Springer-Verlag, Berlin, 234-238.
23. Carlier, G. (1975) Physiol. Veg. Vol. 13, 445-454.

Regulation of Cell Membrane Activities in Plants
E. Marrè and O. Ciferri eds.
© *1977, Elsevier/North-Holland Biomedical Press, Amsterdam*

LEAKAGE OF α-AMYLASE FROM GA$_3$-INDUCED WHEAT ALEURONE LYSOSOMES

R.A. Gibson and L.G. Paleg
Department of Plant Physiology, University of Adelaide,
Waite Agricultural Research Institute, Glen Osmond, S.A. 5064.

Summary

The properties of membranes of GA$_3$-induced wheat aleurone lysosomes were examined. Permeability as measured by leakage of α-amylase, was increased by extremes of pH, high temperature and calcium ions, and was decreased by low temperatures and phosphate ions. GA$_3$ was found to have no in vitro effect on α-amylase leakage from isolated lysosomes, but lysosomes from tissue treated with high levels of GA$_3$ (100, 1 µg/ml) released more α-amylase at elevated temperatures than lysosomes from tissue treated with lower (0.1, 0.01 µg/ml) hormone levels.

INTRODUCTION

In previous papers (1,2,3) many characteristics were described of the enzyme-containing particulate system derived from GA$_3$-treated wheat aleurone. The sedimentable organelles contain α-amylase and proteinase and exhibit many of the conceptual and functional features of animal lysosomes (4). We concluded that they were, in fact, lysosomes, and this report deals with properties of the membrane of this organelle.

In spite of the work devoted to the identification and characterization of the contents of lysosomes in plants (5,6), virtually nothing is known about the membrane of the plant lysosome. The properties of the lysosomal membrane are important with respect to possibly identifying its site of synthesis, controlling the intracellular separation of enzymes and their substrates, regulating fusion between the lysosome and the plasma membrane (in the case of extracellular secretion), etc. The latter point is of added significance since Varner and Mense (7) suggested that the release of α-amylase from aleurone cells must occur before further enzyme synthesis can occur.

Work with steroid hormones and animal lysosomes highlights a further area of ignorance with respect to plant lysosomes. Several reports (8,9) describe an in vivo change in the permeability of lysosomal membranes to the enclosed acid hydrolases as a result of steroid hormone treatment of the intact animal. The work suggested that hormone treatment altered either the composition or the physical characteristics of the lysosomal membrane. Since plant hormones influence processes like senescence, abscission, nutritional reserve utilization, etc., in which lysosomal enzymes may be involved, the influence of hormones on lysosomal properties may be of major importance.

MATERIALS AND METHODS

Lysosome isolation: Unless otherwise stated, the words lysosome or lysosome preparation are used to describe the α-amylase-containing particles in a 60,000g pellet obtained by the methods described in previous publications (1,2,3). The pellets were resuspended in 1 ml of medium with the aid of a plastic piston device having 0.1 mm clearance. Unless otherwise stated, the resuspension medium contained 0.4 M sucrose and 0.05 M Tris-HCl (pH 7.0). The pellets were resuspended by maintaining a twisting pumping action with the piston device until a creamy, homogeneous suspension was obtained (no more than 30 sec). Any additions were made at this stage in a volume of 0.1 ml. GA_3, when added, was included in the resuspension medium.

Leakage of α-amylase: During resuspension of the 60,000g pellets, the samples were kept cold. The resuspended lysosomes were incubated for 30 min. at the required temperature. The volumes were then made up to 10 ml with cold 0.4 M sucrose and again centrifuged at 60,000g. Calcium acetate was added to the supernatant to give a final concentration of 5 mM with respect to calcium. When phosphate was used as the buffer, sufficient (twice the molarity of phosphate) calcium was added to overcome its tendency to precipitate the calcium present. Just prior to the α-amylase assay, and after heating for 20 min at 70°C, the supernatant fractions were adjusted to pH 5.0 and made to a known volume with distilled water.

The pellets obtained from the second centrifugation were treated for 5 min with 1 ml of 0.1% Triton-100 during which time they were stirred with a glass rod. After adding 4 ml of 5 mM calcium acetate, the contents were transferred to a glass homogenizer (7 ml volume), and the membranous material thoroughly ground by hand. The original centrifuge tube and the glass homogenizer were washed thoroughly with 5 ml of 5 mM calcium acetate and the combined sample (10 ml) heated for 20 min at 70°C and assayed for α-amylase activity.

The amount of α-amylase detected in the supernatant fraction was calculated as a percentage of the total α-amylase for that treatment (i.e. supernatant + pellet = total). Since the α-amylase in the supernatant represents enzyme not enclosed by membranes, it is referred to as "% free" enzyme.

Because incubations under various conditions were potentially capable of inactivating α-amylase, a control sample was included in each experiment. The control sample was resuspended in 5 mM calcium acetate and set aside for later assay. The total α-amylase detected in the incubated samples was calculated as a percentage of the total enzyme measured in the control sample and is referred to as "% recovered".

α-Amylase assays: This assay was described earlier (2).

RESULTS

To examine the effects of various treatments on a membrane, at least one property is required that is indicative of known membrane behaviour, is specific for the particular membrane, and is relatively easy to measure. The crude lysosomal preparations used in this study (60,000g pellet) contain differing amounts of most of the organelles present in aleurone cells. However, the assumption was made that by assaying the leakage of α-amylase, the effect of different treatments was on the membranes that enclose the enzyme (i.e., the lysosomal membrane) regardless of the effects the same treatments may have had on other membranes or components.

Osmotic influences on the passage of enzymes through lysosome membranes are frequently observed and, thus, the effect of different osmotica were compared (Table 1). Leakage of α-amylase was inversely proportional to the concentration of sucrose, though the effect was not pronounced. Very slightly higher leakage rates were obtained when sorbitol or mannitol were used, but the recovery of α-amylase was not influenced by any of the osmotic agents.

Calcium ions are important not only for the full expression of α-amylolytic activity, and the stabilization of the enzyme against proteolytic attack, but also in the control of membrane structure and function (10,11). It was of interest to note that addition of calcium ions, over a wide concentration range, had relatively little effect on the lysosomes (Table 1). Calcium did produce a small (7%) concentration-dependent influence on enzyme recovery, and a slightly greater (14%) increase in leakage, but it was concluded that most of the calcium binding sites were probably filled during initial incubation of the aleurone tissue, or had been rendered inaccessible to the cation by some subsequent technique.

TABLE 1

EFFECT OF OSMOTICUM AND CALCIUM ON LEAKAGE OF α-AMYLASE FROM LYSOSOMES

Lysosomes, isolated from GA_3(10 μg/ml)-treated aleurone, resuspended in medium containing 0.01 M Tris-HCl (pH 7.0) and 0.4 M sucrose or sugar indicated, and incubated for 30 min at 30°C

Treatment	% α-Amylase free	% α-Amylase recovered
0.4 M Sucrose (control)	41	95
0.8 M Sucrose	34	101
0.4 M Mannitol	47	101
0.4 M Sorbitol	45	101
1 mM $Ca(NO_3)_2$	47	98
5 mM $Ca(NO_3)_2$	52	101
10 mM $Ca(NO_3)_2$	55	102

The permeability of the lysosomal membrane to enclosed hydrolytic enzymes has been shown to be responsive to pH (12), and with tris-acetate buffer it was found that leakage of α-amylase was minimal between pH 6.5 and 7.0, which was also the range over which greatest recoveries of the enzyme were observed (Fig. 1). α-Amylase appeared to be most stable (judging from recovery values) at or near neutrality and this may not be too unexpected since acid protease is also present in the lysosome (3).

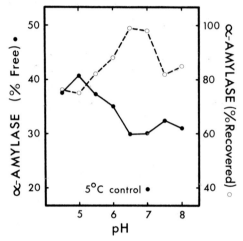

Fig. 1. Effect of pH on the leakage of α-amylase from lysosomes isolated from GA₃-treated (10 μg/ml) aleurone. Lysosomes resuspended in 0.05 M Tris-acetate buffer at the pH indicated and 0.4 M sucrose and incubated for 30 min at 30°C.

The effect of phosphate was also examined. Whether it was added in the presence or absence of tris buffer, the rate of α-amylase leakage was reduced (Table 2), and neither arsenate nor the organic phosphates (ADP and ATP) could reproduce the effect.

TABLE 2

EFFECT OF TRIS AND PHOSPHATE BUFFERS ON LEAKAGE OF α-AMYLASE FROM LYSOSOMES

Lysosomes, isolated from GA₃ (10 μg/ml)-treated aleurone, resuspended in 0.4 M sucrose, material indicated (at pH 7.0), and incubated for 30 min at 30°C

Treatment	% α-Amylase free	% α-Amylase recovered
Phosphate (0.01 M)	27	90
Tris (0.05 M)	40	100
Tris + phosphate	28	98
Tris + arsenate	36	100
Tris + ATP (1 mg)	36	91
Tris + ADP (1 mg)	40	97

The possibility that GA₃ might exert an in vitro effect on the lysosomal membrane was tested. No evidence was obtained that suggested a change in any enzyme or membrane parameter as a result of the addition of pH-adjusted solutions of GA₃ to suspensions of the lysosomes. However, since lysosomes are essentially only present as a result of treating aleurone tissue with GA₃, the possibility existed that all of the GA₃ binding sites (if indeed there were any) were already filled

before isolation of the lysosomes. An attempt was made to circumvent this difficulty by extracting and testing lysosomes isolated from aleurone treated with
optimal (1 µg/ml) or sub-optimal (0.1 µg/ml) amounts of the hormone. The lysosomes were incubated with various concentrations of GA_3 at several temperatures.
The results (Table 3) indicated that GA_3 did not affect the recovery of α-amylase
under any of the conditions. Furthermore, the <u>in vitro</u> treatment of lysosomes
with GA_3 at any of the levels produced no effect, regardless of the incubation
temperature. However, there was a significant difference in the response of lysosomes isolated from 0.1 µg/ml GA_3-treated tissue and those obtained from 1.0 µg/ml
-treated aleurone. Leakage of α-amylase from the 1.0 µg/ml GA_3-treated tissue at
$30°$ was 40% more than leakage from the lysosomes from aleurone treated with the
lower GA_3 concentration.

TABLE 3

EFFECT OF GA_3 AND TEMPERATURE ON LEAKAGE OF α-AMYLASE FROM LYSOSOMES

Lysosomes, isolated from aleurone tissue incubated 24 hr in 0.1 µg/ml or 1.0 µg/ml
GA_3 at 30°C, resuspended in 0.4 M sucrose, 0.05 M Tris-HCl (pH 7.0) and GA_3 at the
concentrations indicated, and incubated for 30 min at temperatures indicated.

Lysosome incubation temperature	Lysosome treatment (µg GA_3/ml)	Leakage from lysosomes from aleurone treated with			
		0.1 µg/ml GA_3		1.0 µg/ml GA_3	
		% Free	% Recovered *	% Free	% Recovered *
5°C	0	12	102	12	98
	1	11	102	10	104
	10	15	99	11	100
	100	<u>14</u>	103	<u>10</u>	100
	Ave.	13.0	n.s.d. from	10.75	
15°C	0	16	97	16	97
	1	16	101	16	99
	10	17	101	16	102
	100	<u>19</u>	99	<u>16</u>	101
	Ave.	17.0	n.s.d. from	16	
30°C	0	22	102	35	105
	1	27	97	34	97
	10	22	100	33	98
	100	<u>23</u>	98	<u>35</u>	99
	Ave.	23.5**		34.25**	

* 100% recovery = 37.5 units (low GA_3), 38.3 units (high GA_3).
** Significantly different at 1% level.

The experiment was expanded and repeated several more times, each producing
essentially the same results (Fig. 2). Lysosomes from aleurone treated with high
concentrations of GA_3 behaved differently, in that they leaked more α-amylase

304

particularly at warmer incubation temperatures, than lysosomes from tissue treated with low GA_3 levels.

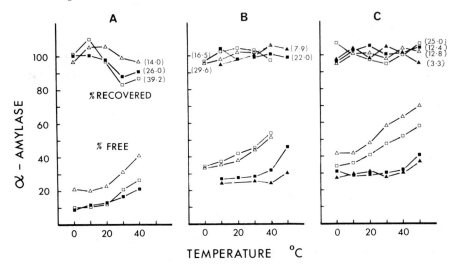

Fig. 2.   Results of three experiments (A, B and C) on the effect of temperature on leakage of α-amylase from lysosomes isolated from aleurone tissue treated with 100 ($\triangle$), 1 ($\square$), 0.1 ($\blacksquare$) or 0.01 ($\blacktriangle$) μg GA_3/ml. Lysosomes resuspended in 0.01 M Tris-HCl (pH 7.0) and 0.4 M sucrose and incubated for 30 min at the temperature indicated. Figures in parenthesis are α-amylase units constituting 100% recovery.

DISCUSSION

Previous experiments with lysosomes from this tissue were designed to show that there was latent enzyme activity associated with this fraction of the cell contents, and the results demonstrated the efficacy of various treatments in liberating it (1). Thus, freezing and thawing, Triton X-100, different phospholipases, etc., were all shown to be effective to a greater or lesser extent.

In the present work, however, and notwithstanding the previous experiments, it seems that the lysosomes are remarkably resistant to most of the treatments applied. For instance, a four-fold change in sucrose concentration produced only about a 10% change in leakage, and the use of sorbitol or mannitol also had little effect. Changing the calcium concentration over a wide range altered leakage by only about 15%, and neither the osmotic agents nor calcium seemed to alter the amount of enzyme recovered.

Phosphate, which unexpectedly decreased leakage, had about a 13% effect. It may also be of interest that phosphate, in combination with temperatures above 30°C, seemed to cause a decrease in enzyme recovery (unpublished data). The effect of pH, contrary to other treatments, was manifest on both leakage and recovery. The magnitude of the effects was not large although the results are strikingly similar to those obtained with mouse liver lysosomes by Mego (13); he says that there is no proteolysis in lysosomes at neutral pH, whereas, because of the presence of acid proteases, intra-lysosomal proteolysis is high at pH 5. On the other hand, Mego also reports that lysosomal breakage at pH 7 is at a minimum, as is leakage of enzymes from the lysosome.

Although the effect of the other in vitro treatments on α-amylase leakage and/ or recovery from lysosomes was slight, exogenous GA$_3$ had no effect on either para-meter. It was totally unexpected, therefore, to find the "carry-over" effect of GA$_3$-treatment of the aleurone influencing the behaviour of the subsequently-isolated lysosomes. However, it seems quite clear that the temperature-dependent leakage of lysosomal enzymes is, at least partially, a function of the GA$_3$ concentration used to treat the tissue.

The results above could be brought about by either a physical or a compo-sitional change in the lysosomal membrane. For example, de Duve et al. (14) and others (8,9) reported changes in the permeability of lysosomal membranes following in vivo application of hormones. The difficulty with this interpretation for the present results is the problem of why in vitro treatment of the lysosomes doesn't have the same effect. Of course, there is the possibility that an in vitro effect is both conceptually unlikely and practically impossible to demonstrate because GA$_3$ must be applied to the aleurone cells to induce lysosomes, and, thus, any hormone-membrane interaction would have already occurred.

An alternative explanation is that GA$_3$ induces the synthesis of different mem-brane components of the lysosome and these changes are reflected by an altered thermal behaviour of the isolated lysosomes. This explanation is perhaps more reasonable, though it is difficult to understand why lysosomal membranes synthe-sized as a response to one hormone concentration should differ in composition from membranes synthesized as a response to a different concentration of the hormone.

Finally, in spite of our ignorance of the reasons that lysosomal membranes dis-play different thermal behaviour patterns as a result of different concentrations of hormone triggering their synthesis, the fact that such differences occur may be of importance in terms of the in vivo secretion of hydrolytic enzymes from the aleurone. The altered thermal behaviour suggests that membranes of lysosomes for-med in high GA$_3$-treated aleurone are more flexible and more fluid at any given temperature than the membranes of lysosomes formed in response to lower GA$_3$ levels. Since active secretion from the cell must involve fusion of lysosomes with the plasmalemma, and, as the ability of one membrane to fuse with another must be dependent on the physical state (fluidity) of the membranes involved, it may be that hydrolytic enzyme synthesis in and secretion from aleurone cells is con-trolled by the physical state of the secretory (lysosomal) system rather than, or in addition to, protein synthesis per se. Several reports (7,15) indicated that release of aleurone lysosomal enzymes must occur before further enzyme synthesis can occur.

ACKNOWLEDGMENTS

The authors gratefully acknowledge the financial support received from the Australian Research Grants Committee, the Barley Improvement Trust Fund and the C.S.I.R.O. They also thank Dr. A. Wood for helpful and stimulating discussions.

REFERENCES

1. Gibson, R.A. and Paleg, L.G. (1972) Biochem J. 128, 367-375.
2. Gibson, R.A. and Paleg, L.G. (1975) Aust. J. Plant Physiol. 2, 41-49.
3. Gibson, R.A. and Paleg, L.G. (1976) J. Cell Sci. (in press).
4. Dingle, J.T. (1969) in Lysosomes in Biology and Pathology, Vol. 1, North-Holland, Amsterdam.
5. Gahan, P.B. (1973) in Lysosomes in Biology and Pathology, Vol. 3, North-Holland, Amsterdam.
6. Matile, P. (1969) in Lysosomes in Biology and Pathology, Vol. 1, North-Holland, Amsterdam.
7. Varner, J.E. and Mense, R.M. (1972) Plant Physiol. 49, 187-189.
8. Szego, C.M. (1972) Advances in Cyclic Nucleotide Research 1, 541-564.
9. Szego, C.M. and Seeler, B.J. (1973) J. Endocr. 56, 347-360.
10. Chrispeels, M.J. and Varner, J.E. (1967) Plant Physiol. 42, 398-406.
11. Triggle, D.J. (1972) in Progress in Surface and Membrane Sciences, Vol. 5, Academic Press, New York.
12. Sawant, P.L. et al. (1964) Arch. Biochem. Biophys. 105, 247-253.
13. Mego, J.L. (1973) in Lysosomes in Biology and Pathology, Vol. 3, North-Holland, Amsterdam.
14. de Duve, C. et al. (1962) Biochem. Pharmacol. 9, 97-116.
15. Jones, R.L. and Armstrong, J.E. (1971) Plant Physiol. 48, 137-142.

Regulation of Cell Membrane Activities in Plants
E. Marrè and O. Ciferri eds.
© 1977, Elsevier/North-Holland Biomedical Press, Amsterdam

CALCIUM-ENHANCED ACIDIFICATION IN OAT COLEOPTILES

Bernard Rubinstein
Department of Botany
University of Massachusetts
Amherst, MA 01002
U.S.A.

Kenneth D. Johnson and David L. Rayle
Department of Botany
San Diego State University
San Diego, CA 92182
U.S.A.

SUMMARY

We have compared the acidification responses initiated by auxin and calcium ions in oat coleoptiles. An exogenously-supplied source of calcium ions is not required for auxin-induced acidification, and calcium alone can evoke a rapid acidification response. Since calcium (but not auxin) can cause the release of acid from metabolically-inactive, frozen-thawed coleoptile sections, the main site of calcium action may be the cell wall where calcium ions exchange off protons which then diffuse into the external medium.

INTRODUCTION

Auxin-induced growth of excised grass coleoptiles and young stem segments is accompanied by, and may be dependent upon, acidification of the cell wall region[1]. If the acidification response to IAA[*] involves the extrusion of protons across the plasmalemma, then maintenance of the cell potential could be accomplished by the simultaneous expulsion of an anion or import of a cation. Marrè et al.[2] and Haschke and Lüttge[3] have shown that the release of acid equivalents is balanced by K^+ uptake during auxin treatment of pea epicotyl segments and oat coleoptiles, respectively. No data on auxin-induced anion release is available.

Recently, Cohen and Nadler[4] have claimed that oat coleoptiles require Ca^{2+} for auxin-induced acidification, and have suggested that a H^+-Ca^{2+} exchange process may be occurring. Such an interpretation of this Ca^{2+} effect is in direct conflict with the H^+-K^+ exchange hypothesis, and is yet unsubstantiated by any Ca^{2+} uptake studies.

To investigate the possibility of an interaction between Ca^{2+} and IAA in the acidification response, we have attempted to characterize the Ca^{2+}-enhanced acidification in oat coleoptile sections. We will show that the initial Ca^{2+} effect on

---

[*]Abbreviations: IAA, indoleacetic acid; DNP, dinitrophenol; CCCP, carbonyl cyanide m-chlorophenylhydrazone.

acidification may be larger than that observed for auxin alone, and the protons released by Ca^{2+} treatment originate in large part from the cell wall.

MATERIALS AND METHODS

Preparation of coleoptiles: Seeds of Avena sativa L. cv. "Rodney" were sown in moist vermiculite and grown in the dark (with one or two exposures to weak white light) at 27 C for 4.5 days. On the fifth day coleoptiles from 2 to 4 cm in length were excised, the cuticular-epidermal layer was stripped away using a fine jeweler's forceps[5], and the "peeled" coleoptiles were then cut to 15 mm at a 3-mm distance from the apical tip. The sections were then rinsed, and subsequently transferred to a preincubation medium containing 1 mM KCl (pH 6.5) and placed on a reciprocating shaker at 25 C.

Frozen-thawed sections were prepared by spraying the peeled coleoptiles on both sides with liquid freon ("Cryokwik", Damon Scientific), thawing rapidly in dis-tilled water (25 C), then repeating this treatment. The sections were then rolled flat with a wooden applicator, rinsed extensively with water, and preincubated like fresh sections.

Measurement of acidification: After 15 min in the 1 mM KCl preincubation medium, the coleoptiles were transferred to 10-ml beakers (usually 10 sections per beaker) containing 2.0 ml of fresh 1 mM KCl (pH 6.5). Five min thereafter, 20 μl of a 100-fold concentration of the desired solution was added, and the pH of the treatment medium was adjusted to 6.40 $\pm$ 0.05. At various intervals thereafter, measured amounts of 3.0 mM KOH were added to return the pH to 6.40 (the pH never fell below 5.8 before titration). pH Determinations were made on a Beckman "Ex-pandomatic" meter equipped with either a Markson "Pencil" pH electrode or an In-gold flat-surface electrode. All solutions were agitated vigorously before each reading.

Measurement of $^{86}Rb^+$ uptake: Groups of 20 15-mm peeled coleoptile sections were placed in 10-ml beakers containing 2.0 ml of 2 mM MES-tris buffer, pH 6.5. After a 45-min preincubation at 24 C on the shaker, 20 μl of 0.1 M $CaCl_2$ and/or 1 mM IAA were added; H_2O was added to make the final volume 2.04 ml. After a 60-min treatment period, 20 μl of 0.1 M KCl and 10 μl of $^{86}RbCl$ (40 μCi/ml; 524 mCi/ g Rb^+; ICN Pharmaceuticals) were added. A 30-min uptake period was terminated by aspirating the radioactive medium and rinsing the coleoptiles twice with 3 ml of 1 mM KCl, followed by a 6-min chase period in 3 ml of 1 mM KCl. The chase medium was aspirated and any solution trapped within the inner hollow space of the coleop-tile was forced out with a stream of air. The sections were then transferred to vials containing 8 ml of "Aquasol" (New England Nuclear) for scintillation counting using ^{14}C window settings.

Measurement of $^{14}CO_2$ incorporation: Groups of 60 15-mm peeled coleoptile sec-tions were placed into each of 4 50-ml beakers containing 10 ml of 2 mM MES-tris buffer (pH 6.5) and 1 mM KCl. After a 45-min preincubation with shaking at 24 C, 0.1 ml of 0.1 M CaCl$_2$ and/or 0.1 ml of 1 mM IAA were added. After 60 min, the sections were transferred to 10-ml serum vials containing 5 ml of fresh treatment medium ($\pm$ CaCl$_2$, $\pm$ IAA), and the vials were stoppered with vaccine caps. Twenty-five μl of a H^{14}CO$_3^-$ solution (25 μCi/ml; 50.3 mCi/mmol; ICN Pharmaceuticals) were then added for a 15-min uptake period. The medium was then aspirated and the sections were rinsed twice with 10 ml of 10 mM NaHCO$_3$, chased for 6 min and rinsed again in 10 mM NaHCO$_3$; 10 sections were then transferred to scintillation vials for determination of total ^{14}CO$_2$ incorporated (all the radioactivity remaining in the tissue after the extensive rinse and chase procedure was in an organic form). The remaining 50 sections from each treatment were extracted with boiling 80% ethanol without grinding the tissue. The ethanol-soluble fraction was rendered slightly alkaline (pH 10) with KOH, partitioned against ethyl ether and the labeled malate present in the aqueous phase was estimated by paper chromatography as described by Johnson and Rayle[6].

RESULTS

When oat coleoptile sections (cv. "Rodney") were floated on 1 mM CaCl$_2$, a rapid decrease in pH of the surrounding medium was observed (Fig. 1). In contrast to the IAA effect, Ca^{2+}-enhanced acidification proceeded with little or no lag period and with a rate which exceeded that seen for IAA during the initial 90 min of treatment. When IAA and Ca^{2+} were present together, the rate of acidification was greater than that seen for either agent alone. Similar results were obtained with oat coleoptiles of the "Victory" cultivar. The measured reduction in pH caused by the various treatments was not due to increases in the levels of dissolved CO$_2$, for when the acidic media taken after the 4-hr pH measurements were bubbled with N$_2$, the pH did not rise more than 0.1 unit.

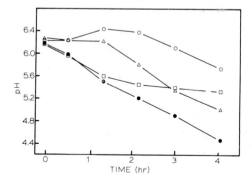

Fig. 1. Time-course of acidification in the presence of 1 mM KCl (o——o); and 1 mM CaCl$_2$ (□——□), or 10 μM IAA (△——△), or IAA and CaCl$_2$ (●——●).

In order to make a better quantitative assessment of the Ca^{2+} effect, the medium upon which the coleoptile sections were floated was titrated back to the initial pH at regular intervals. Figure 2A shows a marked and rapid acidification caused by Ca^{2+} as compared to control conditions (1 mM KCl alone). The pH change effected by control sections was almost negligible for the first 90 min, but then a new, more rapid rate of acidification became apparent. Frozen-thawed sections, with or without Ca^{2+}, also acidified the medium in a time-dependent manner, but the acidification rates tapered off after 2 - 3 hr. The Ca^{2+}-enhanced acidification in frozen-thawed tissue also proceeded with little or no lag period.

Figure 2B shows that titrating at more frequent intervals led to larger amounts of apparent acid released from coleoptile sections, particularly when Ca^{2+} was present. By 3 hr, control rates were stimulated 37% by 15-min titration intervals as compared to hourly intervals; the rate of Ca^{2+}-enhanced acidification was stimulated 88%. Furthermore, when the accumulation of external acid was minimized (as would be the case with frequent back titrations) by the inclusion of dilute buffer (0.5 or 1.0 mM Na-HEPES, pH 6.4; 1 mM Na-phosphate, pH 6.3; or 2 mM MES-tris, pH 6.5) and hourly titrations made, the rate of acidification (in the presence of Ca^{2+}) was substantially greater than that measured in the absence of buffer (data not shown). In contrast to the situation with living sections, increasing the frequency of titration (from hourly to every 15 min) of the medium surrounding Ca^{2+}-treated, frozen-thawed sections had only a slight (25%) positive effect.

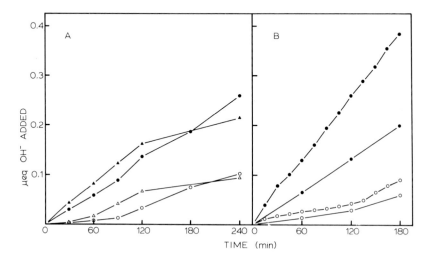

Fig. 2. A. Acidification measured by titration method in living tissue plus (●——●) or minus (o——o) 1 mM $CaCl_2$, or frozen-thawed tissue plus (▲——▲) or minus (△——△) 1 mM $CaCl_2$. B. Effect of titration frequency (hourly or 15-min intervals) on acidification rates using living tissue plus (●——●) or minus (o——o) 1 mM $CaCl_2$. Each point represents cumulative μeq NaOH added.

A concentration curve for Ca^{2+}-enhanced acidification is presented in Figure 3A. Using living coleoptile sections, the optimum $CaCl_2$ concentration was 1 mM (other experiments indicated an optimum between 0.3 and 1 mM). $CaCl_2$ at 10 mM was clearly supraoptimal. On the other hand, frozen-thawed sections exhibited no optimal response at concentrations up to 10 mM $CaCl_2$. The nitrate and sulfate salts of calcium yielded data similar to those obtained with $CaCl_2$.

The effect of varying the KCl concentration on acid release is shown in Figure 3B. There was an almost linear increase in acidification with logarithmic increases in KCl concentration, but even at 20 mM, KCl was not as effective as 1 mM $CaCl_2$. When 1 mM $CaCl_2$ was included with the various KCl concentrations, the resulting acidification was not additive, suggesting that a saturation point had been reached.

TABLE 1

RELATIVE EFFECTIVENESS OF DIVALENT
CATIONS IN ENHANCING ACIDIFICATION

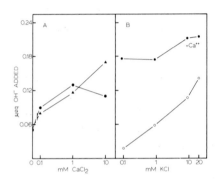

Fig. 3. A. Effect of $CaCl_2$ concentration on acidification by living (●——●) or frozen-thawed (▲——▲) tissue.
B. Effect of KCl concentration on acidification by living tissue in the absence (o——o) or presence of (●——●) 1 mM $CaCl_2$.

Cation	neq OH^- Added/3 hr*	
	Living	Frozen-thawed
Mn^{2+}	176	180
Mg^{2+}	115	---
Ca^{2+}	105	92
Co^{2+}	46	---

*Total μeq added hourly over a 3-hr period; background (1 mM KCl alone) acidification subtracted from each.

The effects of other divalent cations on the acidification response are shown in Table 1. Coleoptile sections incubated on each of the divalent salts (1 mM) acidified more than control tissue (1 mM KCl alone). The relative effectiveness of the divalent cations in eliciting acidification is: $Mn^{2+} > Mg^{2+} \geq Ca^{2+} > Co^{2+}$. Time-course studies (not shown) indicated that the rate of acidification caused by Co^{2+} was not much different than that of Ca^{2+} for the first hr, but thereafter, the rate in Co^{2+} rapidly fell to below that of the KCl control. The greater stimulatory effect of Mn^{2+} relative to Ca^{2+} was also apparent with frozen-thawed coleoptile sections (Table 1).

To evaluate the role of energy metabolism in Ca^{2+}-stimulated acidification, the effects of the respiratory inhibitors, NaN_3 and CCCP, were monitored using living

TABLE 2

EFFECTS OF INHIBITORS ON Ca^{2+}-ENHANCED ACIDIFICATION

Treatment	neq OH^- Added*		
	$- CaCl_2$	$+ CaCl_2$	Ca^{2+} Enhancement
1 mM KCl	53	180	127
1 mM KCl + 1 mM NaN_3	- 44	- 33	11
1 mM KCl + 10 μM CCCP	- 20	12	32

*Total μeq added hourly over a 3-hr period.

sections. Both azide and CCCP caused the control sections to effect an increase in pH of the medium; the presence of Ca^{2+} depressed the inhibitor-induced alkalization (Table 2). Nevertheless, the amounts of OH^- required for titration of the Ca^{2+} plus inhibitor-treated samples were less than that needed for the 1 mM KCl controls. Cycloheximide (10 μg/ml) was similar to the respiratory poisons in its effect on Ca^{2+}-enhanced acidification when added simultaneously with the Ca^{2+}, or 1 hr after Ca^{2+} treatment (data not shown).

A time-course of the effect of 0.1 mM DNP on living and frozen-thawed sections is shown in Figure 4. The inhibition of acidification in living tissue was evident by 1 hr and, as seen with azide and CCCP, even the addition of Ca^{2+} did not raise the DNP-inhibited acidification rate up to that observed with the 1 mM KCl control sample. The Ca^{2+}-enhanced acidification seen with frozen-thawed tissue was not inhibited by DNP (Fig. 4).

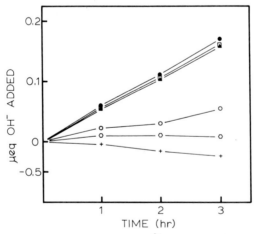

Fig. 4. Effect of DNP on Ca^{2+}-enhanced acidification. $CaCl_2$ and/or DNP were added simultaneously after the normal 5-min preincubation on fresh 1 mM KCl, and the pH was adjusted to 6.40 (zero time). Living sections: no additions (o——o); plus 1 mM $CaCl_2$ (●——●); plus 0.1 mM DNP (+——+); or plus $CaCl_2$ and DNP (O——O). Frozen-thawed sections: plus $CaCl_2$ (□——□) or plus $CaCl_2$ and DNP (▲——▲).

If Ca^{2+} effects the release of acid from the symplast of living cells as is believed to be the case for IAA- and fusicoccin-enhanced acidification[1,2,3,6,7,8], then like these growth regulators, one might predict that Ca^{2+} would enhance both K^+ uptake in exchange for H^+, and $^{14}CO_2$ incorporation into organic acids. Table 3 shows that Ca^{2+} and IAA are almost equally effective in stimulating $^{86}Rb^+$ uptake (taken as a measure of K^+ influx) and $^{14}CO_2$ incorporation into organic acids. When applied together at optimal concentrations for either alone, the Ca^{2+} plus IAA enhancements of $^{86}Rb^+$ uptake and $^{14}CO_2$ incorporation displayed approximate additivity. Lag periods for Ca^{2+} effects on both responses were about 15 min (data not shown).

A majority of the ^{14}C-label found in the sections after a 15-min incubation with $NaH^{14}CO_3$ was in the form of malate. Greater than 90% of the total radioactivity in the tissue was extractable with hot 80% ethanol, and of the alcohol-soluble material, 75 - 80% chromatographed as ^{14}C-malate (data not shown). These percentages did not vary with the treatment, so the Ca^{2+} and/or IAA enhancements of $^{14}CO_2$ incorporation is essentially synonymous with enhanced rates of malate labeling.

TABLE 3

EFFECTS OF Ca^{2+} AND IAA ON $^{86}Rb^+$ UPTAKE AND $^{14}CO_2$ INCORPORATION

Addition	$^{86}Rb^+$ Uptake*	$^{14}CO_2$ Incorporation**
None	127	3687
1 mM $CaCl_2$	185	4630
10 μM IAA	200	4770
$CaCl_2$ + IAA	289	5324

*nmol/hr/10 sections.

**cpm/15 min/10 sections.

DISCUSSION

When peeled oat coleoptile sections were incubated in 1 mM KCl, the addition of Ca^{2+} stimulated a drop in pH of the surrounding medium. This Ca^{2+} effect is more rapid than IAA-induced acidification in the absence of Ca^{2+}. Ca^{2+}-enhanced acidification occurs optimally at about 0.3 - 1.0 mM $CaCl_2$ with little difference in the accompanying anion. Nevertheless, it might be more accurate to say that Ca^{2+}-evoked acidification is a divalent cation effect rather than a specific Ca^{2+} response. Mn^{2+} was more effective than Ca^{2+}, while Mg^{2+} elicited similar acidification rates. The initial Co^{2+}-enhanced acidification rate compared favorably to that seen in the presence of Ca^{2+}, but the decline in rate with longer Co^{2+} incubations may reflect a poisonous side effect on cellular metabolism.

Concerning the mechanism by which Ca^{2+} enhances acidification, it is apparent from studies using frozen-thawed sections that Ca^{2+} (and other divalent cations) can interact with structural components in the tissue (eg., cell walls) to elicit the release of acid. These frozen-thawed sections are metabolically inactive[9] as a consequence of membrane disruption, and do not release acid when treated with IAA or fusicoccin (unpublished results). Nevertheless, Ca^{2+} effects a drop in the external pH of frozen-thawed sections which is equal to or greater than the corresponding Ca^{2+} effect on living tissue, at least over an initial 2 - 3-hr period. Indeed, even when isolated, washed cell walls prepared from oat coleoptiles (adjusted to pH 6.5) were subjected to Ca^{2+}, a rapid drop in the pH of the suspension was observed (unpublished results). We also observed that the added acidification caused by Mn^{2+} compared to Ca^{2+} in living tissue occurred on frozen-thawed sections as well. These results confirm the long-held view that plant cell walls are good cation exchangers, and that Ca^{2+} and other divalent cations are effective proton exchangers.

Since the cell walls of living coleoptile sections would act similarly as a cation exchanger, then at least part of the measured acid released from living sections treated with Ca^{2+} must originate from the cell walls. It would be difficult to assess, however, the quantitative contribution of cell-wall-bound protons exchanged for Ca^{2+} in living tissue because one cannot experimentally eliminate the cytoplasmic source of acid. A direct comparison of living to frozen-thawed tissue indicates that Ca^{2+} has similar quantitative effects on acid released over the first 2 - 3-hr period (see Fig. 2A), but the freeze-thaw process may increase the permeability of the sections to Ca^{2+} and/or expose cytoplasmic-based cation exchangers (eg., proteins) which did not escape the cell wall matrix during rinsing. Therefore, measurement of acidification in frozen-thawed sections may be an overestimate of Ca^{2+} exchange of cell wall protons.

Like the IAA- and fusicoccin-induced acidification responses in oat coleoptile sections[5,10], the Ca^{2+} effect in living tissue is sensitive to respiratory inhibitors. Since such inhibitors actually cause an alkalization of the pH 6.3 medium surrounding coleoptiles, inhibitor action apparently results in a general disturbance of the normal acid extrusion mechanism to the point that the intact symplast actually absorbs acid (leaks out basic substances?) from (to) the medium. Although Ca^{2+} can partially overcome this inhibitor-effected alkalization, the level of acidification with the inhibitors in the presence of Ca^{2+} is still below that of control (minus both Ca^{2+} and inhibitor) tissue. It is important to realize that these results do not necessarily mean that the Ca^{2+}-stimulated acidification process in living cells requires metabolic energy. Rather it seems more likely that when living tissue is treated with respiratory inhibitors, maintenance of a pH gradient between the wall and cytoplast is impossible, and the released protons are reabsorbed into the tissue.

In light of the above-mentioned results, a possible mode of Ca^{2+}-enhanced acidification which must be considered is that Ca^{2+} may act to intensify a proton gradient between the cytoplasm and cell wall in peeled, living tissue when floated on aqueous medium. It is known that Ca^{2+} treatment of root tissue allows monovalent cation uptake without influence from any apparent binding to the negative charges within the cell wall[11]. Calcium may be playing an analogous role in peeled coleoptile tissue by facilitating passage of excreted protons out of the tissue and into the external medium. The possibility of a purely diffusional model is suggested by the apparent increased acidification rates obtained when buffers were applied or when more frequent titration regimes were employed.

This hypothetical model for Ca^{2+}-enhanced release of acid from living coleoptile tissue would imply two sources of acid with one site of Ca^{2+} action. Ionic binding of Ca^{2+} to negatively-charged residues within the wall (and possibly, the outer surface of the plasmalemma) would effect the release of protons into the medium from the first source: cell wall. This displacement of wall protons would create a favorable gradient for the release of protons from the cytoplasm (the second source).

The participation of cytoplasmic proton extrusion in the Ca^{2+} acidification response is suggested by two types of evidence. First, since frozen-thawed sections have lost the metabolic capacity to generate acid equivalents, the Ca^{2+}-induced acidification measured should slow down as the Ca^{2+} approaches saturation of the proton-replacement sites. Such a tapering off of acidification was observed in frozen-thawed tissue after 2 - 3 hr of Ca^{2+} treatment (see Fig. 2A). Second, available evidence strongly suggests that protons extruded outside the protoplasm are exchanged for K^+, and the loss of cytoplasmic acid is compensated by the fixation of CO_2 into malic acid[3,6,12]. Ca^{2+} treatment leads to enhanced rates of $^{86}Rb^+$ uptake and $^{14}CO_2$ fixation into malate (see Table 3).

Such a model would imply that protons are the acidic agents released across the plasmalemma. Although we cannot eliminate the possibility that, in living tissue, hydroxide ions are taken up into the cytoplasm during acidification, we are in a position to argue against some other mechanisms. CO_2 release into the medium cannot account for the observed pH drop (with or without Ca^{2+}), since bubbling the acidified medium with N_2 to remove dissolved CO_2 does not significantly alter the acid content. Active absorption of bicarbonate ions from extracytoplasmically-hydrated CO_2 cannot be an alternate acidification mechanism because HCO_3^- is only sluggishly taken up by coleoptile tissue[6], and because the amount of HCO_3^- uptake at pH's below 6.3 (pK_a of $CO_2 \rightleftharpoons HCO_3^-$ equilibrium) would have to be exceedingly large to generate correspondingly little external H^+. Finally, the release of an organic acid(s) as opposed to protons is not indicated since titration curves for the acidified media show no significant buffering capacity. Thus, a proton extrusion (or OH^- uptake) mechanism seems most likely.

It remains possible that beyond first interacting with the cell wall, Ca^{2+} may

316

also act directly on the cytoplasm either by affecting membrane structure[13], by activating an enzyme (eg., a Ca^{2+}-activated ATPase), or by exchanging across the plasmalemma with extruded protons[4]. It should be noted, however, that any future experiments designed to assess the role of Ca^{2+} _in vivo_ with respect to the latter acidification mechanisms must take care to eliminate the possibility of secondary effects deriving from an earlier Ca^{2+}-cell-wall structural interaction.

We have no evidence that Ca^{2+} plays a direct role in the mechanism of IAA-enhanced acidification. When optimal levels of IAA and Ca^{2+} were applied to living coleoptile sections, the rates of acidification, $^{86}Rb^{+}$ uptake, and $^{14}CO_2$ fixation were found to be greater than the corresponding rates when either was applied singly. Such an additivity provides indirect evidence for the notion that IAA and Ca^{2+} have different sites of action, and also argues against the possibility that one of the acid-generating agents serves to increase the level of the other (ie., IAA would not act to increase the effective level of Ca^{2+}, or vice versa)[14].

ACKNOWLEDGEMENTS

This work was supported by National Science Foundation Grant BMS73-07110 A01 to K.D.J. and D.L.R. We are also grateful for the technical assistance provided by Thomas P. Walters.

REFERENCES

1. Rayle, D. L. and Cleland, R. (1976) Curr. Top. Develop. Biol. 11: in press.
2. Marrè, E., Lado, P., Rasi-Caldogno, F., Colombo, R. and De Michelis, M. I. (1974) Plant Sci. Lett. 3: 365-379.
3. Haschke, H. P. and Lüttge, U. (1975) Plant Physiol. 56: 696-698.
4. Cohen, J. D. and Nadler, K. D. (1976) Plant Physiol. 57: 347-350.
5. Rayle, D. L. (1973) Planta 114: 63-73.
6. Johnson, K. D. and Rayle, D. L. (1976) Plant Physiol. 57: 806-811.
7. Hager, A., Menzel, H. and Krauss, A. (1971) Planta 100: 47-75.
8. Cleland, R. (1976) Biochem. Biophys. Res. Commun. 69: 333-338.
9. Rayle, D. L., Haughton, P. M. and Cleland, R. (1970) Proc. Nat. Acad. Sci. 67: 1814-1817.
10. Marrè, E., Colombo, R., Lado, P. and Rasi-Caldogno, R. (1974) Plant Sci. Lett. 2: 139-150.
11. Epstein, E., Rains, D. W. and Schmidt, W. E. (1962) Science 136: 1051-1052.
12. Stout, R., Johnson, K. D. and D. L. Rayle (1976) Plant Physiol. 57: 2 (Suppl).
13. Epstein, E. (1961) Plant Physiol. 36: 437-444.
14. Lockhart, J. A. (1965) Ann. Rev. Plant Physiol. 16: 37-52.

Regulation of Cell Membrane Activities in Plants
E. Marrè and O. Ciferri eds.
© *1977, Elsevier/North-Holland Biomedical Press, Amsterdam*

ACTION OF TRANSPORT ANTIBIOTICS ON CYTOKININ-DEPENDENT

BETACYANIN SYNTHESIS

Daphne C. Elliott

School of Biological Sciences, Flinders University of South Australia,
Bedford Park, South Aust. 5042

Summary

Potassium ions have been found to stimulate the cytokinin-dependent betacyanin synthesis in *Amaranthus* seedlings. Fusicoccin, which stimulates K^+ uptake, increases endogenous betacyanin accumulation and acts synergistically with benzyladenine. The calcium ionophore, A23187, was found to be inhibitory in this system. These results lend support to a general theory of cytokinin action based on selective effects on ion transport.

Introduction

Cytokinins have been shown to have selective effects on ion uptake, increasing the K^+/Na^+ ratio associated with cell expansion in sunflower leaf discs and cotyledons (1,2). The cytokinin-induced increase in fresh weight in cucumber cotyledons is also accompanied by increased K^+ uptake (3).

In this paper it is shown that the induction of betacyanin synthesis in *Amaranthus* seedlings is enhanced by K^+ ions and by an agent which controls K^+ transport across membranes. The synthesis of betacyanin is a cytokinin-dependent response which involves gene activation and new enzyme synthesis (4). It is now suggested that a primary general action of these growth substances may be on ion transport.

Materials and Methods

The growing conditions for *Amaranthus* seedlings, light sources and measurement of pigment produced were as previously reported (5) except for changes in the test buffer as recorded in the results.

318

Ionophores.   I thank Dr. R. Hamill, Eli Lilly Research Laboratories, for a gift of A23187. Fusicoccin was the gift of Professor A. Ballio and Professor E. Marré of the University of Milan.

Results

Effect of potassium and sodium ions on betacyanin synthesis.   *Amaranthus*

seedlings were germinated in the dark at $37^{\circ}C$ for 46 hr and then

transferred to the test solutions at $25^{\circ}C$ for the 24 hr induction period.

The amount of betacyanin produced in the presence of 5 μM benzyladenine

under the conditions of the standard assay (5) was taken as 100%. The

effect of potassium and sodium ions on the induction given by half-maximal

benzyladenine (0.5 μM) is shown in Table 1. Increasing the potassium ions

in a phosphate buffer while keeping the pH constant affects both

endogenous pigment production and benzyladenine-dependent synthesis.

Maximum effect on benzyladenine-dependent synthesis is obtained with a

low concentration of K^+ (2 mM K_2HPO_4; 4 mEq/1). Of the chloride and

nitrate salts KNO_3 enhances endogenous pigment most but both KCl and

KNO_3 increase synthesis in the presence of benzyladenine. The $NaNO_3$

was much less effective than KNO_3.

Effect of transport antibiotics on betacyanin synthesis.   In low K^+

buffer (10 mM NaH_2PO_4 - K_2HPO_4, containing 2 mM K_2HPO_4) and half-maximal

benzyladenine fusicoccin stimulated betacyanin accumulation (Table 2)

as well as enhancing endogenous colour production without benzyladenine.

A23187 inhibited both endogenous and cytokinin-dependent synthesis.

Both benzyladenine-stimulated and fusicoccin-stimulated induction

were completely prevented by cycloheximide (10 μg/ml).

Discussion

The evidence presented in this paper supports a hypothesis that the

primary action of cytokinins in enzyme induction is to increase membrane

Table 1.

Effect of sodium and potassium phosphates, chlorides and
nitrates on the induction of betacyanin synthesis by half-
maximal benzyladenine in *Amaranthus* seedlings.

Betacyanin production is related to the amount produced in the presence of
5 µM benzyladenine, 10 mM Na_2HPO_4-KH_2PO_4, pH 6.3, and 0.1% tyrosine. This
amount was taken as 100% (5). Assays were performed without and with 0.5 µM
benzyladenine, the concentration giving half-maximal (46%) induction in
these standard conditions. Either 10 mM buffer, pH 6.3, containing 0.1%
tyrosine, or aqueous unbuffered solutions of the salts, containing 0.1%
tyrosine, were used as indicated. Controls (with all components except
benzyladenine) were subtracted in each case before calculating % induction.
All treatments were assayed in triplicate. Standard deviation did not
exceed ± 15%.

Addition	Betacyanin content nmol/seedling			Induction %
	-Benzyladenine	+0.5 µM Benzyladenine	Benzyladenine-dependent synthesis	
10 mM buffer,pH6.3				
Na_2HPO_4-NaH_2PO_4 (22:78, v/v; Na^+, 12 mEq/l)	0.070	0.216	0.146	33
K_2HPO_4-NaH_2PO_4 (22:78, v/v; K^+, 4 mEq/l; Na^+, 8 mEq/l)	0.105	0.331	0.226	51
Na_2HPO_4-KH_2PO_4 (22:78, v/v; K^+, 8 mEq/l; Na^+, 4 mEq/l)	0.096	0.300	0.204	46
K_2HPO_4-KH_2PO_4 (22:78, v/v; K^+, 12 mEq/l)	0.109	0.286	0.177	40
Salt				
None	0.077	0.185	0.108	24
2 mM KCl	0.082	0.247	0.165	37
2 mM NaCl	0.073	0.180	0.107	21
2 mM KNO_3	0.110	0.297	0.187	42
2 mM $NaNO_3$	0.085	0.218	0.133	30

Table 2.   Effect of transport antibiotics on the induction of
           betacyanin by benzyladenine.

Low K^+ buffer (K_2HPO_4-NaH_2PO_4, 10 mM, pH 6.3, containing 0.1% tyrosine)
and 0.5 μM benzyladenine were used. See Table 1 for explanation of %
induction.

Ionophore	Betacyanin content nmol/seedling			Induction %
	-Benzyladenine	+0.5 μM Benzyladenine	Benzyladenine-dependent synthesis	
None	0.117	0.385	0.268	51
1 μM A23187	0.097	0.291	0.194	37
10 μM A23187	0.090	0.263	0.173	33
1 μM fusicoccin	0.225	0.665	0.440	84
10 μM fusicoccin	0.242	0.771	0.529	101

permeability specifically to potassium ions. KCl and not NaCl stimulates
the benzyladenine-dependent betacyanin synthesis in *Amaranthus* seedlings.
Previously published betacyanin assays in *Amaranthus* have included Na-K
phosphate buffer (6) and KNO_3 (7). Köhler and Birnbaum (8) found no
effect of KCl and concluded that stimulation depended on nitrate as well
as potassium ions. KNO_3 is indeed better than KCl but not as good as
low K^+ phosphate buffer in promoting benzyladenine-dependent induction
(Table 1). In all cases the K^+ salt is better than the corresponding
Na^+ salt.

    Fusicoccin activates in the cell membrane an electrogenic proton
pump and increases K^+ uptake into the cell (9). Its action on extension
growth is enhanced by K^+ and its effects on increased fresh weight have
been observed in tissues specific for both auxin (10) and cytokinin (11).
It enhances endogenous betacyanin synthesis (Table 2), possibly a

reflection of its obligatory vectorial nature. Since fusicoccin can substitute for benzyladenine in this induction the question may be asked, do they both act on the one protein receptor. The fact that fusicoccin acts synergistically with benzyladenine (Table 2) rather than additively may appear to deny this proposition. The nature of this interaction however will depend on the response curve to K^+ in the cell and has not been studied yet in sufficient detail to give a definite answer about the identity of the fusicoccin-stimulated system and the benzyladenine-stimulated system. Marré *et al.* (11) suggest that since fusicoccin can not substitute for cytokinin in the prevention of senescence in isolated leaf discs then the primary receptors for the two classes of compounds are different. Fusicoccin can also not substitute for benzyladenine in stimulating accumulation of anthocyanin in sunflower cotyledons (12). These data do not, however, rule out the possibility of secondary effects of fusicoccin in these systems.

The results with A23187 are interesting in relation to recent work showing interactions in biological systems between cytokinins and calcium. In some cases calcium has enhanced the cytokinin response (e.g. enlargement of *Xanthium* cotyledons (13); stimulation of ethylene production in mung bean hypocotyl (14)). In red-cabbage seedlings calcium has been found to inhibit kinetin stimulation of anthocyanin production (15). Cytokinins have also been shown to enhance calcium uptake into cells of the fungus *Achlya*, and to allosterically regulate the binding of calcium to a cell surface glycoprotein from these cells (16,17,18). The inhibitory action of A23187 in the present work may indicate the involvement of a divalent cation in the action of benzyladenine. A23187 has been shown to discharge calcium and magnesium from membranes (19) and it may be inhibitory in the *Amaranthus* system by generally destabilising the membrane or more specifically by removing Ca^{++} or Mg^{++} from a specific Ca^{++} or Mg^{++} dependent protein involved in potassium uptake.

In the work of Ilan *et al.* (1,2) both auxin and kinetin were found to accelerate growth as measured by increases in fresh weight and these increases were correlated with K^+ uptake, each hormone being specific for different parts of the plant. Ilan (2) draws attention to one of three suggestions made by Trewavas (20) for a possible mechanism for auxin control of net RNA synthesis, that growth hormones increase the permeability of the nuclear membrane to inorganic ions. In dipteran chromosomes, Kroeger (21,22) has shown that a graded series of genetic loci can be activated by changing the K^+/Na^+ ratio in the medium. There is also a considerable body of work in plant tissue culture showing correlations between high K^+/Na^+ and the capacity for rapid growth (23). Indeed K^+ ions function as a cation activator for a wide variety of important enzymes including protein synthesising machinery and enzymes involved in nucleotide metabolism (24). The evidence reported in the present paper indicates induction of a specific enzyme synthesis by means of interaction of benzyladenine with an agent which controls K^+ transport across membranes.

A general working hypothesis for cytokinin action depending on control of ion uptake would fit the available facts both for effects involving cell enlargement and increases in fresh weight (2,11) and for effects depending on gene activation such as induction of betacyanin synthesis.

Acknowledgements

This work was supported by grants from the Flinders University and from the Australian Research Grants Committee. The technical assistance of Mr. Marcus Tippett is gratefully acknowledged.

References

1. Ilan, I., Gilad, T. and Reinhold, L. (1971). Physiol. Plant *24*, 337-341.
2. Ilan, I. (1971). Physiol. Plant *25*, 230-233.

3. Sastry, K.S.K., Udayakumar, M. and Rao, S.R. (1973). Current Sci. *42*, 830-832.
4. Piatelli, M., Guidici de Nicola, M. and Castiogiovanni, V. (1971). Phytochem. *10*, 289-293.
5. Elliott, D.C. and Murray, A.W. (1975). Biochem. J. *146*, 333-337.
6. Köhler, K-H and Conrad, K. (1966). Biol. Rdsch. *4*, 36-37.
7. Conrad, K. and Köhler, K-H (1967). Wiss. Z. Univ. Rostock *16*, 657-659.
8. Köhler, K-H and Birnbaum, D. (1970). Biol. Zbl. *89*, 201-211.
9. Lado, P., DeMichelis, M.I., Cerana, R. and Marrè, E. (1976). Plant Sci. Lett. *6*, 5-20.
10. Marrè, E., Lado, P., Rasi-Caldogno, F. and Colombo, R. (1973). Plant Sci. Lett. *1*, 179.
11. Marrè, E., Colombo, R., Lado, P. and Rasi-Caldogno, F. (1974). Plant Sci. Lett. *2*, 139-150.
12. Servettaz, O., Castelli, D. and Longo, C.P. (1975). Plant Sci. Lett. *4*, 361-368.
13. Leopold, A.C., Pooviah, B.W., dela Fuente, R.K. and Williams, R.J. (1974). In Plant Growth Substances, 1973, pp. 780-788, Hirokawa Publishing Co., Tokyo.
14. Lau, O-L. and Yang, S.F. (1975). Plant Physiol. *55*, 738-740.
15. Bassim, T.A.H. and Pecket, R.C. (1975). Phytochem. *14*, 731-733.
16. LeJohn, H.B. and Cameron, L.E. (1973). Biochem. Biophys. Res. Commun. *54*, 1053-1060.
17. LeJohn, H.B. and Stevenson, R.M. (1973). Biochem. Biophys. Res. Commun. *54*, 1061-1066.
18. LeJohn, H.B., Cameron, L.E., Stevenson, R.M. and Meuser, R.V. (1974). J. Biol. Chem. *249*, 4016-4020.
19. Reed, P.W. and Lardy, H.A. (1972). In The Role of Membranes in Metabolic Regulation, pp. 111-131, Academic Press, New York.
20. Trewavas, A. (1968). Progress in Phytochem. *1*, 113-160.
21. Kroeger, H. (1963). Nature *200*, 1234-1235.
22. Kroeger, H. (1966). Proc. Symp. Endoc. Genet. p. 55.
23. Steward, F.C. and Mott, R.L. (1970). Int. Rev. Cytol. *28*, 275-370.
24. Evans, H.J. and Sorger, G.J. (1966). Ann. Rev. Plant Physiol. *17*, 47-76.

AUTHOR INDEX

SUBJECT INDEX